GUIDELINES FOR INTEGRATING PROCESS SAFETY INTO ENGINEERING PROJECTS

**PUBLICATIONS AVAILABLE FROM THE
CENTER FOR CHEMICAL PROCESS SAFETY
of the
AMERICAN INSTITUTE OF CHEMICAL ENGINEERS**

GUIDELINES FOR INTEGRATING PROCESS SAFETY INTO ENGINEERING PROJECTS

CENTER FOR CHEMICAL PROCESS SAFETY OF THE AMERICAN INSTITUTE OF CHEMICAL ENGINEERS

120 Wall Street, 23rd Floor • New York, NY 10005

GUIDELINES FOR
INTEGRATING PROCESS SAFETY
INTO ENGINEERING PROJECTS

CENTER FOR CHEMICAL PROCESS SAFETY OF THE
AMERICAN INSTITUTE OF CHEMICAL ENGINEERS

This book is one in a series of process safety guidelines and concept books published by the Center for Chemical Process Safety (CCPS). Please go to www.wiley.com/go/ccps for a full list of titles in this series.

It is sincerely hoped that the information presented in this document will lead to an even more impressive process safety record for industry; however, neither the American Institute of Chemical Engineers, its consultants, CCPS Technical Steering Committee and Subcommittee members, their employers, their employers officers and directors, nor BakerRisk® and its employees warrant or represent, expressly or by implication, the correctness or accuracy of the content of the information presented in this document. As between (1) American Institute of Chemical Engineers, its consultants, CCPS Technical Steering Committee and Subcommittee members, their employers, their employers officers and directors, and BakerRisk® and its employees, and (2) the user of this document, the user accepts any legal liability or responsibility whatsoever for the consequence of its use or misuse.

This edition first published 2019

© 2019 the American Institute of Chemical Engineers

A Joint Publication of the American Institute of Chemical Engineers and John Wiley & Sons, Inc.

Registered Office
John Wiley & Sons, Inc., 111 River Street, Hoboken, NJ 07030, USA

Editorial Office
111 River Street, Hoboken, NJ 07030, USA

For details of our global editorial offices, customer services, and more information about Wiley products visit us at www.wiley.com.

Wiley also publishes its books in a variety of electronic formats and by print-on-demand. Some content that appears in standard print versions of this book may not be available in other formats.

Library of Congress Cataloging-in-Publication Data is available.

Hardback ISBN: 9781118795071

Cover Images: Silhouette, oil refinery © iStock.com/manyx31; Stainless steel © Creativ Studio Heinemann/Getty Images, Inc.; Dow Chemical Operations, Stade, Germany / Courtesy of The Dow Chemical Company

Printed in the United States of America

V10005725_110218

CONTENTS

List of Figures   xvii

List of Tables   xviii

Acronyms and Abbreviations   xx

Glossary   xxviii

Acknowledgments   xxxii

Files on the Web   xxxiv

Preface   xxxvi

1   INTRODUCTION   1

    1.1   Background and Scope   2

    1.2   Why Integrating Process Safety is Important   3

    1.3   What Type of Projects Are Included?   5

    1.4   Project Life Cycle   7

    1.5   Relationship to Other Programs   10

    1.6   Structure of this Document   13

2   PROJECT MANAGEMENT CONCEPTS AND PRINCIPLES   16

    2.1   Common Principles and Structure   16

        2.1.1   Statement of Requirements   16

        2.1.2   Project Scope   17

        2.1.3   Basis of Design   17

        2.1.4   Project Budget   18

        2.1.5   Project Plan   18

        2.1.6   Project Life Cycle   19

    2.2   Project Management   20

    2.3   Project Governance   21

    2.4   Types of Project   22

        2.4.1   Greenfield Projects   22

        2.4.2   Brownfield Projects   22

	2.4.3	Retrofit / Expansion Projects	23
	2.4.4	Control System Upgrade Projects	24
	2.4.5	Demolition Projects	24
	2.4.6	Management of Change Projects	24
	2.4.7	Mothballing Projects	25
	2.4.8	Re-Commissioning Projects	25
	2.4.9	Restarting a Project	25
	2.4.10	Post-Incident Projects	26
2.5		Project Organization	26
	2.5.1	Pre-Project Team	26
	2.5.2	Typical Project Team	27
	2.5.3	Unit Based Team	28
	2.5.4	Equipment Based Team	28
	2.5.5	Site Based Team	28
	2.5.6	Small Projects	28
	2.5.7	Roles and Responsibilities	29
2.6		Strategies for Implementation	32
	2.6.1	Contractor Selection	35
	2.6.2	Engineering Only	36
	2.6.3	Engineering and Procurement	36
	2.6.4	Engineering, Procurement and Construction	36
	2.6.5	Operation	37
	2.6.6	Contractor Oversight	37
2.7		Risk Management	38
2.8		Project Controls	40
	2.8.1	Planning and Progress	40
	2.8.2	Estimates, Budgets and Cost Control	41
	2.8.3	Reporting	41
	2.8.4	Metrics	41
	2.8.5	Action Tracking	42
	2.8.6	Change Management	42
2.9		Other Considerations	43
	2.9.1	Materials Management	43
	2.9.2	Quality Management	43
	2.9.3	Lessons Learned	43
	2.9.4	Post-Project Close-Out	44
2.10		Stage Gate Reviews	44

3   FRONT END LOADING 1   46

3.1   Preliminary Hazard Identification   48
3.2   Preliminary Inherently Safer Design Review   49
3.3   Concept Risk Analysis   51
3.4   Other Activities   52
        3.4.1   Process Safety and EHS Plan   52
        3.4.2   Risk Register   52
        3.4.3   Action Tracking   52
3.5   Stage Gate Review   53
3.6   Summary   54

4   FRONT END LOADING 2   56

4.1   Evaluation of Development Options   58
        4.1.1   Hazard Identification   59
        4.1.2   Preliminary Inherently Safer Design Review   59
        4.1.3   Concept Risk Analysis   60
        4.1.4   Selection of the Development Option   60
4.2   Further Definition of the Selected Option   63
        4.2.1   Design Hazard Management Process   63
        4.2.2   Preliminary Inherently Safer Design (ISD)   68
        4.2.3   Hazard Identification and Risk Analysis (HIRA)   68
        4.2.4   Engineering Design Regulations, Codes, and Standards 69
        4.2.5   Design Philosophies/Strategies   70
        4.2.6   Preliminary Facility Siting Study   71
        4.2.7   Preliminary Fire and Explosion Analysis   72
        4.2.8   Transportation Studies   72
        4.2.9   Preliminary Blowdown and Depressurization Study 74
        4.2.10 Preliminary Fire & Gas Detection Study   74
        4.2.11 Preliminary Fire Hazard Analysis   74
        4.2.12 Preliminary Firewater Analysis   75
        4.2.13 Preliminary Security Vulnerability Analysis   75
        4.2.14 Other Engineering Design Considerations   75
4.3   Other Activities   76
        4.3.1   EHS and Process Safety Plan   76
        4.3.2   Risk Register   76
        4.3.3   Action Tracking   76
        4.3.4   HIRA Strategy   76
        4.3.5   Documentation   77

| | | 4.3.6 | Stage Gate Review | 77 |
| 4.4 | | Summary | | 79 |

5 FRONT END LOADING 3   80

5.1		Evaluation of Development Options		82
5.2		Further Definition of the Selected Option		82
	5.2.1	Design Hazard Management Process		84
	5.2.2	Inherently Safer Design Optimization		85
	5.2.3	Facility Siting and Layout		86
	5.2.4	Refine Design Safety Measures		93
	5.2.5	Set Performance Standards		94
	5.2.6	Hazard Identification and Risk Analysis (HIRA)		95
	5.2.7	Safety Assessments		102
	5.2.8	Re-Evaluate Major Accident Risk		113
	5.2.9	Finalize Important Safety Decisions		113
	5.2.10	Finalize Basis of Design		113
5.3		Other Engineering Considerations		114
	5.3.1	Asset Integrity Management		114
	5.3.2	Quality Management		115
	5.3.3	Contractor Selection		115
	5.3.4	Brownfield Developments		116
5.4		Other Activities		116
	5.4.1	EHS and Process Safety Plans		116
	5.4.2	Risk Register		116
	5.4.3	Action Tracking		116
	5.4.4	Change Management		116
	5.4.5	Documentation		117
	5.4.6	Preparation for Project Execution		117
5.5		Case for Safety		119
5.6		Stage Gate Review		119
5.7		Summary		120

6 DETAILED DESIGN STAGE   121

6.1		Detailed Design		124
	6.1.1	Design Hazard Management Process		124
	6.1.2	Inherently Safer Design Optimization		125
	6.1.3	Site Layout		126
	6.1.4	Design Safety Measures		126

	6.1.5	Set Performance Standards	127
	6.1.6	Hazard Identification and Risk Analysis (HIRA)	127
	6.1.7	Safety Assessments	128
	6.1.8	Re-Evaluate Major Accident Risk	129
	6.1.9	Other Design Reviews	129
6.2		Procurement	130
6.3		Asset Integrity Management	131
6.4		Other Process Safety Activities	132
	6.4.1	Case For Safety	132
6.5		Other Project Activities	133
	6.5.1	EHS and Process Safety Plans	133
	6.5.2	Risk Register	134
	6.5.3	Action Tracking	134
	6.5.4	Change Management	134
	6.5.5	Documentation	135
	6.5.6	Constructability	136
	6.5.7	Contractor Selection	137
6.6		Preparation for Construction	138
6.7		Preparation for Pre-Commissioning, Commissioning, and Startup	139
6.8		Stage Gate Review	140
6.9		Summary	141

7 CONSTRUCTION   143

7.1		Planning	146
7.2		Pre-Mobilization	147
7.3		Mobilization	149
7.4		Execution	150
	7.4.1	Procurement	151
	7.4.2	Fabrication	151
	7.4.3	Safety Culture	151
	7.4.4	Workforce Involvement	152
	7.4.5	Stakeholder Outreach	152
	7.4.6	Contractor Management	152
	7.4.7	Transportation	153
	7.4.8	Equipment and Materials Handling	153
	7.4.9	Hazard Evaluation	154
	7.4.10	Engineering Design	156

	7.4.11	Safe Work Practices	157
	7.4.12	Operating, EHS and Process Safety Procedures	159
	7.4.13	Training and Competence Assurance	159
	7.4.14	Asset Integrity Management	160
	7.4.15	Change Management	161
	7.4.16	Emergency Response	162
	7.4.17	Incident Investigation	163
	7.4.18	Auditing	164
	7.4.19	Performance Measurement	164
	7.4.20	Operations Case for Safety	165
	7.4.21	Pre-Commissioning	166
	7.4.22	Mechanical Completion	170
	7.4.23	Documentation	171
7.5	Other Project Activities		172
	7.5.1	EHS and Process Safety Plans	172
	7.5.2	Risk Register	172
	7.5.3	Action Tracking	172
	7.5.4	General Construction Management	172
7.6	De-Mobilization		173
7.7	Preparation for Commissioning and Startup		174
7.8	Final Evaluation and Close-out		175
7.9	Stage Gate Review		175
7.10	Summary		177

8   QUALITY MANAGEMENT   178

8.1	Design/Engineering	183
8.2	Procurement	186
8.3	Fabrication	187
8.4	Receipt	189
8.5	Storage and Retrieval	190
8.6	Construction and Installation	191
8.7	Operation	193
8.8	Documentation	194
8.9	Summary	194

9   COMMISSIONING AND STARTUP   196

9.1	Preparation		199
	9.1.1	Planning	199
	9.1.2	Safety	201

9.2	Operational Readiness	202
	9.2.1 Pre-Startup Stage Gate Review	203
	9.2.2 Operational Readiness Review	204
	9.2.3 Start-Up Efficiency Review	207
9.3	Commissioning	208
	9.3.1 Equipment Testing	209
	9.3.2 Commissioning Procedures	211
9.4	Startup	213
	9.4.1 Preparation for Startup	213
	9.4.2 Calibration of Instruments and Analyzers	213
	9.4.3 Startup with Process Chemicals/Fluids	214
9.5	Common Process Safety Elements	215
	9.5.1 Hazard Evaluation	215
	9.5.2 Safe Work Practices	216
	9.5.3 Procedures	217
	9.5.4 Training and Competence Assurance	217
	9.5.5 Management of Change	218
	9.5.6 Incident Investigation	219
	9.5.7 Emergency Response	220
	9.5.8 Auditing	221
	9.5.9 Documentation	221
	9.5.10 Performance Measurement	222
9.6	Other Project Activities	222
	9.6.1 EHS and Process Safety Plans	222
	9.6.2 Risk Register	223
	9.6.3 Action Tracking	223
9.7	Performance Test Runs	223
9.8	Handover	224
9.9	Preparation for Ongoing Operation	225
9.10	Project Close-Out	226
	9.10.1 Close Out Report	226
	9.10.2 Post-Project Evaluation	226
9.11	Summary	227
10	**OPERATION**	**228**
10.1	Process Safety Management System	231
	10.1.1 Process Safety Culture	233
	10.1.2 Compliance with Standards	233

10.1.3   Process Safety Competency   233
10.1.4   Workforce Involvement   234
10.1.5   Stakeholder Outreach   234
10.1.6   Process Knowledge Management   234
10.1.7   Hazard Identification and Risk Analysis   234
10.1.8   Operating Procedures   235
10.1.9   Safe Work Practices   235
10.1.10 Asset Integrity and Reliability   236
10.1.11 Contractor Management   238
10.1.12 Training and Performance Assurance   239
10.1.13 Management of Change   239
10.1.14 Operational Readiness   239
10.1.15 Conduct of Operations   240
10.1.16 Emergency Management   240
10.1.17 Incident Investigation   240
10.1.18 Measurement and Metrics   241
10.1.19 Auditing   242
10.1.20 Management Review and Continuous Improvement   242
10.1.21 EHS and Process Safety Procedures   243
10.2   Other Project Activities   243
10.2.1   EHS and Process Safety Plans   243
10.2.2   Risk Register   243
10.2.3   Action Tracking   243
10.3   Technical Support   243
10.4   Performance Test Runs   244
10.5   Operation Stage Gate Review   244
10.6   Post-Operational Review   245
10.7   Project Close-Out   246
10.8   Summary   246

11   END OF LIFE   247

11.1   Design for Decommissioning   249
11.2   Planning for Decommissioning   250
11.2.1   Engineering Survey   251
11.2.2   Hazard Evaluation   254
11.2.3   Hazardous Materials   254
11.2.4   Process Safety Plan   255
11.2.5   Utilities   255

		11.2.6	Re-Engineering	256
	11.3		Decommissioning Procedures	256
		11.3.1	Late-Life Operations	257
		11.3.2	Cessation of Production	258
		11.3.3	Cleaning and Decontamination	258
		11.3.4	Mothballed Facilities And Equipment	258
	11.4		Deconstruction and Demolition	259
		11.4.1	Deconstruction	259
		11.4.2	Demolition	260
	11.5		Process Safety for Decommissioning	261
		11.5.1	Contractor Management	261
		11.5.2	Safety Culture	262
		11.5.3	Workforce Involvement	263
		11.5.4	Stakeholder Outreach	263
		11.5.5	Hazard Evaluation	263
		11.5.6	Safe Work Practices	264
		11.5.7	EHS and Process Safety Procedures	265
		11.5.8	Training and Competence Assurance	266
		11.5.9	Asset Integrity Management	267
		11.5.10	Change Management	267
		11.5.11	Operational Readiness Review	267
		11.5.12	Emergency Management	268
		11.5.13	Incident Investigation	269
		11.5.14	Auditing	269
		11.5.15	Disposal	270
		11.5.16	Remediation	270
	11.6		Other Project Activities	271
		11.6.1	EHS and Process Safety Plans	271
		11.6.2	Risk Register	271
		11.6.3	Action Tracking	271
		11.6.4	General Decommissioning Management	272
		11.6.5	Stage Gate Reviews	272
	11.7		Summary	274
12	**DOCUMENTATION**			**275**
	12.1		Document Management	275
	12.2		Process Knowledge Management	278
		12.2.1	Front End Loading 1 Stage	278

12.2.2	Front End Loading 2 Stage	279
12.2.3	Front End Loading 3 Stage	280
12.2.4	Detailed Design Stage	281
12.2.5	Construction Stage	283
12.2.6	Commissioning and Startup Stage	286
12.2.7	Handover	287
12.2.8	Operation Stage	288
12.2.9	End of Life Stage	291
12.3	Summary	292

APPENDIX A. TYPICAL PROCESS SAFETY STUDIES OVER PROJECT LIFE CYCLE   **293**

APPENDIX B. PROJECT PROCESS SAFETY PLAN   **295**

APPENDIX C. TYPICAL HAZARD & RISK REGISTER   298

APPENDIX D. SAFETY CHECKLIST FOR PROCESS PLANTS   301

APPENDIX E. EXAMPLE OF SITE-SPECIFIC DECOMMISSIONING CHECKLIST / QUESTIONNAIRE   **315**

APPENDIX F. TYPICAL PROJECT DOCUMENTATION   **322**

APPENDIX G. STAGE GATE REVIEW PROTOCOL FOR PROCESS SAFETY   **337**

REFERENCES   **365**

INDEX   **379**

LIST OF FIGURES

Figure 1.1. Front End Planning Process Map (CII 2012).   7

Figure 1.2. Capital Project Stages   9

Figure 2.1. Simplified Diagram for Typical Project Organization   27

Figure 2.2. Risk Assessment Cycle   39

Figure 3.1. Front End Loading 1   46

Figure 4.1. Front End Loading 2   56

Figure 5.1. Front End Loading 3   80

Figure 5.2. Example of Overpressure Contour Plot   97

Figure 6.1. Detailed Design   121

Figure 7.1. Construction   143

Figure 7.2. Improperly Installed Electrical Cables   169

Figure 7.3. Damaged Instrument Cable   169

Figure 7.4. Improperly Installed Tubing   170

Figure 7.5. Improper Handling of Pressure Safety Valve   170

Figure 8.1. Corroded Solenoid   190

Figure 8.2. Wrapped Equipment with Expired Desiccant   190

Figure 9.1. Startup   196

Figure 10.1. Operation   228

Figure 11.1. End of Life   247

LIST OF TABLES

Table 1.1. Types of Projects Covered by these Guidelines   6

Table 1.2. Relationships between Projects and Risk-Based Process Safety
          Elements   11

Table 1.3. Chapters Addressing Project Life Cycle Stages   14

Table 2.1. Important Interpersonal Skills for Project Managers (PMI 2013)   30

Table 2.2. Impact of Contracting Strategy   34

Table 3.1. Simplified HAZID Checklist   49

Table 3.2. FEL-1 Stage Gate Review Scope   53

Table 4.1. Economically Feasible Platform Concepts vs. Water Depth   62

Table 4.2. Typical Steps in a Design Hazard Management Process   64

Table 4.3. Hierarchy of Risk Reduction Measures   65

Table 4.4. FEL-2 Stage Gate Review Scope   79

Table 5.1. Typical Deliverables in a FEED Package   83

Table 5.2. Typical Examples of Safety Critical Equipment / Elements   105

Table 5.3. FEL-3 Stage Gate Review Scope   120

Table 6.1. Detailed Design Stage Gate Review Scope   141

Table 7.1. Typical Planning Activities at Pre-Mobilization   148

Table 7.2. Typical Pre-Commissioning Activities   168

Table 7.3. Typical Punch-List Categories   169

Table 7.4. Construction Stage Gate Review Scope   176

Table 8.1. Typical Human Errors That Occur in Projects   180

Table 8.2. Typical Project Activities Involving Quality Management   180

Table 8.3. Typical Quality Activities During FEL and Detailed Design   185

Table 9.1. Typical Commissioning and Startup Plan   200

Table 9.2. Pre-Startup Stage Gate Review Scope   204

Table 9.3. Typical Operational Readiness Review Checklist Categories   206

Table 10.1. Risk-Based Process Safety Elements   232

Table 10.2. Operation Stage Gate Review Scope   245

Table 11.1. Typical Content of Engineering Survey Report   252

Table 11.2. Example of Safety Stop Checklist   268

Table 11.3. End of Life Stage Gate Review Scope   273

ACRONYMS AND ABBREVIATIONS

ACC	American Chemistry Council
AIA	American Insurance Association
AIChE	American Institute of Chemical Engineers
AIHA	American Industrial Hygiene Association
AIM	Asset Integrity Management
ALARP	As Low As Reasonably Practicable
ANSI	American National Standards Institute
API	American Petroleum Institute
APM	Association of Project Management
ASME	American Society of Mechanical Engineers
ASSE	American Society of Safety Engineers
AST	Aboveground Storage Tank
ATEX	Appareils destinés à être utilisés en ATmosphères Explosibles (94/9/EC Directive)
BDL	Building Damage Level
BEP	Basic Engineering Package
BM&M	Benchmarking and Metrics program
BOD	Basis of Design
BPCS	Basic Process Control System
BSI	British Standards Institution
BST	Baker-Strehlow-Tang blast model
CAD	Computer-Aided Design

CAPEX	Capital Expenditure
CCPS	Center for Chemical Process Safety
CFR	United States Code of Federal Regulations
CII	Construction Industry Institute
CMMS	Computerized Maintenance Management System
CO/CO2	Carbon Monoxide/Carbon Dioxide
COMAH	Control of Major Accident Hazards
CPT	Client Project Team
CRA	Concept Risk Analysis
CSB	United States Chemical Safety Board
DCN	Design Change Notice
DCS	Distributed Control System
DHA	Dust Hazards Analysis
DHM	Design Hazard Management
DHS	United States Department of Homeland Security
DIN	Deutsches Institut für Normung (German standard)
DOT	United States Department of Transportation
DSP	Decision Support Package
EER	Evacuation, Escape, and Rescue study
EHS	Environment Health & Safety
EI	Energy Institute
EN	European Norm standard maintained by CEN (European Committee for Standardization)
EPA	United States Environmental Protection Agency
EPC	Engineering, Procurement and Construction
EPCM	Engineering, Procurement, Construction and Management

ERPG	Emergency Response Planning Guidelines (AIHA)
ESD	Emergency Shutdown
ESDS	Emergency Shutdown System
ESDV	Emergency Shutdown Valve
EU	European Union
F&G	Fire and Gas
FAT	Factory Acceptance Test
FEED	Front End Engineering Design
FHA	Fire Hazard Analysis
FID	Final Investment Decision
FEL	Front End Loading
FMEA	Failure Modes and Effects Analysis
FSA	Functional Safety Assessment
FSS	Facility Siting Study
GB	Chinese national standard
GTR	Guarantee Test Run
HAC	Hazardous Area Classification
HAZID	Hazard Identification Study
HAZOP	Hazard and Operability Study
HCA	High Consequence Area
HIPS	High Integrity Protection System
HIPPS	High Integrity Pressure Protection System
HIRA	Hazard Identification & Risk Analysis
HF	Hydrofluoric Acid
HP	High Pressure
HFA	Human Factors Analysis
HMI	Human-Machine Interface

HR	Human Resources
HSE	United Kingdom Health and Safety Executive
HVAC	Heating, Ventilation and Air Conditioning
I/O	Input/Output
ICC	International Code Council
IChemE	Institution of Chemical Engineers
IEC	International Electrotechnical Commission
IOGP	International Association of Oil and Gas Producers
IPL	Independent Protection Layer
IRI	Industrial Risk Insurers
ISA	International Society of Automation
ISD	Inherently Safer Design
ISO	International Organization for Standardization
ITP	Inspection and Test Plan
ITPM	Inspection, Testing, and Preventive Maintenance
JHA	Job Hazard Analysis
JIT	Just-in-Time
JSA	Job Safety Analysis
JV	Joint Venture
KPI	Key Performance Indicator
LHG	Liquefied Hazardous Gas
LNG	Liquefied Natural Gas
LOC	Loss of Containment
LOPA	Layer of Protection Analysis
LOTO	Lock Out / Tag Out
LP	Low Pressure
LPG	Liquefied Petroleum Gas

MAH	Major Accident Hazard
MAWP	Maximum Allowable Working Pressure
MEL	Master Equipment List
MOC	Management of Change
MODU	Mobile Offshore Drilling Unit
MTI	Materials Technology Institute
N2	Nitrogen
NACE	National Association of Corrosion Engineers
NDT	Non-Destructive Testing
NFPA	National Fire Protection Agency
NGO	Non-Governmental Organization
NIST	National Institute of Standards & Technology
NORM	Naturally Occurring Radioactive Material
NOx	Mono-nitrogen oxides: NO and NO2 (nitric oxide and nitrogen dioxide)
OEM	Original Equipment Manufacturer
OM	Operations Manager
OPEX	Operating Expenditure
ORR	Operational Readiness Review
OSHA	United States Occupational Safety and Health Administration
P&ID	Process and Instrumentation Drawing/Diagram
PCB	Polychlorinated Biphenyl
PED	Pressure Equipment Directive
PEP	Project Execution Plan
PERT	Program Evaluation Review Technique
PFD	Process Flow Diagram
PLC	Programmable Logic Controller

PM	Project Manager and Preventive Maintenance
PMBOK	Project Management Body of Knowledge
PMI	Positive Material Identification and Project Management Institute
PMT	Project Management Team
PPA	Post-project Appraisal
PPE	Personal Protective Equipment
PQP	Project Quality Plan
PRA	Project Risk Assessment
PS	Process Safety
PSI	Process Safety Information
PSM	Process Safety Management
PSSR	Pre-startup Safety Review
PSV	Pressure Safety Valve
PreHA	Preliminary Hazard Analysis
QA	Quality Assurance
QC	Quality Control
QM	Quality Management
QMS	Quality Management System
QRA	Quantitative Risk Analysis
RACI	Responsible, Accountable, Consulted, Informed matrix/chart
RAGAGEP	Recognized and Generally Accepted Good Engineering Practices
RAM	Reliability, Availability, and Maintainability study
RBI	Risk Based Inspection program

RBPS   Risk Based Process Safety

RCM   Reliability Centered Maintenance

RFC   Ready for Commissioning

RFI   Request for Information

RMP   Risk Management Program

ROV   Remotely Operated Vehicle

RP   Recommended Practice (i.e., API guidance)

RV   Relief Valve

SAR   Search and Rescue

SAT   Site Acceptance Test

SCADA   Supervisory Control And Data Acquisition

SCAI   Safety Controls, Alarms, and Interlocks

SCBA   Self-Contained Breathing Apparatus

SCE   Safety Critical Equipment/Element

SDS   Safety Data Sheet (formerly MSDS)

SGIA   Smoke and Gas Ingress Analysis

SIF   Safety Instrumented Function

SIL   Safety Integrity Level

SIMOPS   Simultaneous Operations

SIP   Shelter in Place

SIS   Safety Instrumented System

SME   Subject Matter Expert

SOR   Statement of Requirements

SOW   Statement of Work

SOx   Sulfur oxides: sulfur monoxide (SO), sulfur dioxide
            (SO2), sulfur trioxide (SO3), disulfur monoxide
            (S2O), disulfur dioxide (S2O2), etc.

SRS   Safety Requirements Specification

SUE	Start-up Efficiency review
SVA	Security Vulnerability Analysis
THA	Task Hazard Analysis
TQM	Total Quality Management
TR	Temporary Refuge
UFD	Utility Flow Diagram
UK	United Kingdom
UKOOA	United Kingdom Offshore Operators Association
UPS	Uninterruptible Power Supply
US	United States
UST	Underground Storage Tank
UV/IR	Ultra Violet/Infrared
VCE	Vapor Cloud Explosion
VOC	Volatile Organic Compound
WSA	Waterway Suitability Assessment

GLOSSARY

This Glossary contains the terms specific to this Guideline and process safety related terms from the CCPS Process Safety Glossary. The specific CCPS process safety related terms in this Guideline are current at the time of publication; please access the CCPS website for potential updates to the CCPS Glossary.

Term	Definition
Basis of Design	Technical specifications and documentation that identify how the design meets the performance and operational requirements of the project.
Change Management	The process of incorporating a balanced change culture of recognition, planning, and evaluation of project changes in an organization to effectively manage project changes. These changes include: scope, error, design development, estimate adjustments, schedule adjustment, changed condition, elective, or required.

Term	Definition
Commissioning	The process of assuring that all systems and equipment are tested and operated in a safe environment to verify the facility will operate as intended when process chemicals are introduced
Constructability	Optimum use of construction knowledge and experience in planning, design, procurement, and field operations to achieve overall project objective.
Facility	A portion of or a complete plant, unit, site, complex or any combination thereof. A facility may be fixed or mobile.
Functional Safety	Part of the overall safety relating to the process and its control system which depends on the correct functioning of the safety controls, alarms, and interlocks (SCAI) and other protection layers
Gatekeeper	Person responsible for evaluating the project deliverables at each stage gate
Inherently Safer Design	A way of thinking about the design of chemical processes and plants that focuses on the elimination or reduction of hazards, rather than on their management and control.
Lessons Learned	Knowledge gained from experience, successful or otherwise, for the purpose of improving future performance.
Mechanical Completion	Construction and installation of equipment, piping, cabling, instrumentation, telecommunication, electrical and mechanical components are physically complete, and all inspection, testing and documentation requirements are complete.

Term	Definition
Pre-Commissioning	Verification of functional operability of elements within a system, by subjecting them to simulated operational conditions, to achieve a state of readiness for commissioning.
Project Governance	Management framework within which project decisions are made
Project Life Cycle	The series of phases that a project passes through from its initiation to its closure.
Project Risk	An event or set of circumstances that, should it occur, would have a material effect, positive or negative, on the final value of the project.
Project Scope	Work performed to deliver a product, service, or result with specified features and functions.
Quality	The degree to which a set of inherent characteristics fulfills requirements.
Quality Assurance	Activities performed to ensure that equipment is designed appropriately and to ensure that the design intent is not compromised, providing confidence throughout that a product or service will continually fulfill a defined need the equipment's entire life cycle
Quality Control	Execution of a procedure or set of procedures intended to ensure that a design or manufactured product or performed service/activity adheres to a defined set of quality criteria or meets the requirements of the client or customer.
Quality Management	All the activities that an organization uses to direct, control and coordinate quality.

Term	Definition
Safety Critical Equipment / Element	Equipment, the malfunction or failure of which is likely to cause or contribute to a major accident, or the purpose of which is to prevent a major accident or mitigate its effects.
Scope Creep	Uncontrolled changes or continuous growth in a project's scope
Site Acceptance Test	The system or equipment is tested in accordance with client approved test plans and procedures to demonstrate that it is installed properly and interfaces with other systems and equipment in its working environment.
Startup	The process of introducing process chemicals to the facility to establish operation.
Statement of Work	Narrative description of products, services, or results to be delivered by a project.
System	Section of a facility that can be pre-commissioned independently, but in parallel with other sections of the facility under construction.

ACKNOWLEDGMENTS

The Chemical Center for Process Safety (CCPS) thanks all of the members of the Guidelines for Integrating Process Safety into Engineering Projects Subcommittee for providing technical guidance in the preparation of this book. CCPS also expresses its appreciation to the members of the Technical Steering Committee for their advice and support.

The chairman of the Subcommittee was Eric Freiburger of Praxair. The CCPS staff consultant was David Belonger. Acknowledgement is also given to John Herber, who was the CCPS staff consultant at the beginning of this project.

The Subcommittee had the following key contributing members:

Ignacio Jose Alonso	Consejo de Seguridad de Procesos
Christopher Buehler	Exponent
Donnie Carter	Retired (formerly BP)
Robert Dayton	Chevron
Dr. S. Ganeshan	Adjunct Professor of Chemical Engineering, Bombay
Andrew Goddard	Arkema
Anil Gokhale	CCPS
Emmanuelle Hagey	Nova Chemicals, Inc.
Kevin Watson	Chevron

The following members also supported this project: Susan Bayley (Linde); Jack Brennan (BASF); Phil Bridger (Nexen); Jessica Chen (Diageo); Sean Classen (Shell); Jonas Duarte (LANXESS, formerly DuPont and Chemtura); Marisa Pierce (DNV); and Robert Wasileski (formerly NOVA Chemicals).

AIChE and CCPS wishes to acknowledge the many contributions of the BakerRisk® staff members who contributed to this edition, especially the principal author Michael Broadribb and his colleagues who contributed to portions of this manuscript: Joe Zanoni (FEL2) and Chuck Peterson (Commissioning /startup). Editing assistance from Moira Woodhouse, BakerRisk®, is gratefully acknowledged, as well.

Before publication, all CCPS books are subjected to a thorough peer review process. CCPS gratefully acknowledges the thoughtful comments and suggestions of the peer reviewers. Their work enhanced the accuracy and clarity of these guidelines.

Peer Reviewers:

Anne Bertelsmann	Marathon Petroleum
Denise Chastain-Knight	Exida
Marlon Harding	Merck
Patti Jones	Praxair
Pamela Nelson	Solvay
John Remy	LyondellBasell
Steven Thomas	Chevron

FILES ON THE WEB

The following files are available to purchasers of *Guidelines for Integrating Process Safety into Engineering Projects*. They are accessible from the AIChE/CCPS website below using the password *P250-files*.

www.aiche.org/ccps/publications/EngineeringProjects

Typical Process Safety Studies Over Project Life Cycle

Project Process Safety Plan

Typical Hazard & Risk Register

Safety Checklist for Process Plants

Example of Site-Specific Decommissioning Checklist / Questionnaire

Typical Project Documentation

Stage Gate Review Protocol for Process Safety

PREFACE

The American Institute of Chemical Engineers (AIChE) has been closely involved with process safety, environmental and loss control issues in the chemical, petrochemical and allied industries for more than four decades. Through its strong ties with process designers, constructors, operators, safety professionals, and members of academia, AIChE has enhanced communications and fostered continuous improvement between these groups. AIChE publications and symposia have become information resources for those devoted to process safety, environmental protection and loss prevention.

AIChE created the Center for Chemical Process Safety (CCPS) in 1985 soon after the major industrial disasters in Mexico City, Mexico, and Bhopal, India in 1984. The CCPS is chartered to develop and disseminate technical information for use in the prevention of accidents. The CCPS is supported by more than 200 industry sponsors who provide the necessary funding and professional guidance to its technical steering committees. The major product of CCPS activities has been a series of guidelines to assist those implementing various elements of the Risk Based Process Safety (RBPS) approach. This book is part of that series.

Process safety should be a major consideration during the development of engineering projects within the chemical, petroleum and associated industries. Whether the project is a major capital project or a modification governed by management of change, incorporating process safety activities throughout the project life cycle will reduce risks and help prevent and mitigate incidents. In particular, the adoption of process safety early in the project life cycle can achieve levels of inherent safety that becomes more difficult and expensive in later design development. The CCPS Technical Steering Committee initiated the creation of this guideline to assist companies in integrating process safety into engineering projects.

This guideline book addresses process safety activities that are appropriate for a range of engineering projects, although not all activities will applicable to a specific project. It is not the intent of this guideline book to explain methodologies for the activities as these are covered in other CCPS publications. The guideline book also provides an introduction to project terminology so that process safety engineers and others can articulate the recommended process safety activities in a language that project management teams can understand.

1 INTRODUCTION

This chapter introduces the integration of process safety activities throughout the life cycle of an engineering project. The discipline of process safety has evolved to prevent fires, explosions, and accidental releases of hazardous materials from chemical process facilities. This involves effective management systems comprising practices, procedures, and responsible human performance and behaviors to ensure proper equipment design and installation, and to maintain the integrity of the facility during operations.

Projects are a temporary endeavor undertaken to create a unique product, service, or result. In the case of engineering projects in the process industry, the result is usually a new or modified facility. Engineering projects can vary widely in scope and size, so these guidelines present the broad objectives and considerations for process safety that are appropriate at different stages of the life cycle.

Project Life Cycle

The series of phases that a project passes through from its initiation to its closure.

(from PMBOK Glossary (PMI, 2013)

In oil and gas, and chemical companies in the process industry, the term "stages" is also used in reference to the phases of a project.

Facility

A portion of or a complete plant, unit, site, complex or any combination thereof. A facility may be fixed or mobile.

(from AIChE/CCPS Glossary)

The temporary nature of a project means that its closure corresponds to a point in time when its objectives (i.e. commissioning of a new or modified facility) have been achieved or when the project is terminated because the objectives will not be met. Most projects are undertaken to create a lasting product or result, in this case a facility.

After the project has ended, the facility will continue to operate for a number of years until it is retired, disposed, or dismantled/demolished. During this time the facility will likely be subject to startup/shutdown, periodic inspection, maintenance, and turnarounds. Therefore the facility has its own life cycle, which may partially overlap with the project life cycle. For example, the project may not be closed until the new facility has met production and/or product quality targets, or later the facility may be debottlenecked to increase production or modified, which will involve another project.

The main focus of these guidelines is on *proactively* implementing process safety activities at the optimum timeframe, but also addresses *reactively* conducting "cold eyes" reviews to provide assurance that nothing significant has been missed. This approach ensures that, if the right process safety activities are conducted at the right time, project leadership will have the right (process safety) information in order to be able to make the right risk management decisions regarding safety.

The intent of this book is not to describe in detail *how* to perform specific process safety activities, but rather to identify *what* needs to be addressed at each stage of a project. Other CCPS publications, together with industry codes, standards and recommended practices, describe methods for specific process safety activities and are referenced throughout the book. For example, the design and management of functional safety is covered in great detail in: *Guidelines for Safe Automation of Chemical Processes*, 2nd edition (CCPS 2017b), and *Functional Safety - Safety instrumented systems for the process industry sector - Part 1: Framework, definitions, system, hardware and application programming requirements*, IEC 61511-1 (IEC 2016), which are both referenced in multiple chapters of this book.

Process safety in engineering projects involves leadership, managers, engineers, operating and maintenance personnel, contractors, vendors, suppliers and support staff. Therefore, these guidelines were prepared for a wide audience and range of potential users. The chapter concludes by introducing the structure of this document.

1.1   BACKGROUND AND SCOPE

Process safety management systems have been widely credited for reductions in major accident risk within the onshore process industries, such as oil refineries and chemical plants, and some offshore regions like the North Sea. Most companies have had practices for various process safety elements, such as operating procedures and emergency response, for many years, although the scope and quality of these practices was sometimes inconsistent until specific process safety regulations were promulgated.

Some international process safety regulations, such as the Seveso Directive and its various national implementations in Europe (Seveso 1982), and the Offshore Installation (Safety Case) regulations (HM Government 1992), set goal-setting or performance-based requirements for major project facility design and operation. In the United States, the Occupational Safety and Health Administration (OSHA) introduced the Process Safety Management (PSM) standard (OSHA 1992). This was followed by the Environmental Protection Agency (EPA) Risk Management Program (RMP) rule (U.S. EPA 1996). However, the focus of these relatively prescriptive U.S. regulations was primarily on operations rather than engineering projects, although they did address some basic practices for small Management of Change (MOC) projects.

Historically, project managers have been focused on managing the risks and performance indicators related to costs, schedules, and, in some cases, technological risks, i.e. will the facility work and meet production and quality targets. Often safety concerns, from a project manager's perspective, were primarily focused on the construction stage and the *occupational* safety of a contractor's workforce. Increasingly major operating companies have recognized the need to more comprehensively address process safety in their engineering projects as a means of optimizing the residual safety risk that operations teams are required to manage for the life of the facilities. However, despite growing awareness in certain quarters, some project managers have resisted change and remain focused on cost and schedule, almost to the exclusion of process safety.

These guidelines were written primarily for engineering projects within the process industries, and outline effective approaches for integrating process safety into both large and small projects, including small management of change (MOC) works. Some content may be applicable to other industries. Many engineering and operating companies have their own practices, with differing terminologies, for managing capital projects. The guidance in this book follows the general approach for project management advocated by the Construction Industry Institute (CII) (CII 2012), although some of the terminology varies by industry sector. Although written in the United States, a conscious effort has been made to offer guidance applicable to projects worldwide.

1.2   WHY INTEGRATING PROCESS SAFETY IS IMPORTANT

As Trevor Kletz was fond of saying "… if you think safety is expensive, try an accident. Accidents cost a lot of money. And, not only in damage to plant and in claims for injury, but also in the loss of the company's reputation."

Certainly, process safety activities can incur significant resource requirements. However, several major incidents that involved newly commissioned projects with a range of inherent weaknesses bear testimony to the need for building process safety systematically into future engineering projects.

Case Study: T2 Laboratories

T2 Laboratories was a small facility in Jacksonville, Florida that produced specialty chemicals. On December 19, 2007, a chemical reactor ruptured, causing an explosion that killed four employees, injured another 32, including 28 members of the public, and hurled debris up to a mile from the plant. The batch reactor was producing methylcyclopentadienyl manganese tricarbonyl (MCMT), a gasoline additive, at the time of the rupture.

In their report (CSB 2009), the U.S. Chemical Safety and Hazard Investigation Board (CSB) determined that the immediate cause was due to failure of the reactor cooling water system, which led to a runaway exothermic reaction. CSB further determined the root cause was that T2 Laboratories did not fully understand the reactivity hazards, especially those associated with MCMT runaway reactions. No evidence was found that indicated a Hazard and Operability (HAZOP) study had ever been conducted, which would likely have identified the need for more thermodynamic data.

CSB also identified two contributory factors: inadequate overpressure protection, and lack of redundancy in the cooling water system. No data on the sizing and relief pressure of the reactor rupture disk could be found, although it is believed to have been sized based on normal operations, without considering potential emergency conditions. The cooling water system was susceptible to single-point failures, such as an inadvertently closed valve, blockage and faulty thermocouple, and lacked design redundancy. Operating procedures did not address loss of reactor cooling.

The plant was destroyed and T2 Laboratories has ceased all operations. An understanding and implementation of fundamental process safety principles and practices (e.g. layers of protection and HAZOP) during design would have prevented this tragic incident.

1.2.1   Risk Management

No matter how good the process safety input is into any engineering project, the newly installed and commissioned facility has a residual safety risk that the operations team must manage through an effective process safety management system for the life of the facility. This is true for all projects. Therefore, one of the main benefits of successfully integrating process safety into a project is to reduce this residual safety risk. Inevitably, project managers have several competing priorities to consider, such as financial, political, and practical factors, in addition to safety, so that the final solution may be a compromise. Nevertheless, project management should seek to optimize residual risk to as low as reasonably practicable through careful selection of the final development concept and good engineering design. This goal infers an inherently safer design (ISD) approach that should place fewer demands on operations personnel, while also limiting potential for major incidents.

The adoption of an ISD approach requires project management to introduce the appropriate ISD policies and practices as early as possible in the project life cycle, although opportunities for risk reduction continue, albeit diminish, throughout the project life cycle. Therefore, ISD policies and practices should ideally be integrated into a company's capital project management system. The successful implementation of ISD practices throughout a company's portfolio of engineering projects can reduce major incidents, and contribute to long-term business success. Companies that experience major incidents also experience significant business interruption and reputation damage, and often struggle to survive in a competitive industry. Indeed, this is consistent with the CCPS Business Case for Process Safety (CCPS 2006), which identifies four benefits involving demonstration of corporate responsibility, greater business flexibility, improved risk reduction, and creation of sustained value.

Another benefit of conducting the right process safety activities at the right time is the avoidance of costly change orders during project execution, or even more costly modifications to the facility after startup. It is much more efficient and inexpensive to iteratively develop and change the design on paper during the early stages of the project.

To successfully integrate process safety into projects and achieve the full benefits described above strong and consistent leadership from company executives and project management is required. This implies that these same individuals need to understand basic process safety principles and practices. It is important that project managers know when and which process safety activities to request in order to reduce risks and add value, or, at the very least, know they can trust and rely on an experienced process safety engineer to advise and make the correct calls. Project managers should also know which challenging process safety questions to ask across the multiple interfaces that they have to manage. This level of informed leadership, knowing that the right activities are occurring in the correct order, will have the ability and confidence to assure executives and other stakeholders that a fully functional process safety management system will be delivered to Operations when the facility is ready to startup.

1.3   WHAT TYPE OF PROJECTS ARE INCLUDED?

Engineering projects for the process industries come in all shapes and sizes – from management of change (MOC) works to large capital projects for new facilities. These projects cover a wide range of facilities including, but not limited to, research and development, exploration, production, transportation and storage of oil and gas, chemicals, and pharmaceuticals, as illustrated in Table 1.1.

The objectives of the relevant process safety activities at each stage of the project are broadly consistent irrespective of the nature of the project, although the scope and level of detail may vary. For example, hazard evaluation for a relatively simple modification covered by MOC may use checklists or a What If approach, whereas a complex capital project may warrant HAZID, HAZOP, LOPA and QRA. Nevertheless, both examples share a common objective of identifying hazards and evaluating whether safeguards are adequate to manage the hazards and their risk.

Table 1.1. Types of Projects Covered by these Guidelines

Types of Projects
Greenfield and Brownfield
Onshore and Offshore
Continuous and Batch Operations
Indoors and Outdoors
Modifications (covered by MOC)
Modular and Stick-built
Pilot Plants and Full-scale Process Units
Chemical Complexes and Refineries
Fixed and Semi-Submersible Production Platforms
Drilling Rigs and MODUs
Debottlenecking
Control Systems (DCS, SCADA, SIS, HIPS, etc.)
Tankage and Storage
Utility Systems (Electrical Power, Fuel Gas, Cooling Water, Nitrogen, Compressed Air, etc.)
Buildings (Control Rooms, Offices, Workshops, Warehouses, etc.)
Loading and Offloading Systems (Road, Rail, Marine)
Pipelines (Cross-Country, Intra-Plant, Subsea)
Other Infrastructure

1.4 PROJECT LIFE CYCLE

Previous publications have described the life cycle of projects within the chemical industry, and the requirement to integrate EHS activities, including process safety (CCPS 1996a, CCPS 2001b). However these publications focus more on the integration of the individual EHS disciplines rather than their integration into the project. Furthermore, much of the focus on early conceptual design was related to laboratory experimentation and pilot plant scale operation.

The CII places much emphasis on Front End Planning, which is a process that involves developing sufficient information early in the project's life cycle to allow companies (i.e. owners) to address risk and make decisions to commit resources in order to maximize the potential for a successful project (CII 2012). The front end of a project is a phase when the ability to influence changes in design is relatively high and the cost to make those changes is relatively low.

Front End Planning is divided into three main phases:

- Feasibility

- Concept

- Detailed Scope

This is illustrated in CII's Front End Planning Process Map (see Figure 1.1).

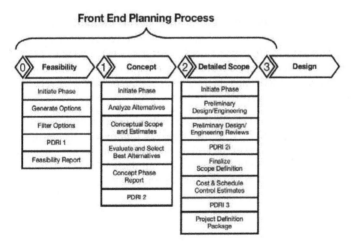

Figure 1.1 Front End Planning Process Map[1] (CII 2012).

---

[1] PDRI: Project Definition Rating Index is a comprehensive checklist of scope definition elements to enable evaluation of the status of an industrial project (CII 1996). A.k.a. FEL Index.

Front End Planning is also known as pre-project planning, front-end engineering design (FEED), feasibility analysis, and conceptual planning. However the most popular terminology in many oil and gas, and chemical companies in the process industry is Front End Loading (FEL). For the purposes of these guidelines, the terminology of FEL will be used.

Under FEL, the three phases or stages are commonly referred to as:

- FEL 1   Appraise, Appraisal or Visualization

- FEL 2   Select, Selection, or Conceptualization

- FEL 3   Define or Definition

After FEL and the completion of all planning activities, projects usually move into execution, where the plan(s) developed in FEL are put into action. In the process industry, this typically involves at least three phases or stages:

- Detailed Design or Detailed Engineering

- Construction

- Commissioning and Startup

Pre-commissioning activities are normally included in the construction phase, but some companies may address them as a separate phase or include them in the commissioning phase.

After project execution, the project life cycle moves into the Operation phase, which generally lasts until stable production is achieved at which point the project is closed. The facility life cycle continues for a number of years. Some facilities commissioned in the mid-twentieth century remain in operation today. However, eventually the facility will enter the final phase of the facility life cycle, End of Life, when its useful life is at an end.

Therefore the typical stages in the life cycle of a capital project and its resulting facility in the process industry are illustrated in Figure 1.2. The project typically closes during the early phase of the facility operation. Thereafter, small projects and management of change modifications may occur during facility operation. Finally the facility reaches its end of life and a new project is initiated for decommissioning the facility.

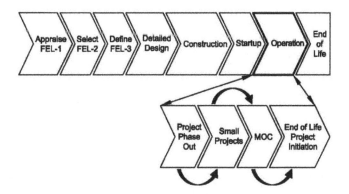

Figure 1.2   Capital Project Stages

The objectives of each stage from a business and project management perspective are as follows:

Appraise (FEL-1)

A broad range of development options is identified, and the commercial viability of the project is evaluated. A technical and commercially viable case plus alternatives should be identified for the project to proceed.

Select (FEL-2)

The alternative concept options are evaluated seeking to identify the optimum project by maximizing opportunities, while reducing threats and uncertainties to an acceptable level. Upon completion of technical and commercial studies, a single concept is selected.

Define (FEL-3)

The technical definition and execution plan for the project are improved to confirm the conceptual design, cost and schedule. A basic design is developed with plot plan, preliminary process flow diagrams, material and energy balances, and equipment data sheets. Timing varies between companies/projects, but sanction for financial investment usually occurs at the end of this stage, if sufficient confidence in the project is achieved.

Detailed Design

Detailed engineering of the defined scope from the front end loading (FEL) process is completed, scope changes managed, and materials and equipment procured.

Construction

Fabrication, construction, installation, quality management, and pre-commissioning activities are completed. Operational readiness activities are performed in preparation for commissioning, startup and operation.

Commissioning

The project is commissioned, and the facility and documentation handed over to the operations team for normal operation.

Operation

Test runs may be required to confirm that performance specifications are met before the project is closed. The project team may conduct a lessons learned review to aid future projects. At this point the facility is handed over completely to the client Operations team, the project team phases out, and the project is closed.

End of Life

When a business decision is taken to cease operations, the facility is de-commissioned. Depending upon local circumstances and regulations, the facility may be dismantled, diposed and/or demolished, or modified for future use. End of facility life typically involves a new project.

Although small modification type projects covered by MOC may not follow these stages in a formal manner, each MOC should address similar objectives. Small capital projects or identical repeat projects may elect to combine two or more stages to streamline efficiencies, while meeting the overall objectives.

Each stage of a project has specific process safety activities in support of the overall project management objectives. These process safety activities are described below.

1.5   RELATIONSHIP TO OTHER PROGRAMS

Successful engineering projects usually have a Safety Plan, often comprising Health and Environment into an EHS Plan, which lays out a strategy and schedule of process safety and occupational safety activities over the project life cycle. Starting from early feasibility (FEL 1), these plans tend to be *living* documents that evolve over time as more detail is added as the project definition is established. Effective integration of process safety into a project makes use of process safety elements routinely employed in day-to-day process plant operations.

Although *Guidelines for Risk Based Process Safety* (RBPS) (CCPS 2007b) was developed primarily for operations, its elements are appropriate at various stages of a project. For example, all four pillars of RBPS are involved, as follows:

- Commit to Process Safety

  Project EHS Plans and engineering standards demonstrate commitment.

- Understanding Hazards and Risks

  Design of new facilities requires process knowledge, hazard identification and risk analysis.

- Manage Risks

  New facilities require integrity, operability and maintainability by competent personnel.

- Learn from Experience

  Lessons learned from similar facilities should be built into new facilities.

Significant relationships with process safety elements are shown in Table 1.2. As can be seen from this table, nearly all elements of a risk-based process safety management system have some bearing on project development. However, reliance on integrating RBPS alone may not be sufficient for many projects. Other process safety practices are likely to be relevant, such as inherently safer design (ISD), and other engineering design practices.

Table 1.2. Relationships between Projects and Risk-Based Process Safety Elements

RBPS Pillar	RBPS Element	Project Activities Related to RBPS Element
Commit to Process Safety	Process Safety Culture	Present in all project activities
	Compliance with Standards	Use standards and RAGAGEP
	Process Safety Competency	Involve competent employees and contractors
	Workforce Involvement	Safety responsibilities in design, construction, and operations for employees and contractors
	Stakeholder Outreach	Consult and inform on potential risks during project planning and execution
Understand Hazards & Risk	Process Knowledge Management	Incorporate knowledge on materials, technology and equipment
	Hazard Identification and Risk Analysis	Identify hazards and assess associated risks. Identify measures for risk reduction

RBPS Pillar	RBPS Element	Project Activities Related to RBPS Element
Manage Risk	Operating Procedures	Develop procedures for commissioning and operations
	Safe Work Practices	Develop procedures for construction activities Plan and perform installation and pre-commissioning
	Asset Integrity & Reliability	Ensure maintainability and reliability, especially SCE Ensure quality of design, procurement and construction
	Contractor Management	Pre-qualify candidate contract firms Ensure contracted services meet safety goals
	Training and Performance Assurance	Train employees and contractors Certifications for engineers, inspectors and technicians
	Management of Change	Evaluate post-HAZOP design changes Evaluate field changes
	Operational Readiness	Confirm assets as installed meet design specifications Confirm no outstanding actions and/or documentation
	Conduct of Operations	All project activities Promptly address unsafe activities / conditions
	Emergency Management	Develop ERP Plans for construction and operations
Learn from Experience	Incident Investigation	Incorporate lessons learned from similar facilities Investigate incidents promptly
	Measurement & Metrics	Collect, analyze and archive data
	Auditing	Conduct independent technical / stage gate reviews
	Management Review and Continuous Improvement	Evaluate if all RBPS elements performing as intended and producing desired results

A well-designed facility should start by addressing ISD principles from an early stage (FEL-1). CCPS provides guidance through their publication, *Inherently Safer Chemical Processes: A Life Cycle Approach, 2nd edition* (CCPS 2009d). As the project definition progresses, guidance from the CCPS publication *Guidelines for*

Engineering Design for Process Safety, 2nd edition (CCPS 2012a) is available for further reference.

Depending upon the scope and magnitude of the engineering project, a vast array of process safety studies and activities may be appropriate at various stages of the project life cycle. Table 1.2 represents a matrix of some of the key process safety activities at each stage of a typical project. Some of these activities may be conducted by experienced process safety engineers, while other multi-discipline engineering studies would benefit from input by process safety expertise.

Appendix A presents an overview of typical process safety studies at each stage of a project life cycle.

1.6 STRUCTURE OF THIS DOCUMENT

These guidelines begin with a chapter that sets the groundwork for engineering projects. Chapter 2 discusses the management and organization of capital projects, and introduces the project structure and terminology promoted by the Project Management Institute (PMI) and the Construction Industry Institute (CII). The characteristics of various types of projects and strategies for their implementation are discussed. Finally, the management and objectives of process safety risk are introduced.

Once this basic understanding of projects is established, the life cycle of an engineering project is addressed in terms of the process safety objectives, scope and activities of each stage. These include:

- Front End Loading 1 (FEL-1)
- Front End Loading 2 (FEL-2)
- Front End Loading 3 (FEL-3)
- Detailed Design
- Construction
- Commissioning/Startup
- Operation
- End of Useful Life

Each of these stages is addressed in turn in Chapters 3 through 7, and 9, 10, and 11, as illustrated in Table 1.3.

Chapter 3 covers the feasibility of proceeding with a new project to produce a specific product(s) in a certain location, employing various process technologies. This initial phase of Front End Loading (FEL-1) involves preliminary Hazard Identification and Risk Analysis (HIRA) of multiple development options, from which a range of viable options are identified.

Chapter 4 deals with the next phase of FEL (FEL-2) where the various development options are reduced through a concept selection process involving more detailed HIRA, including offsite major accident risk. The site, process technology, facilities, and infrastructure requirements are determined considering an ISD approach, and a preliminary EHS and Process Safety plans developed.

Chapter 5 addresses the final phase of FEL (FEL-3) during which the technical scope of a single development option is defined. Increasingly more detailed HIRA studies are used to determine the site layout, spacing, grading and other siting concerns as a result of potential fires, explosions and toxic releases. The front-end engineering and design (FEED), including assumptions, philosophies, and engineering codes and standards, is completed, as well as the detailed EHS and Process Safety plans.

Table 1.3. Chapters Addressing Project Life Cycle Stages

Project Stages	New Equipment	Procurement	Quality Management	Documentation
FEL-1 Appraisal	Chapter 3	-	-	Chapter 12
FEL-2 Selection	Chapter 4	-	-	Chapter 12
FEL-3 Definition	Chapter 5	-	-	Chapter 12
Detail Design Detail Engineering	Chapter 6	Chapter 6	Chapter 8	Chapter 12
Construction PreCommissioning	Chapter 7	Chapter 7	Chapter 8	Chapter 12
Commissioning Startup	Chapter 9	-	-	Chapter 12
Operation ITPM	Chapter 10	-	-	-
End of Useful Life Decommissioning	Chapter 11	-	-	-

Chapter 6 covers the first stage of project execution, detailed design, involving layout and detailed engineering of individual items of equipment. Change management is introduced following the final HIRA study, and process safety information documented and compiled. The procurement of long-lead items of equipment are also covered.

Chapter 7 addresses the construction phase of the project, involving construction plans and management, procurement of equipment and materials, fabrication, installation, and management of engineering and integrity baseline documentation.

Chapter 8 covers quality management activities to ensure that the new facilities are designed, procured, fabricated and installed according to the technical specifications.

Chapter 9 deals with commissioning and startup activities, commencing with pre-commissioning, shakedown, check-out and resolution of problems, and hand-over to Operations before proceeding with startup. Operations readiness activities such as training and pre-startup safety reviews are performed in preparation to operate and startup.

Chapter 10 addresses post-project operation, when the facility is running with acceptable product quality. The project has been closed out and the facility, data, and documents have been handed over to Operations. Technical safety projects are performed periodically throughout the operational phase to ensure performance specifications are met, maximize return to shareholders, and protect license to operate.

Chapter 11 covers decommissioning, abandonment, demolition/dismantling and other end-of-useful-life issues from a process safety perspective.

Chapter 12 reviews the essential design files and process safety information that must be compiled by the project team for hand-over to Operations.

2 PROJECT MANAGEMENT CONCEPTS AND PRINCIPLES

In this chapter, some general concepts around projects and their management, and underlying principles of structure and execution are discussed. This is not intended as an in-depth guide to project management; rather it is a basic introduction to some aspects and terminology that are common to many projects. Further more detailed information and guidance is available from the following publications:

- A Guide to the Project Management Body of Knowledge (PMBOK Guide), 5th edition (PMI 2013)
- CII Best Practices Guide: Improving Project Performance, (CII 2012)

This chapter also introduces terminology that is common to most projects. Process safety engineers should familiarize themselves with this terminology, so that they may have effective communications with project personnel and ensure that process safety issues are fully considered.

A project represents an original idea or concept that when given resources, time and effort, becomes a reality. The most accomplished project management team will not deliver a successful project if the concept is inadequate or if the project is denied adequate resources. However, by addressing project concept, organization, and control issues, and understanding the potential risk areas, the pitfalls can generally be avoided.

From a process safety perspective, it is important that the project team has a strong process safety focus from the earliest stages of the project. It is only by starting early that the residual risk inherent in the completed project may be reduced in a cost effective and efficient manner. It is this residual risk that the Operations team will have to live with and manage for the life of the facilities. Process safety should be built into the common principles and structure of a project, as discussed below.

2.1 COMMON PRINCIPLES AND STRUCTURE

2.1.1 Statement of Requirements

The concept must be restated as an objective(s) through a process of refinement. This objective is frequently termed the Statement of Requirements (SOR), and also known as the Statement of Work (SOW). The principle of establishing an SOR applies, irrespective of the size of the project.

Statement of Work

Narrative description of products, services, or results to be delivered by a project.

(PMI, 2013)

An example of a company's definition of SOR is: "Description of the fundamental business requirements and success factors for a project, and forms the basis upon which project objectives, technical definition and execution planning are developed."

It is important that the SOR is known and understood by the project team, including the main contractors. As the design and execution of the project evolves through multi-disciplinary input, the design should be checked against the SOR on a regular basis.

2.1.2   Project Scope

The multi-disciplinary ideas and inputs to the evolving project result in execution strategies (for design, procurement, construction, commissioning, and operations), and a technical design that meets user requirements. This set of information forms the basis for translating intellectual ideas into hardware; i.e. turning the project into reality, and is commonly known as the Project Scope or Scope of Work. The Project Scope may be integrated into the SOR as a single document.

Project Scope

Work performed to deliver a product, service, or result with specified features and functions.

(from PMBOK Glossary (PMI, 2013)

The Project Scope allows identification of the resources necessary to deliver the project, and determination of the project duration. Various resources will be needed in design, procurement, construction, commissioning, and overall project management. The addition of increased resources may speed up one or more stages of the project.

2.1.3   Basis of Design

The business requirements in the SOR and project scope need to be defined in terms of the technical and safety standards and design basis for the project. This is commonly known as the Basis of Design (BOD) or Design Basis.

Basis of Design

Technical specifications and documentation that identify how the design meets the performance and operational requirements of the project.

An outline BOD is often developed during FEL-2, and then updated and frozen in FEL-3. The BOD is one of the principal inputs to development of the project's cost and schedule estimates.

2.1.4   Project Budget

The resources needed to deliver the project will require financing. A cost estimate for the project can be developed based on historical experience and any future trends. Additional financial provisions may be necessary to cover specific project risks (including safety risks) should they be realized, and to cover any areas of uncertainty in the scope. The net result is an estimate of expenditure required for the project to proceed. As the project definition evolves during FEL, the confidence in the cost estimate should improve. These costs should be progressively evaluated during the early FEL stages to determine the viability of the project.

2.1.5   Project Plan

Assuming the project is viable, a logical sequence of tasks utilizing the resources needs to be established for delivering the project. This Project Plan uses planning tools that range from Gantt charts to more sophisticated tools such as logic networks, especially for complex projects with multiple interactions and links between resource and time elements. The sequence of tasks should aim to deploy the available resources in the most efficient manner to complete the project scope within the approved budget and schedule. The timing, duration and resource requirements for the process safety tasks and activities discussed in later chapters should be included in the plan. The development of the plan is an iterative activity due to the potential for change, and is steadily elaborated throughout the project life cycle.

Following front end loading (FEL) and financial sanction, a Project Execution Plan (PEP) is normally developed as a high-level plan focused on the main strategies through the execution stages of the project (i.e. detailed design, construction, and startup) up to full production. The PEP establishes the means to execute, monitor, and control a project by addressing the most effective methods and maximizing efficiency in the project execution. The PEP is usually developed by the key project participants (i.e. client, project team, contractors) led by the project manager, and approved by company management. It should be updated as future plans and procedures change.

2.1.6 Project Life Cycle

It is common practice in the oil & gas and chemical process industries to divide the life cycle of a capital project into a number of discrete stages. Typical stages were illustrated in Chapter 1 (Figure 1.2), and fall into two groups: FEL and project execution. FEL involves the development of sufficient strategic information with which the company can address risk and make decisions to commit resources in order to develop a plan for the project. Project execution (or implementation) is when the plan designed in FEL is put into action to deliver the project.

The terminology and objectives of each stage are described in detail in the chapters that follow. Smaller projects and management of change (MOC) works may not clearly delineate into discrete stages or may combine several stages. However, these small projects should still address the overall objectives of each stage.

Depending upon the scope of a project, a range of process safety activities are applicable at each stage of the project, and these activities are discussed in detail in later chapters. Some of these activities are applicable to even small projects and modifications.

As part of the project governance process (see Section 2.3) a virtual gate is placed between successive stages. When the project reaches a gate at the end of a particular stage, an impartial *gatekeeper* judges whether the project still meets the business needs, has adequately delivered the stage objectives, is adequately managing the project and safety risks, and should continue to the next stage.

Gatekeeper

Person responsible for evaluating the project deliverables at each stage gate.

To assist the gatekeeper, it is common for a capital project team to schedule a technical peer review(s) and develop some form of Decision Support Package (DSP) at each stage gate[2]. Process safety should feature prominently in the stage gate reviews (see Appendix G for example process safety questions) and DSP, addressing the technical risks that are relevant to the project and the actions required to ensure that they are properly managed. Small MOC works should receive an independent technical review prior to approval for implementation. Process safety should be included in the scope of the technical review.

---

[2] The term "stage gate" is being used in its common and generic form as it is used throughout industry. Stage-Gate® is also a registered trademark of the last listed owner, Stage-Gate International. CCPS and BakerRisk® disclaim any proprietary right or interest in the registration of the mark.

2.2   PROJECT MANAGEMENT

As can be seen in Table 1.1 (Chapter 1), the scope and size of projects can vary greatly from small management of change (MOC) to large capital projects. Most projects are unique, although some repetitive elements may be present in some projects. For example, a second essentially identical process unit can be added to an existing site. Nevertheless, each process unit is unique with a different location, different feedstock and product pipeline routing, etc. that may have unique technical and process safety implications. Another difference between repetitive projects may involve the project organization with different team members, contractors and vendors.

The unique characteristics of projects require a systematic and disciplined application of good project management practices irrespective of the type or size of project. All projects have a structure and an execution plan, and involve some form and degree of risk that needs to be understood and minimized where possible.

A key project management responsibility is balancing the competing project constraints, which include, but are not limited to:

- Scope
- Quality
- Schedule
- Budget
- Resources, and
- Risks (PMI 2013).

Moreover, the client is likely to impose requirements for process safety, EHS, regulatory compliance, and stakeholder outreach that may add further constraints. The process safety risks, in particular, require careful management to reduce the residual risk, that Operations will have to manage, to a level that meets corporate tolerance criteria.

If any of these factors change during the project, there will be a knock-on effect impacting another factor(s). For example, schedule changes can increase (or decrease) resource and budget requirements. For this reason, most projects implement a Change Management process (see Section 2.8.6 below) to control change, especially scope creep, i.e. uncontrolled or unapproved expansion of the project scope that can occur gradually without adjustments to time, cost, and resources. This can also add safety risks that will need to be identified, assessed, and managed.

Scope Creep

Uncontrolled changes or continuous growth in a project's scope

---

Any changes in scope once detailed engineering has commenced should be resisted and challenged. Only those approved by appropriate senior management should be allowed. The effect of scope changes on process safety needs evaluation for potentially adding risk, and should be subject to review at the next stage gate.

Project managers tend to measure project success in terms of cost and schedule (timeliness vs. percentage completion), although other factors such as quality and EHS performance may also feature. Nevertheless, process safety needs to be recognized as a critical discipline and incorporated into the scope, budget and schedule, including the time required to properly address actions as a result of process safety studies and activities.

Other activities undertaken by successful project managers involve aligning the project team and interaction with stakeholders. To align the project team, clearly understood objectives for all personnel should be developed, and their commitment obtained to work toward those goals, so that each member is focused on the same set of project objectives (CII 2012). Stakeholders include project sponsor, corporate executives, corporate functions (e.g. process safety, EHS, engineering, etc.), the project team, Operations management, contractors, vendors, regulators, and local communities. They have various needs, concerns and expectations, require regular collaborative communications, and require managing in order to meet project requirements and deliverables (PMI 2013).

2.3   PROJECT GOVERNANCE

A senior executive(s) is likely to be the project sponsor or "client". The sponsor endorses the project objectives, and, if he/she is satisfied with the commercial viability, feasibility studies and implementation strategies, recommends the project for corporate sanction. The project governance process under which the final decision is made to proceed with the project will vary from one company to another, but is likely to comprise an oversight function and some internal financial and technical policies and standards.

Project Governance

Management framework within which project decisions are made

After the project is sanctioned, the project team is responsible to the sponsor for implementing the approved strategies for design, procurement, and construction. The project team will report various key performance indicators (KPIs) to the sponsor on a regular basis. These KPIs are likely to include cost, schedule, and process safety and EHS performance (i.e. injuries, spills/loss of containment, open action items from studies and reviews, etc.). The project governance process usually continues throughout the project life cycle. Approaching project completion, the new facilities will be progressively handed over to the Operator for acceptance and eventual operation when construction and commissioning are complete.

2.4   TYPES OF PROJECT

An example of the range of types of project covered by these guidelines was illustrated in Table 1.1 (Chapter 1). Each type of project will have its own characteristics and technical and process safety challenges. Some of the more significant types of projects are discussed below.

2.4.1   Greenfield Projects

Greenfield projects are, as the name implies, located in a completely undeveloped area, i.e. a "green field." While there will be few if any constraints on the project due to previous development, the challenges may include, but are not limited to:

- Limited infrastructure, such as roads, rail, utilities, emergency services, hospitals, etc.
- Limited local workforce, support services, and logistics
- Limited accommodation for first workers / construction workforce
- Green preservation and environmental footprint
- Potential decommissioning constraints (i.e. revert to greenfield)
- Acceptance by local community, if any

In addition, the new facility will need to establish a complete management system of policies, standards and procedures prior to commissioning. This management system will need to address financial, human resources, legal, technical/engineering, and EHS as well as the elements of process safety.

2.4.2   Brownfield Projects

Brownfield projects are different to greenfield in that the location has had some development. There may be existing facilities and buildings on previously cleared land, which may be operated by the company, partner, or others, such as a chemical park. Brownfield could also apply to offshore projects as well as to an expansion, revamp or upgrade on an existing facility.

The challenges of a greenfield project above may actually be the opposite in the case of brownfield. For example, there is likely to be existing infrastructure and local workforce and services. The challenges of a brownfield project may include, but are not limited to, several constraints:

- Limited area for the new project
- Proximity of local community, neighboring facilities
- Demolition of old facilities (see Section 2.4.5 below)
- Interface of new components and existing facilities (e.g. different standards, technology)
- Locating engineering documentation, including process safety information
- Locating underground utilities
- Upgrades to existing utility and firewater systems
- 'Hot tap' tie-in to existing process and utility piping
- Simultaneous operations (e.g. construction, operations, maintenance, drilling)
- Disruption to construction and/or operation of existing facilities (e.g. full or partial shutdown)
- Adjusting scope to existing environmental, community or legal requirements

2.4.3   Retrofit / Expansion Projects

Retrofit, expansion, debottlenecking, upgrade, optimization, and revamp are types of brownfield projects, and share many of the same challenges. Some projects may merely replace or update an existing facility while maintaining production at the existing capacity.

As an example, the upstream oil and gas industry has a high demand for these types of project. A new production facility may maintain a plateau of peak production for a few years, but eventually recovery rates decline. Major projects, such as gas re-injection and/or water injection, can stabilize or even reverse declining production. Other retro-fit and revamp projects may be required to handle increased volumes of produced water or structural strengthening of aging facilities due to the corrosive nature of the offshore environment.

Such projects can introduce the challenge of new or greater process safety hazards to be managed. Inherently safer design (ISD) approaches can sometimes be limited by brownfield challenges (see 2.4.2 above) and may require the compromise of a combination of engineering and greater reliance on procedural measures. This will put more responsibility on Operations leadership to maintain robust process safety management and strong operating discipline.

2.4.4   Control System Upgrade Projects

A control system and/or safety system upgrade project is a sub-set of a brownfield project, and again shares the same challenges. One of the drivers for a control system upgrade can be advances in technology that make existing hardware obsolete, and within 10-20 years result in difficulty obtaining spares. While the main focus may be on upgrading the HMI, controllers, and I/O, other challenges can involve interface connections with existing field equipment and wiring, network infrastructure and connectivity, ancilliary systems (e.g. power and UPS) and space requirements, and out-of-date documentation for the existing system. From a process safety perspective, control system projects are likely to have significant training requirements for both Operations and Maintenance personnel.

2.4.5   Demolition Projects

Demolition or deconstruction is often required as part of a brownfield project or it may be a standalone project (see Chapter 11). Typical challenges involve, but are not limited to:

- Proximity of neighboring facilities and buildings may require dismantling and prohibit toppling/explosives
- Deconstruction of some equipment for future re-use
- Partial decommissioning of operating facility
- Presence of asbestos and PCBs in older facilities
- Simultaneous operations with adjacent facilities
- Vibration may affect adjacent operations
- Underground cables and piping, and sewers
    - Location unknown
    - Connect with other adjacent facilities
- Environmental remediation

Robust process safety and EHS plans and procedures are required, especially hazard identification and safe work practices. Most clients are unfamiliar with management of this type of project, so contractor selection and oversight is important to ensure the appropriate competencies and behaviors.

2.4.6   Management of Change Projects

While large projects are invariably managed using a capital project governance approach, management of change (MOC) projects tend to be relatively small, and are a sub-set of brownfield projects. These smaller modification projects may not follow the stages of a capital project, but are managed under local MOC procedures. Nevertheless, each MOC project should address similar objectives of identifying hazards, assessing risks associated with those hazards, and managing the risks to prevent and/or mitigate process safety and EHS incidents.

Some modification projects may be initiated by the site maintenance department, for example when replacement-in-kind original equipment manufacturers (OEM) equipment and spare parts are not readily available. It is important that these projects are subject to MOC procedures to ensure that any hazards are properly managed.

Further information and guidance is available from the following CCPS publication: *Guidelines for Management of Change for Process Safety* (CCPS 2008c).

2.4.7 Mothballing Projects

Facilities that are temporarily shutdown for a period of time and require some form of preservation are often referred to as *mothballed*. The main challenge is preventing deterioration so that the facility may be put back into production or the project completed at a later date. Preservation techniques will depend upon the characteristics of the facility, such as type of equipment, metallurgy, and local environment. Common practices include periodically rotating motors, capping vents/flares, maintaining nitrogen blankets, coating or filling machinery with oil, and use of desiccants/biocides. However, a multi-discipline project team including process safety should determine the appropriate preservation measures to ensure asset integrity in consultation with the OEM.

Further information and guidance on asset integrity of mothballed facilities is available from the following publications: *Guidelines for Asset Integrity Management (CCPS 2017a); Guidelines for Mothballing of Process Plants (MTI 1989).*

2.4.8 Re-Commissioning Projects

Re-commissioning of a mothballed facility will to some extent depend upon how long the facility has been mothballed and how meticulous the prevention measures were maintained. If the facility has been shutdown for longer than a few months, it is likely that a multi-discipline project team will be required to inspect and test equipment to determine its integrity. For even short-term shutdowns, a team may be required to reverse any preservation measures taken. An operations readiness review should be conducted irrespective of the length of shutdown.

2.4.9 Restarting a Project

It may be expedient to stop a project at a certain stage of development due to commercial reasons, such as a significant drop in market prices for a product. In these circumstances, the project may restart after a year or two when market conditions improve. The main issues to be managed in a project restart are a potential loss of continuity in terms of team members (and possibly project manager), document control, decision-making history, and changes in technology. The PMT should consider these issues if and when a project is temporarily stopped. There may also be concerns if there is pressure to compress the project schedule to take advantage of favorable market conditions. In addition to competency and continuity issues, the project plan, including plans for process safety and EHS, should be updated. Any changes to the SOR or BOD will also require new HIRA studies.

2.4.10   Post-Incident Projects

Rebuilding a facility after a major incident, such as a fire, explosion, flood or hurricane, represents a special case. There may be considerable urgency to re-establish production due to commercial pressures as the company is likely to have commitments to supply customers. In these circumstances, several of the good practices (e.g. ISD) described in these guidelines may be deliberately omitted. The company may decide to merely copy the original design specifications or procure whatever materials are available on short delivery even if the specification is not identical to the original plant. The company may also sole-source construction contracts to expedite the rebuild. Nevertheless, the company should carefully evaluate and manage the risks associated with rebuilding the facility (i.e. demolition and construction risks), and incorporate findings from the incident investigation into the rebuild.

2.5   PROJECT ORGANIZATION

Each stage of the project life cycle will have a project team that includes a project manager and a multi-disciplinary group of individuals who perform the work necessary to achieve the project's objectives. As the project definition and execution progresses, the size, structure and organization of the project team will continuously change throughout the life cycle.

Some project characteristics are common, such as finite life, multi-disciplinary teams, a progressive environment, and the requirement to control cost and schedule. Nevertheless, each project is somewhat unique, and therefore project organization is likely to vary from one project team to another. For example, the complexity of the project will determine the size of the project team, and the project BOD will require specific technical expertise. The project's strategy for engineering, procurement and construction will also influence the organization. However, some common approaches are described below.

2.5.1   Pre-Project Team

A corporate pre-project team may conduct the feasibility studies and develop the concept options in FEL-1, and possibly FEL-2 also. If so, the team is likely to be a small, highly experienced multi-disciplinary group. Ideally the pre-project team includes a process safety specialist. If not, the team should be able to access and involve the appropriate expertise, such as process safety. Without this specialist input to apply inherently safer design principles early, the pre-project team is unlikely to identify the optimum development options, which could ultimately impact the cost and residual risk of the project. At the conclusion of the pre-project stage(s), the team hands over further development to a project team under the leadership of a project manager.

2.5.2   Typical Project Team

A typical project organigram is shown in Figure 2.1. This simplified diagram illustrates some of the basic principles. The basic structure identifies a number of managers of sub-teams in support of the project manager. The roles and responsibilities of each sub-team should be clearly understood to facilitate good teamwork with minimal possibility for conflict to arise between sub-teams. The roles and responsibilities of each sub-team are discussed below in Section 2.5.7.

Figure 2.1 Simplified Diagram for Typical Project Organization

In a small project, the project team may directly manage all tasks, whereas in a large capital project, contractors are likely to be employed to carry out some or all of the work scope. If contractors are employed, they will be contractually responsible for performing their scope of work under the control of a corporate management team, sometimes referred to as the Project Management Team (PMT) or Client Project Team (CPT). Selection and management of contractors should incorporate CCPS guidance: *Guidelines for Risk Based Process Safety* (CCPS 2007b).

While the format of the PMT may be similar to the organization in Figure 2.1, their roles and responsibilities will be significantly influenced by the contracting strategy. The process safety discipline may be stand alone, or merged with design engineering or EHS. Other management branches may be relevant such as procurement, quality, and contracts depending upon whether there is a direct project involvement or outsourcing. No matter how much the management team expands the basic project organization is followed leading to hierarchical structures requiring good links between groups.

Process safety may reside within the Design sub-team, within an engineering design contractor, and/or a corporate function. It may also be contracted out to a specialist consultancy for services, such as facilitating a hazard and operability (HAZOP) study, quantitative risk analysis (QRA), facility siting study, and other process safety studies identified in Appendix A. There may also be a requirement to brief the project manager and his team on the fundamental principles of process safety, and relevant corporate policies, standards and practices.

2.5.3   Unit Based Team

Within the overall simplified project organization in Figure 2.1, the engineering design, procurement, and/or construction functions of a large capital project may be divided on a unit basis between different contractors. For example, Contractor A may be licensor for a specific process technology, while Contractor B may specialize in cross-country pipelines or co-generation power plants. The client company may also be competent to handle part of the overall work scope, such as debottlenecking an existing process unit. Each contractor may be required to have a process safety subject matter expert (SME).

2.5.4   Equipment Based Team

Another approach sometimes favored for smaller projects is to organize the project team on an equipment basis. In this case, different technical disciplines are responsible for the engineering design of specific equipment, such as pressure vessels, piping, rotating machinery, and control systems. However, a process engineer should develop the overall design with input from the other disciplines, including process safety.

2.5.5   Site Based Team

It is possible that a very large project may have a number of work sites and facilities, in which case the project may be handled by several design and construction teams working in parallel. For example, an offshore project in a greenfield location may comprise one or more offshore production platforms, sub-sea pipelines, and an onshore terminal. The project organization is then expanded to reflect multiple parallel teams, but still follows the same basic principles. Process safety SMEs are likely to be required at each site.

2.5.6   Small Projects

While the characteristics and principles of organization are similar for small projects, some or all of the necessary multi-disciplinary support, including process safety, may be on a part-time ad-hoc basis from elsewhere within the corporate organization or on contract. Nevertheless, the project manager has overall responsibility for delivery of the project within cost and schedule.

2.5.7 Roles and Responsibilities

Organizational capability is essential for the success of any project. Adequate resources alone are insufficient to meet the project schedule, **and** ensure a safe, reliable, and efficient operation when the facilities are handed over to Operations. The project human resources must possess the relevant technical and administrative competencies, and have and understand clear roles and responsibilities. A common practice is the development of a set of Project Co-ordination Procedures that clearly state the duties, responsibilities, authority, and reporting relationships (including a RACI chart) within the project team. The roles and responsibilities of key project personnel are discussed below.

Project Manager

The Project Manager's (PM) fundamental responsibility is to complete the project to an agreed specification within budget and schedule. The PM is ultimately accountable for all technical aspects of the specification, including process safety. In this respect, the PM should have a basic level of understanding of many technical disciplines, and a fundamental knowledge of process safety and its importance is imperative. To meet these objectives, the PM must be supported by a task force style team that is strongly goal orientated and capable of dealing with all aspects of the project. Therefore one of the first responsibilities for the PM is to appoint competent (knowledge, skill, experience) personnel in all of the key positions within the organization. Given the progressive environment of projects, the PM will need to adjust his team to ensure the best match of competencies to the project stage. He or she will also need to manage team dynamics by motivating personnel and ensuring cross functionality between different disciplines and sub-teams.

Another key responsibility for the PM is interface management. Most projects have a large number of stakeholders and interfaces both internally and externally. Some of the most important interfaces are with the contractor(s) and any sub-contractors, where the PM needs to be seen to be in control of contractors' management. The PM should monitor each contractor's performance, and influence remedial actions if performance is unsatisfactory. Another important interface is with the project sponsor, to whom the PM should report progress on a regular basis.

To fulfill these and other responsibilities, the PM should possess a number of important interpersonal skills (see Table 2.1).

**Table 2.1. Important Interpersonal Skills for Project Managers
(PMI 2013)**

Interpersonal Skill
Leadership
Team Building
Motivation
Communication
Influencing
Decision Making
Political and Cultural Awareness
Negotiation
Trust Building
Conflict Management
Coaching

Sometimes it is necessary to change the project manager at an intermediate stage of the project. This is quite common when a pre-project team conducts feasibility studies to decide whether a commercial project exists. In these circumstances, the PM may not be appointed until a formal project is established. If a new PM is required to assume responsibility for a project, the transition should be regarded as another risk that needs to be managed.

Project Management Team

The staff that make up the Project Management Team (PMT) are shown in Figure 2.1. These managers have key roles in support of the PM, who may wish to appoint staff that he has worked with before to achieve a good mix of skills and experience. Depending on the scope of the project and its execution strategy, several other sub-team managers (e.g. procurement, quality, contracts) may also be on the PMT.

Normally the PMT is organized on a functional basis, and each manager is responsible to the PM for his/her function's deliverables at each stage of the project. For example, the Design Manager (a.k.a. Engineering Manager) is responsible for the basic engineering of the project, including, but not limited to, managing engineering risk, technical integrity, design safety, compliance with local regulations and industry standards, change management, and engineering documentation (including process safety information).

The PMT may also be responsible for evaluating any deviation from established engineering codes and standards.

Technical Staff

The required technical staff should be based upon the characteristics of the project. For example, a control system upgrade will require instrument/control and electrical engineering with input from a process engineer and Operations. Other disciplines that most, if not all, projects require are process safety, EHS, procurement, and quality management.

The responsibilities of the technical staff are generally to contribute to the project deliverables within their function. This may involve a combination of establishing engineering codes and standards, technical studies, design activities such as calculations, technical reviews, and ad-hoc input.

Contractors

If some or all tasks are contracted out, the responsibilities and interfaces between the various contractors must be effectively defined at the outset. Each contract should clearly state the deliverables, but Project Co-ordination Procedures may add further clarity. This is vitally important because the client and contractor have differing objectives. For example, the client wants a facility that operates as intended, on/before schedule, and below budget; whereas the contractor wants to win the contract and then assess cost, maximize profit but win repeat work, and minimize responsibility and risk. Various contracting strategies are discussed in Section 2.6.

Support Staff

The complexity of the project is likely to determine the sophistication of the administrative and control systems. Key functions for most projects are project controls (i.e. planning and cost control) that provide essential information to the PM on a regular basis. In particular, they have a responsibility to identify and advise any deviations from the intended plan. This allows the PM to minimize the disruptive effects of change, such as delayed delivery of procured equipment and materials, or a sudden rise in global commodity prices (e.g. steel, copper, etc.).

Operations

An Operations Manager (OM) or representative(s) plays an important role interfacing with the Design Manager's team to make sure operability is fully supported in the design. The OM should be appointed at an early stage of the project to ensure that a good engineering design is not difficult to operate, as it will be expensive to change the design later.

Key responsibilities for the OM are developing an operations and maintenance strategy/philosophy for the project, providing input on operational lessons learned, and preparing the Operations team for handover. In addition to participating in hazard and risk studies, and design reviews (e.g. P&ID, model, stage gate), Operations should also provide ad-hoc input throughout the project life cycle.

Partners

A common strategy for companies to share the cost and risks associated with a project, especially a very large project, is to jointly pursue the development with one or more partners. Equity interests in joint ventures (JV) may vary between the partners. One partner may be responsible for managing the engineering design, procurement, construction, and the same or different partner will operate the completed JV after handover.

While the responsibilities of the managing partner (Operator) are similar to the concepts presented in this chapter, the responsibilities of the other partner(s) are somewhat different. Companies that have an equity interest in a non-operated project are a co-sponsor, and may expect to influence performance and manage relationships with the Operator. Oversight should be strong enough to protect the company's investment. A few key personnel may even be seconded to the project team and/or participate in stage gate reviews.

2.6   STRATEGIES FOR IMPLEMENTATION

There are a variety of strategies for the design and construction of projects, and some situations will favor the use of in-house resources as opposed to using contractors. The choice is governed primarily by a question of resources. Most companies do not maintain a large cadre of in-house resources capable of engineering design and construction as project workloads tend to fluctuate. Other factors that influence the decision may include cost (e.g. contractors may be cheaper) and expertise (e.g. contractors may have specialized expertise from previous experience). Nevertheless, small projects are often conducted in-house or with limited contractor support. For example, a large chemical or oil company may develop all the engineering including FEL for smaller projects (i.e. <$200 million). However, in the case of a large project, such as an ethylene cracker, where the company does not have the resources or technology expertise, the company is likely to contract with an experienced technology supplier. In that case the company's project role will focus on oversight.

Before determining the strategy for project implementation, it is important to understand the different objectives of the client company vs. the contractor(s). The client requires: (i) the new project to function as defined in the SOR; (ii) completion on or ahead of schedule; and (iii) minimal cost (i.e. below budget) with an acceptable balance between capital expenditure (CAPEX) and operating expenditure (OPEX). Whereas the contractor's objectives are: (i) to win the contract and then determine the value/man-hours, etc.; (ii) maximize profit while maintaining reputation to win repeat business; and (iii) minimize responsibility and risk exposure. These differing objectives are obviously in conflict and can drive adverse behaviors if not addressed up front. It is important to define roles, responsibilities, rights and risks, and establish a relationship that is fair to both parties when making contractual agreements. Finally, a means of overcoming the conflicting objectives should be found in order to successfully motivate the contractor to work to the benefit of the client.

Typical language in a services contract may include text such as the following:

"...The Company expects the Contractor to provide a facility design that meets the requirements of the Basis of Design and can be safely started, operated and shut down. The Contractor shall provide a facility design that meets Company's qualitative and quantitative risk tolerance criteria (Exhibit XX). After subjecting the design to hazard identification / risk analysis (HIRA) studies, any elevated (i.e. unacceptable, intolerable) risks shall have mitigation measures implemented (study action items / recommendations) that move the risks to a managed (i.e. acceptable, tolerable) level as per Company's risk management process and procedures (Exhibit XX). The mitigation measures proposed to manage any elevated risk issue shall follow an inherently safer design strategy by implementing a risk mitigation / control hierarchy that first considers inherently safer design options before engineering controls, and engineering controls before administrative controls (Exhibit XX). Any elevated risks that remain prior to start up shall be communicated and subject to approval / recycle by the Company ..."

It may be difficult for the contractor to understand and apply the client's risk tolerance criteria, but regular client oversight of the HIRA, ISD, and DHM studies, and participation in engineering design reviews and stage gate reviews can ensure that risks are reduced in line with the client's tolerance criteria. One of the exhibits should include details of the inherently safer design / risk control hierarchy strategy that is likely documented in the risk management program procedures, but is worth highlighting at the highest level in the contract to avoid any later misunderstanding of responsibilities.

Different contracting strategies involve varying degrees of risk and control between the client and the contractor(s). For example, a fast track project with a compressed development and execution timeline may limit application of inherently safer design principles. Division of work scope between multiple contractors is a common method of spreading project risk (see Section 2.7). The final decision on contracting strategy is likely to be based upon a combination of the project objectives, constraints, preferred delivery methods, contract form/type, and the client's contract administrative practices. Table 2.2 indicates the impact of three common contracting strategies; reimbursable, lump sum and turnkey.

A reimbursable contract is a contract where a contractor is paid for all of its allowed expenses to a set limit, plus additional payment to allow for a profit. This type of contract requires the lowest up-front definition of services, gives the client the greatest control and hence a potentially positive impact on controlling process safety activities, but has the highest financial risk exposure for the client.

A lump sum contract is a contract where a contractor is responsible for completing the project within the agreed fixed cost set forth in the contract. This type of contract compared to a reimbursable contract requires more up-front definition of services, gives the client less control and hence a potentially intermediate impact on controlling process safety activities, but has less financial risk exposure. Lump sum contracts are often awarded to a contractor for a single project stage, such as engineering or construction.

Table 2.2. Impact of Contracting Strategy

Type of Contract	Definition Necessary	Risk Client	Exposure Contractor	Client Control	Impact on Process Safety
Reimbursable	Lowest	Highest	Lowest	Highest	Positive
Lump Sum	Intermediate	Intermediate	Intermediate	Intermediate	Intermediate
Turnkey	Highest	Lowest	Highest	Lowest	Negative

In a lump sum contract, a contractor may consider HIRA recommendations as potential change orders, but, if the above suggested language is in the contract, the client can counter by stating that the contractor has not met the requirements of the contract and must provide a design that does.

For both lump sum and reimbursable contracts, the design "freeze" point should be clearly determined. This is typically during detailed design after the design hazard management (DHM) process (see Chapter 6 Section 6.1) has identified and mitigated elevated risks through design safety measures to a managed level. At this point the change management process (see Section 2.8.6 below) should be fully implemented to discourage change and establish a high hurdle for any change to even be considered.

A turnkey contract is a contract where a contractor completes a project, then hands it over in fully operational form to the client, i.e. the client need to do nothing but 'turn a key' to commission/startup. This type of contract requires the highest up-front definition of services, gives the client the least control and hence a potentially negative impact on controlling process safety activities, but has the least financial risk exposure for the client. Turnkey contracts are often awarded to a contractor for multiple project stages, such as engineering, procurement and construction (EPC), and sometimes even include commissioning and startup prior to handover to the client.

For turnkey contracts, it is essential that the process safety competency, procedures, and practices of the contractor are evaluated and agreed as part of the contractor selection process (see Section 2.6.1 below).

The contracting strategy is normally documented in the Project Execution Plan (PEP) (see Section 2.1.5 above) together with applicable oversight monitoring and control practices. A number of contractual arrangements are discussed below.

2.6.1 Contractor Selection

Once a strategy to employ contractors has been decided, there are three main options for selecting contractors, as follows:

- Competitive tendering
- Extension of an existing contract
- Single contractor

Each option has its advantages and disadvantages, but may be appropriate under certain circumstances. Most contracts for major capital projects are placed following competitive tendering. Smaller projects, especially operations and maintenance projects where continuity of services and personnel are important, may elect to extend an existing contract where feasible. Negotiations with a single contractor are more likely in an emergency where immediate mobilization is required, where secrecy is needed, or where the contractor is sole source for a particular technology or expertise.

Competitive tendering requires extensive preparatory work and time to allow prospective contractors to understand the scope of work and then submit their tenders. Further time is required to assess each tender and seek clarification where necessary before awarding the contract. The use of selective tendering or prequalification may be used to reduce the assessment work where the client has previous knowledge of the contractors' resources, technical capability and financial status.

Cost is not the only criterion in determining the contract award. Many companies also assess and rank technical and EHS (including process safety if delegated to a contractor) capability between competing tenders. If the contractor is responsible for process safety, it is essential that the contractor's process safety competency and capability is thoroughly evaluated prior to contract award. Adherence to the client's process safety standards should also be written into the contract. The client and PMT should maintain close oversight of the contractor to ensure that process safety activities are performed properly.

Further information and guidance on contractor management is available from the following CCPS publication: *Guidelines for Risk Based Process Safety* (CCPS 2007b).

2.6.2   Engineering Only

The use of design contractors is fairly widespread within the process industry. The range of work varies from relatively simple modifications to existing facilities using local contractors to major capital projects employing international engineering contractors.

Feasibility studies, FEED studies, and cost estimates are invariably handled on a reimbursable contract basis. Detailed engineering design may be handled as a turnkey or reimbursable contract, although there are a number of variations between these two extremes. Turnkey requires a fully detailed scope of work and offers high contractor risk/low client risk, whereas reimbursable is the reverse.

The choice of design contractor is influenced by technology expertise, prior experience, organization (including key personnel), and cost. Some contractors have developed or acquired knowledge of specialist process technologies, and may offer key engineering personnel familiar with the technology. Where key personnel are essential to the success of the project, the client should seek restrictions on re-assignment of these personnel by the contractor.

2.6.3   Engineering and Procurement

It is common for design contractors to take responsibility for procuring equipment and materials associated with their design and specifications. This is especially true for long lead items, such as some process compressors. Alternatively, the client company may handle procurement in-house or employ an independent procurement contractor. Either way the client will reimburse actual order costs.

The client normally approves vendors, tender lists, bid approvals, and any amendments to purchase orders. The client may also seek assurance that the design contractor has sound, auditable management systems for administration of commercial decisions, purchase orders, commitments, and timely provision of vendor documentation.

2.6.4   Engineering, Procurement and Construction

Major capital projects often employ an engineering, procurement and construction (EPC) contracting strategy, where the contractor is responsible for all three EPC functions. In some cases the client may also outsource project management (i.e. EPCM strategy). Some major engineering contractors may have full EPC capability in-house, employ sub-contractors, and/or employ individuals with relevant expertise on contract (i.e. act as "body shop"). Some modification projects may also adopt an EPC strategy, particularly if the modification is complex or requires certain skills that the client does not have in-house. Nevertheless, it is essential that process safety is not compromised. Therefore, EPC contractors should have in-house capability or subcontract adequate process safety capability for the duration of the project.

Construction of simpler projects and modifications may be undertaken by the client's own engineering and maintenance teams. However, if skills such as piling and civil engineering are required, the client may out-source only these works, while handling the mechanical/electrical/control work directly.

Whether a full EPC or piece-meal strategy to construction is adopted, management of construction activities needs to address overall planning, material control, quality management, management of sub-contractor interfaces, industrial relations, occupational safety, environmental issues, late design changes, and contract administration. The client normally expects to approve all sub-contractors either through a pre-approved list or later when selected by the main construction contractor.

Contractor selection is influenced by relevant experience; construction planning capability; provision of experienced management, supervision, site personnel, and construction equipment; quality control; and financial stability. Clients usually prefer lump sum construction contracts, although frequently have to settle for a "bill of quantities" approach with a fixed fee for profit and overheads.

2.6.5   Operation

Most client companies operate their new facilities upon project completion, although specialist assistance from the design contractor, technology licensor, and vendors may be required during commissioning and early operation. In certain circumstances the client may out-source facility operation to a third party, who on occasion may also be the EPC contractor or a contractor responsible for a portion of the overall scope of work.

The most common out-sourcing contracts cover the operation and maintenance of utility systems, such as cogen and wastewater treatment plants, and other supporting infrastructure. In these circumstances the contractor assumes responsibility for technical and/or commercial operation of the facilities as a service to the client, who retains ownership of the facilities. The contractor often provides similar services to other clients, and therefore has the strength in depth and comprehensive technical knowledge of operations that may not be core to the client's main business. Most contracts are based on a fixed fee for profit and overheads.

While not as common, there is an increasing trend for some companies in the process industry to out-source operation of process plants. The scope of this out-sourcing may cover operation and maintenance of the new facilities or may be restricted to only maintenance and/or warehousing. The contractor may also assume responsibility for recruitment of plant management and the workforce.

2.6.6   Contractor Oversight

Whichever contracting strategy is adopted, it is important to clearly define the scope of work and the roles and responsibilities of the client and the contractor(s). It is also important for the success of the project to develop a good working relationship between the client's overall project management team (PMT) and the contractor(s). Nevertheless, the client's PMT should perform a level of oversight commensurate with the scale of the project to hold contractors accountable for supplying the services and meeting the specifications agreed to under the terms of the contract(s).

As noted above, the client through the PMT should maintain close oversight of the HIRA, ISD, and Design Hazard Management (DHM) studies to ensure that risks are reduced so that they meet the client's risk tolerance criteria. Oversight of other design and procurement activities can be achieved through audits and random checks of documents and calculations. Another significant concern for the client is the continuity of key contract personnel (e.g. project management and technology expertise) essential to the success of the project, and the client should monitor any contractor staff changes.

Although a formal contract may not exist, the same philosophy should be applied to internal service providers, such as a client's engineering function, if they are conducting feasibility studies and/or design work.

An important consideration for construction contractors and sub-contractors is their safety performance. While the contract companies have the responsibility to monitor the actions of their employees and to enforce appropriate safety requirements, the client generally has the ultimate responsibility for ensuring the safety of the worksite. Oversight can be achieved by random, unannounced inspections of contractor activities to monitor adherence to safe work practices, and other safety procedures and safe working conditions. Occasional formal audits should also assess documentation, such as training and qualification records.

Finally, all identified concerns from contractor oversight and monitoring activities should be brought to the attention of the contractor(s) as soon as possible, and a satisfactory resolution agreed. If agreement is not possible, the client should consider dismissing the contractor and terminating the contract.

2.7   RISK MANAGEMENT

Projects can face a diverse set of risks and uncertainties covering political, geographic, markets/commercial, economic, regulatory, technical, security, and cultural issues, in addition to project definition/execution, operational, EHS and process safety risks.

Project Risk

An event or set of circumstances that, should it occur, would have a material effect, positive or negative, on the final value of the project.

Another definition of project risk is "an uncertain event or condition that has a positive or negative effect on a project's objectives" (PMI). As can be seen, unlike safety risks, project risks can be positive as well as negative. Project risks that have a negative impact are generally referred to as threats, while those with a positive impact are called opportunities. All of these risks need to be identified, evaluated, and managed throughout the project life cycle. This can be a complex task requiring assessments that fully investigate potential impacts and their likelihood of occurrence. Action plans to mitigate risks should be established in FEL-1, and then periodically reviewed and updated as the project progresses. Project risks are usually managed at two levels, i.e. discipline level and project level.

At the discipline level, project engineers use established methodologies to undertake detailed risk assessments. Each discipline generally has its tools, standards, and means of communicating risk. For example, process safety engineers frequently use a combination of HAZID, HAZOP, QRA, LOPA, etc. to minimize risks from fire, explosion and toxic hazards, as will be described in detail in later chapters. These activities should be specified in the process safety plan. Whichever tools are used, risks should be assessed early in the project life cycle, and revisited later to review and update findings. The project scope evolves, and may change, during early design stages, and risks that initially appeared acceptable may become significant. For example, initial layout may be satisfactory until more congestion occurs during design that poses a serious blast overpressure risk to a control room. The ultimate goal should be to reduce residual risks to a tolerable level that Operations personnel will have to manage throughout the life of the completed facility.

Project Risk Assessment (PRA) is a process used on capital projects (but can be applied to smaller modification projects) to identify, evaluate, and manage the key risks and uncertainties at the project level. A risk can be either a threat or an opportunity; e.g. a 4-month delay in receiving regulatory approval is a threat if it stretches to 6 months, or it can be an opportunity if approval can be achieved within 2 months. PRA requires strong project management commitment to managing risks holistically, and providing clear roles and responsibilities within the project team to implement PRA. It requires a systematic, documented approach that is adequately resourced, and typically follows a cycle, such as that illustrated in Figure 2.2.

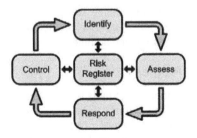

Figure 2.2 Risk Assessment Cycle

The risk assessment cycle in Figure 2.2 involves four steps: identify the risks, assess the risks, respond to the risks, and control the risks. This process should be conducted using holistic techniques to avoid surprises later, and with rigor to reliably prioritize all risks. A combination of checklists and brainstorming by experienced managers, engineers and operations personnel is commonly employed, but also progressively draws on input from detailed assessments at the discipline level. An "owner" should be designated for each significant risk, and is responsible for ensuring effective action plans that are adequately resourced to manage the risk. Given the complexity of risks that capital projects face, a means of regularly tracking the progress of action plans for each risk, and communicating their status to the project team, is necessary for effective risk management. This is an iterative process with the cycle being performed at each stage of the project, and is often embedded within the project's work processes.

An increasingly common approach to documenting project risks is the use of a risk register (Appendix C). They can vary widely in complexity, but generally are a spreadsheet or database tool that contains each identified risk, its description, ownership, assessment (impact, likelihood), and actions taken. The risk register is discussed in later chapters at each stage of the project life cycle, when risks are updated or new identified risks are added. Significant risks are often monetized to express financial impact on project value.

Most projects conduct some form of independent review at various milestones during the project life cycle. These reviews are variously known as "peer review/peer assist," "stage gate review," "cold eyes review," etc. They can vary widely in scope, but are primarily focused on risk management either at project level and/or discipline level.

Further information and guidance on PRA is available from the following publication: *Project Risk Analysis and Management Guide*, (APM 2004).

2.8   PROJECT CONTROLS

Most project managers employ a range of controls to manage the myriad of risks, uncertainties, actions, commitments, interfaces, and other requirements necessary for a successful project, i.e. a safe reliable facility that meets specification within budget and schedule. These controls cover planning, progress, estimates, budgets, cost control, reporting, accounting, and administration.

2.8.1   Planning and Progress

The primary role of the planning function is to carry out and coordinate work in an orderly, efficient manner. This requires tasks, work scopes, and resources to be identified and sequenced in a logical order to reduce time and cost. Progress can then be regularly measured against target dates and slippages analyzed. Plans are often developed at two or more levels of detail, using Gantt charts, logic networks or other sophisticated planning tools. For example, during the construction stage of the project, detailed plans may be broken down into each discipline or craft on a system or area basis to identify any possible improvements in productivity.

A project safety plan should be developed and periodically updated to identify the process safety activities that should be performed throughout the project life cycle. The process safety plan may be merged with the EHS plan, but the tasks within both plans should be included within the overall plan for the project. Appendix A illustrates the process safety activities that, depending on the project scope, may be applicable at each stage of the project life cycle.

2.8.2 Estimates, Budgets and Cost Control

A cost estimate is necessary for all projects to determine whether it is a viable commercial undertaking. Initial estimates may only be accurate to order of magnitude, but as the design develops, and the level of technical information is defined, increasingly accurate estimates are possible. At project sanction approval, the estimate accuracy is typically better than plus/minus 20%, and this becomes the project budget. When detailed design is substantially complete, and firm prices from purchase orders and contracts are available, accuracies of better than plus/minus 15% are possible, allowing for contingencies due to market conditions, currency fluctuations, weather delays, late scope changes, etc.. A typical approach to cost control is to break down the project into manageable components, each further broken down into work packs that mirror the tasks in the project plans.

Projects require an accounting function to pay invoices for goods and services received, and to monitor expenditure vs. the sanctioned budget. While turnkey and lump sum contracts are fixed price, change orders and other reimbursable contracts need to be closely monitored. Regular cost reports show expenditure, commitments, and alert project management to any budget overruns, enabling corrective actions to be taken.

2.8.3 Reporting

Regular project reports inform the client organization of progress, key events and statistics, and future targets. Key items in reports received from contractors and major suppliers may be included in client reports. Where partners are involved in the project, the project manager is likely to regularly brief the partner's technical representatives.

2.8.4 Metrics

In addition to metrics for cost and schedule mentioned above, projects may be required to collect and report other performance indicators, such as local employment vs. expatriates, local contracts for goods and services, process safety, injury, illness and environmental statistics, and other client corporate requirements.

Some projects may also benchmark their performance against recognized leaders (other internal or external projects) for the purpose of determining best practices that lead to superior performance when adapted and utilized.

2.8.5   Action Tracking

Most projects generate a large number of recommendations from not only risk assessments but also multiple different technical, commercial, security, and EHS studies, to which must be added actions and commitments arising from regulatory, planning approval, and other negotiations and meetings. As a result, the project will need a system to prioritize, document decisions, and track progress of each recommendation and action.

2.8.6   Change Management

A system for managing change is vitally important to most projects. During FEL as the design evolves from multiple options to a single concept, many changes are likely to occur. Change management in a project context is a process to evaluate and control changes to the project's scope, design, cost, schedule, etc., whereas management of change (MOC) is generally applied to changes in chemicals, equipment, procedures, and organization of an existing facility. Nevertheless, the basic principles of MOC, i.e. hazard/risk evaluation, technical review, and formal approval, apply equally to change management. Change management is normally implemented after completion of the HIRA studies and DHM process, when changes need to be tightly controlled.

Change Management

The process of incorporating a balanced change culture of recognition, planning, and evaluation of project changes in an organization to effectively manage project changes. These changes include: scope, error, design development, estimate adjustments, schedule adjustment, changed condition, elective, or required.

(from CII)

A scope change almost always requires adjustments to the project budget and/or schedule, and is more likely to result in higher cost and delays to project completion. Project managers need to be alert to any uncontrolled expansion of scope, i.e. scope creep, without the formal agreement of the client. Many projects freeze the design at the final hazard/risk evaluation, and any changes thereafter require a formal approval process involving evaluation of technical justification, hazard/risk, and cost impact. This applies to all disciplines and financial changes. Some projects even introduce a philosophy of transition to *management of no change* as the project moves into execution.

Further information and guidance on management of change is available from the following CCPS publications: *Guidelines for Risk Based Process Safety*, (CCPS 2007b); *Guidelines for the Management of Change for Process Safety* (CCPS 2008c).

2.9 OTHER CONSIDERATIONS

2.9.1 Materials Management

Materials account for a large percentage of a capital project's total cost, and a comprehensive materials management system can contribute to project success by reducing costs, and improving productivity and quality. Materials management involves supplier qualification, sourcing, purchasing, quality management (see below), expediting, transportation, logistics, and handling of equipment and other materials at the project construction site. Correct handling is important to ensure that materials received meet the procured specification, are properly stored, correctly preserved prior to and after installation, and correctly identified as visually similar materials may have different specifications. Documentation for all materials should be collated for inspection and quality management purposes, and final handover to Operations.

2.9.2 Quality Management

Projects require a quality management system (QMS) to set and deliver quality requirements for the complete supply chain from design through operation. The QMS ensures that quality characteristics are incorporated within design specifications, and effective processes are in place to ensure that procurement, fabrication, construction and handover deliver equipment and other materials that meet specification. Quality management is discussed in detail in Chapter 8.

2.9.3 Lessons Learned

Large operating companies that execute multiple projects normally conduct some form of post-project appraisal (PPA) in order to identify lessons learned and opportunities to improve future projects.

Lessons Learned

Knowledge gained from experience, successful or otherwise, for the purpose of improving future performance.

(from CII)

The PPA includes lessons (positive and negative) related to investment decision-making, technology, cost estimating, stakeholder relations, operational input, joint ventures, contractor management, DHM, and general project management. DHM lessons can have significant importance for process safety application to future projects. These lessons are variously compiled in spreadsheets, databases or booklets so that they are readily available to the PMT of future projects.

2.9.4   Post-Project Close-Out

Close-out activities should be considered and adequate planning performed early in the project schedule. This is especially relevant to brownfield developments, where tie-ins to existing plants may need to be scheduled for plant shut-down periods. Early planning for information (including process safety information (PSI)) required to be handed over to the future Operator upon project completion will also allow the PMT to organize their documentation system to facilitate the handover.

Other post-project close-out activities include:

- Coordination of punch listing and the operational readiness review (a.k.a. pre-startup safety review (PSSR)) with the future Operations team,
- Agreement with the client to assume financial responsibility and ownership of the completed facilities, and for the project to continue to provide technical responsibility for a limited timeframe,
- Coordination of test runs with the future Operations team,
- Agreement with the client to assume responsibility of future liabilities for the facilities, where there is a warranty for specific equipment,
- Conduct of the post-project appraisal (as described above) and any other audits required by the client or partners,
- Management of claims from contractors or vendors,
- Preserve and archive project documentation, records, and data for use by future projects,
- Progressive demobilization of project personnel, who may be seconded from within the client organization or contracted externally,
- Preparation of a project close-out report.

2.10 STAGE GATE REVIEWS

Many operating companies within the process industries conduct reviews at key milestones during the life cycle of capital projects. These reviews are variously known as stage gate reviews, 'cold eyes' reviews, peer reviews, project EHS reviews, etc., and are normally conducted by an independent and experienced multi-discipline team familiar with the relevant facility/process and technology.

The objectives, scope and extent of the reviews vary between companies, but typically focus on technical and EHS issues, relevant to the stage of the project development, with a strong emphasis on process safety. One of the primary goals is to assess whether the PMT has adequately identified and evaluated the hazards and their associated risks inherent in the project, and developed (or is capable of developing) plans to effectively manage those risks. The Project's process safety and EHS plans are a major input to the stage gate reviews. The stage gate review team may use a protocol and/or checklist to guide their assessment. An example protocol is in Appendix G.

If the review team identify that the PMT have not completely met their technical, EHS and process safety objectives (activities and deliverables) for the stage, they will make recommendations for any improvements needed. This may also include recommendations in respect of plans for future stages of the project. Successive reviews may also verify whether recommendations from preceding stages have been adequately resolved. Finally, the stage gate review team may recommend to the Gate Keeper whether the project is ready to proceed to the next stage, although the final decision rests with the Gate Keeper.

Additional reading

Rosentrater, G., *Manage Projects Effectively*, Chemical Engineering Progress, November 2001.

Rosentrater G., *Preliminary and Final Engineering Scopes of Work*, Chemical Engineering Progress, December 2001.

Rosentrater G., *Complete Your Capital Project Efficiently*, Chemical Engineering Progress, January 2015.

Walkup G.W., Ligon J.R., *The Good, the Bad, and the Ugly of the Stage Gate Project Management Process in the Oil and Gas Industry*, paper presented at the SPE Annual Technical Conference and Exhibition, San Antonio TX, 2006.

3 FRONT END LOADING 1

The first stage of any project comprises a range of feasibility studies to appraise the commercial and technical viability of a potential project. This Front End Loading (FEL) 1 stage is sometimes known as Appraise, Appraisal or Visualization. Figure 3.1 illustrates the position of FEL 1 in the project life cycle.

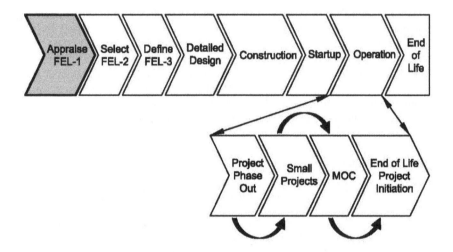

Figure 3.1. Front End Loading 1

As discussed in Chapter 2 (see Section 2.5.1), the early feasibility stages may be conducted by a pre-project team or a formal project team depending upon client company practices and confidence in the likelihood of the project proceeding. For the remainder of this book, the term 'project management team' is used.

Project Management Team

During the FEL-1 stage of a potential project, the project management team (PMT) develops a broad range of development options in line with the company's business strategy and objectives, and evaluates the commercial viability of the project. Each alternative option is assessed in terms of value, risk (threats and opportunities), and uncertainty. Key challenges involve the assessment of alternative technologies, processes, and locations. Examples of alternative options may include fractionation vs. absorption, batch vs. continuous processing, access to infrastructure, such as marine, road, rail, and/or pipelines, and proximity to communities and sensitive areas.

The most promising development options in terms of both technical and commercial viability should be identified in order for the project to proceed. An outline statement of requirements (SOR) for the project is often developed in sufficient detail to act as a basis for further developing and assessing the various options during FEL-2. The PMT should also assess each development option in terms of EHS and process safety, as discussed below.

Environment, Health and Safety

Besides technical and commercial objectives, the PMT needs to identify any significant environmental, health and safety (EHS) aspects of each alternative option that could impact the viability of the project. In particular, for development options in sensitive areas, an adequate risk management plan should be in place to protect company reputation. The PMT should determine whether the information required to assess EHS risks, liabilities, regulatory compliance, commitments, and adverse impacts is either available or measures are in place to obtain it. Finally, the PMT must consider whether the design and operation of each proposed option is capable of meeting corporate EHS policies and standards.

Process Safety

From a process safety perspective, the key objectives of the FEL-1 stage are to:

- Determine whether there are any potential process safety risks associated with the options being considered, such as novel technology or processes,
- Ensure that inherently safer design (ISD) principles are considered when developing each option, and
- Assess the proposed location(s) for any process safety issues, such as potential impacts on local communities, the environment, and other industry.

Other important process safety goals involve identifying any uncertainties with each development option, such as unknown chemical reactivity/stability or corrosivity, and any issues that might significantly influence process safety performance, e.g. project construction on a brownfield site involving heavy lifts over live equipment. Based on these objectives and goals, the main process safety input to this stage will comprise hazard identification and risk analysis (HIRA), which is discussed in detail below.

3.1   PRELIMINARY HAZARD IDENTIFICATION

A preliminary hazard identification (Prelim HAZID), a.k.a. preliminary hazard analysis (PreHA), should be conducted on each of the options being considered by the PMT. At this early stage of the project, only basic information will be available. For example, feed and product throughputs, locations, technologies, processes, concept layouts, existing infrastructure, and means of transportation of feed/products, which may be sufficient to identify other basic information such as process block diagrams, possibly generic PFDs, potential intermediates/by-products, waste streams, emissions, and estimated range of sizes for facility acreage, inventories, etc. Nevertheless, the information should be sufficient to conduct a preliminary HAZID.

Later, when the number of alternative options has been reduced, more detailed information should become available as the concept designs evolve. It may then be appropriate to update, as necessary, the initial preliminary HAZID.

The most popular methodologies for conducting preliminary HAZID studies are checklists and brainstorming, or a combination of the two. Most major, and some smaller, companies have their own internal checklists. Brainstorming should be conducted by a small experienced group with knowledge of the technologies, processes and process safety/loss prevention. HAZID studies are typically very broad in their scope, looking at all possible sources of major hazards to the project by examining each area/unit/module/system in turn. The HAZID should focus on potential impacts to people, environment and the facility.

Other methodologies that are used by some companies for preliminary hazard identification include Dow Fire/Explosion Index, Dow Chemical Index, Mond Index, and What If analysis.

A simplified and generic checklist of concerns and issues to consider in the HAZID is illustrated in Table 3.1.

It may also be appropriate to include potential occupational health and environmental impacts, such as major emissions, hazardous wastes, discharges (with treatment options), and natural resource use, in the HAZID unless separate studies are planned for these.

Further information and guidance on methodologies for hazard identification, including more detailed checklists, is available from the CCPS publication *Guidelines for Hazard Evaluation Procedures, 3rd edition* (CCPS 2008b).

Table 3.1. Simplified HAZID Checklist

Concern / Issue
Material properties (toxicity, flammability/explosivity, reactivity, corrosivity, etc.)
Process conditions (physical state, severity, pressure, temperature, thermodynamics, etc.)
Physical processes (failure modes, cranes, drilling, conveyors, refueling, etc.)
Environmental factors (earthquake, flood, wind, wave, hurricane, snow/ice, etc.)
Loss of containment scenarios (process, utilities, pipelines, tankage, blowout, etc.)
Structural failure (subsidence, scour, corrosion, fatigue, excess weight, etc.)
Loss of stability/buoyancy, mooring/anchor integrity (only floating facilities)
Escalation/domino scenarios
Third party impact (vehicles, aircraft, shipping, farming, excavation, dropped loads, etc.)
Location – impact to/from adjacent industry, community, workforce
Location – site terrain (nature, stability, etc.)
Raw material/product handling and transportation (road, rail, marine, pipeline, etc.)
Logistics (transport, supplies, spares, support services, emergency response/mutual aid, etc.)
Incident history of similar technology/process
Hazard evaluation of similar technology/process

3.2   PRELIMINARY INHERENTLY SAFER DESIGN REVIEW

An inherently safer design is one that avoids hazards instead of controlling them, particularly by reducing the amount of hazardous material, designing equipment for worst case conditions, and reducing the number of hazardous operations in the facility.

Inherently Safer Design

A way of thinking about the design of chemical processes and plants that focuses on the elimination or reduction of hazards, rather than on their management and control.

(CCPS, 2009)

The application of inherently safer design (ISD) is most effective at the earliest stages of a project. Although opportunities to apply ISD exist in later stages, there is likely to be less flexibility or a significant cost impact. Therefore Project should consider ISD principles when developing each of the alternative options.

The four ISD principles are:

- **Intensification / Minimize:** reduce the quantity of hazardous material (e.g. continuous stirred tank reactors are smaller than batch reactors for a given production rate).

- **Substitution / Substitute:** substitute hazardous material with less hazardous material (e.g. use hypochlorite for water treatment instead of chlorine).

- **Attenuation / Moderate:** use less hazardous conditions, less hazardous form of material, or facilities that minimize the impact of a release of hazardous material (e.g. dilution - using aqueous ammonia instead of anhydrous ammonia).

- **Simplification / Simplify:** design facilities that eliminate unnecessary complexity to make operating errors less likely, and that are more forgiving of errors that are made (e.g. avoid SIS by rating LP separator for upstream HP breakthrough).

Case Study: New Offshore Platform

The PMT for a small offshore gas field was challenged to create a commercially viable development option. Standard design practice at the time consisted of a manned platform with processing to separate condensate, two export pipelines for gas and condensate liquids, power generation, accommodation module, and helideck for crew change.

Various manned platform options were considered, but proved uneconomic. By following ISD principles, the PMT was able to develop a novel unmanned design operated remotely from the shore. By opting for a multi-phase export pipeline to a shore processing facility, the platform was simplified by removing process equipment, power generation (supplied by cable from shore), and accommodation. The helideck was also removed as boat access was possible for monthly maintenance visits scheduled during calm weather. Other measures included use of corrosion resistant alloy for topsides piping rated for well head shut-in pressure.

Limiting personnel presence, eliminating helicopter travel, and minimizing hazardous inventories substantially reduced safety risks.

These ISD principles are generally considered more reliable than other strategies or approaches to reduce risk. Other strategies involve the addition of passive, active and procedural risk reduction measures, and are more applicable for addressing residual risk during later detailed engineering.

Some companies have found it worthwhile to conduct preliminary ISD reviews during FEL-1 to better understand the hazards, and find ways to reduce or eliminate the hazards inherent in the proposed development options. The reviews are generally based upon a combination of What If analysis, checklist and/or brainstorming.

Further information and guidance on ISD is available from the following CCPS publications: *Inherently Safer Chemical Processes, A Life Cycle Approach, 2nd edition (CCPS 2009d); Guidelines for Hazard Evaluation Procedures, 3rd edition (CCPS 2008b).*

3.3 CONCEPT RISK ANALYSIS

A Concept Risk Analysis (CRA) evaluates significant safety (and sometimes environmental and health) aspects and any adverse impacts that could affect the viability of the potential project. It typically addresses key issues, such as location (land take, communities, sensitive environments, infrastructure, logistics, etc.) and technology/process (hazards, ISD, etc.). The preliminary HAZID provides the basis for identifying the significant scenarios of interest.

The CRA is a simplified form of quantitative risk analysis (QRA), based upon a combination of generic technology/process data and site specific data. Due to the basic nature of the available information on the technologies and processes at this stage of the project, the analysis uses industry data, such as the likelihood of fires / explosions for similar facilities. The consequence part of the analysis may use either estimates for inventories and process conditions or generic industry data from similar facilities. Site specific data that is available includes meteorological data, and locations of hazardous inventories, local communities, workforce, and other areas of interest.

This simplified QRA is unlikely to be as accurate in *absolute* terms as QRA studies conducted at later stages of the project when the detailed design has evolved. However the application of CRA to multiple options in a *comparison* approach largely overcomes the problem of inaccuracies in the assumptions used in the absence of definitive data. The difference in risk between options is the important factor, not the absolute level of risk. This allows the PMT to compare the safety risks between options and rank them accordingly. It can also provide insights into potential business interruption and property damage related risks between options. All risks should be captured in a risk register (see 3.4.2 below).

The CRA should be conducted by a competent and experienced risk analyst familiar with the sensitivity of using estimates and assumptions in place of definitive data. The risk analyst requires input from personnel familiar with the technologies and processes of each development option.

The PMT should carefully weigh the commercial and technical attributes of each development option together with their process safety and EHS risks. Further analysis of promising options in FEL-2 is likely to be necessary before a preferred option can be selected.

Further information and guidance on QRA is available from the following CCPS publication: *Guidelines for Chemical Process Quantitative Risk Analysis, 2nd edition* (CCPS 2000).

3.4   OTHER ACTIVITIES

There are a number of other activities that support FEL. These activities will continue throughout the project life cycle and will require to be updated periodically.

3.4.1   Process Safety and EHS Plan

Preliminary plans for process safety and EHS should be developed in FEL-1 to identify all the studies required for each development option (Appendix B). These plans may be combined in a single document. The plan(s) should address the level of detail for each study and its timing.

3.4.2   Risk Register

For each development option, hazards and risks identified in the preliminary HAZID, ISD review, and CRA should be recorded in a risk register (Appendix C). This risk register may be separate or a sub-set of the overall project risk register.

3.4.3   Action Tracking

Any actions that are identified in FEL-1 should be recorded in a project database or spreadsheet and tracked to resolution.

3.5   STAGE GATE REVIEW

Within the process industry it is common to conduct a stage gate review(s) towards the end of FEL-1. The extent of the reviews varies between companies, but normally technical and EHS issues are addressed either separately or in a combined review. Process safety is often included in the EHS review. The review is conducted by an independent and experienced multi-discipline team, who assess whether the PMT has fulfilled their process safety (and EHS) objectives (activities and deliverables) for the stage.   At the conclusion of the review, the review team will make recommendations for any improvements needed, and indicate to the Gate Keeper, based on process safety, whether the project is ready to proceed to the next stage, FEL-2.

The stage gate review team may use a protocol and/or checklist. Appendix G includes an example of a stage gate review protocol. A typical process safety scope for a FEL-1 stage gate review is illustrated in Table 3.2.

Table 3.2. FEL-1 Stage Gate Review Scope

Scope Item
Review the technology and process for potential Process Safety risks
Confirm all project options were assessed for inherently safer design
Review all potential locations for possible Process Safety impacts on neighboring facilities, local community and environment
Examine project options for issues that can significantly influence Process Safety performance
Identify Process Safety uncertainties/unknowns of each project option

---

Case Study: Oilfield Expansion

A small oilfield stabilized the crude oil at a gathering center to remove light ends, and then exported the oil in railcars. An exploration drilling program discovered a significant extension to the reservoir, so a project team was formed to develop options to increase production by a factor of 5.

Various development options were evaluated involving new wells and flowlines, an enlarged gathering center with LPG recovery, and additional railcar loading facilities. An option for an oil export pipeline was dropped due to routing concerns through an environmentally sensitive area. At the end of FEL-1, the project team only had a single oil export option (rail) that was considered technically and commercially feasible.

The FEL-1 stage gate review team examined all options including rail export, which required a train of 12 railcars to be loaded every hour, 24 hr/day, 365 days/yr. There was only a single track line connecting the rail terminal to the mainline several miles away, down which empty railcars for the next train, and full railcars needed to be moved in opposite directions every hour. Safety checks of railcars were required prior to loading and on completion of loading, which, if performed properly, left significantly less than an hour for the loading operation. This option was further complicated by the export of LPG by rail in 3 trains per day down the same single track line.

The stage gate review team considered the risks of a major incident involving crude oil and/or LPG railcars was too high. In particular, empty railcars arriving early or late, and any loading equipment problems could potentially impact operating discipline and lead to cursory safety checks and short cuts. The gate keeper agreed, and the project team developed an alternative oil export option involving a pipeline.

3.6   SUMMARY

The best opportunity to make a positive impact on the life-cycle of a major capital project is during the early conceptual and planning stages before capital outlay occurs. FEL-1 involves developing sufficient strategic information on multiple development options with the highest potential of meeting business objectives. From a process safety perspective, this involves understanding the hazards, risks and uncertainties of each option and location when ISD principles are applied. This understanding represents the foundation for further development and selection of a preferred option in FEL-2.

Additional Reading

Amyotte P.R., Goraya A.U., Hendershot D.C., Khan F.I., Incorporation of Inherent Safety Principles in Process Safety Management, Proceedings of 21st Annual International Conference, Center for Chemical Process Safety, Orlando FL, 2006.

Bridges W., Tew R., Controlling Risk During Major Capital Projects, Chemical Engineering Progress, April 2009.

Ebert J.M., Front-end loading for a successful capital project, Inform Magazine, Vol. 27 (6), 2016.

van der Weijde, G.A., Front-end loading in the oil and gas industry: towards a fit front-end development phase, Master Thesis, Delft University of Technology. 2008. Accessed online on May 26, 2017 at: http://repository.tudelft.nl/

4 FRONT END LOADING 2

Assuming that the various development options from FEL-1 are technically and commercially viable, the project moves to the next phase of Front End Loading (FEL-2), sometimes known as Select, Selection, or Concept, which involves refining and evaluating the options to maximize opportunities, while reducing threats and uncertainties to an acceptable level. Figure 4.1 illustrates the position of FEL-2 in the project life cycle.

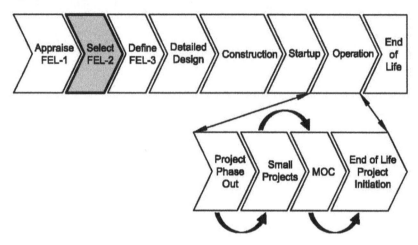

Figure 4.1. Front End Loading 2

Project Management Team

The focus of the project management team (PMT) is on completion of technical and commercial studies to sufficient level to select a single optimum concept and preliminary development plan, including the site, facilities, and infrastructure requirements, to take forward into FEL-3. In order to develop the required level of detailed information for each technically and commercially viable option during FEL-2, the project team requires a high-level of cross-discipline integration and appropriate engagement of functional expertise, including process safety.

Typical deliverables at the end of FEL-2 include:

- Final Statement of Requirements (SOR),
- Technology plan addressing any novel/unproven technologies,
- Basic Engineering Package (BEP) including, but not limited to:
    - o   Outline Basis of Design (BOD),
    - o   Material & energy balance,

      o   Preliminary engineering drawings (Layout, PFDs, P&IDs, etc.),

      o   Process description,

      o   Process control description,

      o   Preliminary lists of equipment and instrumentation,

      o   Equipment and instrument datasheets.

- Procurement plan addressing long lead items,
- Preliminary strategies for project organization, commissioning, and operations and maintenance,
- Updated project risk register, and
- Conceptual cost estimate and preliminary schedule for the project.

The transition from FEL-2 to FEL-3 is a significant decision point in the project life cycle, as FEL-2 is often the last opportunity to modify or terminate the project without incurring major financial and schedule impacts.

Environment, Health and Safety

From an EHS perspective, the project team needs to identify and update all EHS risks addressing the full life cycle of the project, especially those relating to novel technology and the characteristics of the location(s). A project EHS plan is developed to identify how EHS risks will be managed in future project stages. This EHS plan is a living document that will be continually updated and evolve as the development proceeds through later stages of the life cycle. Other FEL-2 activities include the development of a project EHS management system meeting corporate policy, identification of all applicable regulations, standards, and relevant corporate expectations, and confirmation that applicable regulatory and permitting approvals have been obtained or plans are in place to acquire them. Finally, the team needs to ensure that the design and operation of the selected option is capable of meeting corporate EHS policies and standards.

Process Safety

The process safety objectives of FEL-2 generally build on those established in FEL-1, namely:

- Identify all process safety concerns (significant hazards, uncertainties, etc.) relating to the full life cycle of the facility, novel/unproven technology, and characteristics of the location,
- Identify all applicable process safety regulations, standards, and relevant corporate expectations,
- Establish a process safety plan (may be incorporated in the project EHS plan), and

- Establish a risk management strategy, including future Hazard Identification and Risk Analysis (HIRA) studies.

While the various development options must meet certain technical and commercial criteria for the project to be viable, the process safety aspects of the options must also meet established corporate risk criteria. Evaluating the process safety aspects at FEL-2 normally involves more detailed hazard evaluation and risk analysis studies, including the assessment of offsite major accident risk. Early identification and assessment of hazards provides critical input for project decisions at a time when design changes have the minimum cost penalty. The application of an ISD approach is generally prudent in the determination of the optimum site, process technology, facilities, and infrastructure requirements.

The appropriate process safety studies and activities are discussed below for:

  i.   Evaluating the development options and selecting a single development option (Section 4.1),

  ii.   Further definition of the selected development option (Section 4.2), and

  iii.   Other process safety activities (Section 4.3).

4.1   EVALUATION OF DEVELOPMENT OPTIONS

A primary objective during the FEL-2 stage is to select the process technology that will be used. The process safety studies performed for each development option will be dependent upon the project scope. Most studies will focus on HIRA, but depending upon the options being considered, will vary from high level HAZID that identifies major hazards to simplified quantitative risk assessments (QRAs) that will evaluate anticipated risks based on industry data for similar facilities. Inherently Safer Designs and alternate layouts for the different processes may also be evaluated to determine possible risk reductions that can be achieved in each option. Potential incident scenarios may also be evaluated for their impact at each of the possible site locations. Comparative ranking of the risk assessment results will identify the process and location combination that offers the lowest residual risk.

Depending on project scope and the results and recommendations from HIRA studies performed throughout the project life cycle, some projects may require additional specialist studies. For example, an offshore oil development close to a shipping lane may require a study to evaluate the risk of ship collision. Examples of other specialist studies may include risks associated with hydrates, pipeline surge, seismic activity, transportation, etc.

4.1.1   Hazard Identification

A hazard identification (HAZID) study should be conducted on each of the options being considered by the project team. During FEL-2, the HAZID studies are high-level, systematic reviews of potential major accident hazards that are identified to assist selection of the single optimum concept for the project. For smaller projects, a single HAZID may be all that is needed.

At this stage of the project, more information will be available than was available at FEL-1. If preliminary HAZIDs were previously prepared, they should be updated using the available data as the definition of each option evolves. If a preliminary HAZID was not conducted previously, the HAZID should focus on the hazardous materials and major process areas of the facility, where there are potential major accident hazards that can impact people, environment and/or property.

The most popular methodologies for conducting HAZID studies are similar to those described for preliminary HAZID studies in Chapter 3, namely: checklists and brainstorming, or a combination of the two. The 'What If' methodology may also be used. A small experienced group with knowledge of the technologies, processes, and process safety/loss prevention should conduct the HAZID.

For HAZID studies to be effective, a process safety engineer or other competent facilitator will need to develop appropriate checklists that adequately cover the scope of the development options. One checklist may not adequately cover all of the studies needed. The simplified checklist in Table 3.1 may be added to, based upon the specific nature of the development options. Potential occupational health and environmental impacts may also be included in the HAZID.

The HAZID study also provides the initial basis for a Hazard Register that summarizes the hazards present in a facility together with their sources, locations, significance, and controls. The Hazard Register offers a starting point for hazards management and is a regulatory requirement in some jurisdictions.

Further information and guidance on methodologies for HAZID studies, including more detailed checklists, is available from the CCPS publication *Guidelines for Hazard Evaluation Procedures, 3rd edition* (CCPS 2008b).

4.1.2   Preliminary Inherently Safer Design Review

The application of ISD principles is used to reduce the risk of major incidents through eliminating or mitigating hazards rather than applying controls and other safeguards. ISD reviews of each development option should be conducted in FEL-2 to identify ways to reduce or eliminate the inherent hazards. If ISD reviews were previously conducted, they should be updated using the available information as the definition of each option evolves. If not, ISD reviews should be performed in FEL-2, as the importance of applying ISD principles during the early stages of a project cannot be over emphasized. This is a key approach in reducing risks to an acceptable level consistent with corporate policies. The reviews consider ISD opportunities for the site(s), process technologies, facilities, and infrastructure requirements, so that, as the development options are screened, the differences are clear. This will allow the project team to develop risk profiles to compare against cost estimates for the various options to aid in the selection process.

The reviews are generally based upon a combination of What If analysis, checklist and/or brainstorming, and follow the principles outlined in Chapter 3.

If the project strategy is fast track with a compressed development timeline, it may not be possible to adequately incorporate significant ISD content. If so, there should be a mechanism for the PMT to communicate lessons learned for future projects during the project close-out and/or post-project evaluation activities.

Further information and guidance is available from the following CCPS publications: *Inherently Safer Chemical Processes, A Life Cycle Approach, 2nd edition* (CCPS 2009d), and *Guidelines for Hazard Evaluation Procedures, 3rd edition* (CCPS 2008b).

4.1.3 Concept Risk Analysis

If a concept risk analysis (CRA), a type of QRA, was conducted in FEL-1 to evaluate significant safety aspects and adverse impacts of each development option, it may be updated to reflect additional information as the definition of each option progresses. Early in a new project, information is constrained which limits the depth of analysis. As the project advances, the information constraint is gradually reduced. The HAZID may also identify additional hazard scenarios of interest. Differences in risk between options should have a significant input to decision-making relating to the selection of the optimum development option.

While the depth of analysis may increase in FEL-2, the scope and approach for CRA remains as discussed in Chapter 3. Further information and guidance on QRA is available from the following CCPS publication: *Guidelines for Chemical Process Quantitative Risk Analysis, 2nd edition* (CCPS 2000).

After the optimum development option is selected, the definition of that option is raised by the project team to a level consistent with corporate requirements for the FEL-2 stage gate. Some companies perform a more detailed Preliminary Risk Analysis of the selected option at that time. This is discussed below in Section 4.2.

4.1.4 Selection of the Development Option

Selection of the final development option will depend upon many factors. Most importantly, the option must be commercially and technically viable from a business perspective. Otherwise the project will not proceed. However, the selected option must also meet corporate EHS policies and risk criteria. Some jurisdictions also set a risk criterion, in which case the PMT must meet the most onerous criterion.

A company's EHS policies and risk criteria may be qualitative or they may also include quantitative risk criteria. In the latter case, the level of definition of the development options at FEL-2 and their corresponding results from preliminary risk analyses are likely to have quite wide confidence intervals. Therefore any comparison of results against a corporate criterion for individual or societal risk will need careful evaluation, especially with regard to the public.

Further information and guidance on risk criteria is available from the following CCPS publication: *Guidelines for Developing Quantitative Safety Risk Criteria*, (CCPS 2009b).

Other process safety and EHS considerations for the preferred location include:

- Adequate area to allow separation of hazardous inventories from people (including local community), avoid unnecessary congestion, and allow for buffer zone / future expansion. (Brownfield projects may present area and separation distance constraints requiring special design safety measures to reduce risk).

- Adequate area to allow flexibility in the placement and control of access/egress points, safe routing of transportation corridors, and provision of reliable utilities. (Brownfield projects may require elevated piperacks and other compromises, such as interface of new/old control systems).

- Construction may involve lifting heavy process vessels and/or other equipment over existing brownfield process equipment, and the HAZID study should address this risk.

- Availability of skilled workforce and other resources in the region.

- Availability and adequacy of local emergency responders capable of handling the hazards associated with the facility operations, including firefighting, rescue, medical, security and police.

- Neighboring facilities that may impact the site.

- Brownfield projects may need to upgrade existing facilities to current environmental requirements.

The project team will also need to determine which technology to select. If the technology is a licensed process, there should be adequate information for the process safety studies available from the licensor. In addition, visits to operational facilities with the same technology may assist in the selection. If the technology has been developed internally within the company, the available information may range from laboratory research, pilot plant development, or operating experience of similar facilities.

A preliminary constructability review may be appropriate at this stage to support the selection process. Further discussion of constructability reviews is in Chapter 6 Section 6.5.6.

If the technology is novel or unproven, there is likely to be little data available to analyze risks, and the final decision may be somewhat subjective. If the benefits of the novel option are potentially large, the project team may decide to select a second option of more conventional technology for development in parallel. In this eventuality, the final option may not be selected until FEL-3.

The selection of some alternative technologies represents a balance between product quality/yield, onsite/offsite risk, and cost. For example, the choice of alkylation catalyst is usually either sulfuric acid or hydrofluoric acid. The latter has advantages in octane rating and catalyst consumption, but requires significant risk reduction measures (metallurgy, water sprays, dump tanks, etc.), and may not meet corporate risk criteria if close to local communities.

Offshore oil platform design is another example of alternative technologies, which could be fixed on the seabed (steel or concrete legs, compliant tower), an artificial island, jack-up, subsea or floating (semi-submersible, drillship, FPSO, tension leg, spar). Each option has different process safety issues that need to be addressed. The final selection is based upon water depth and cost, but other factors such as mobility, topsides equipment, crew size, pipeline export, and drilling are also considered. Some typical water depths that are economically feasible are illustrated in Table 4.1.

Table 4.1. Economically Feasible Platform Concepts vs. Water Depth

Platform Concept	Water Depth*
Jack-Up	up to 550 ft (170 m)
Compliant Tower	1500 – 3000 ft (450 – 900 m)
Fixed Jacket	up to 1700 ft (520 m)
Tension Leg	600 – 6,000 ft (200 – 2,000 m)
Spar	up to 8000 ft (2,440 m)
Semi-Submersible	200 – 10,000 ft (60 – 3,050 m)
Drillship	up to 12,000 ft (3,660 m)

Note: ongoing technological advances may increase these water depths

The project team will likely apply value engineering in a systematic and structured approach to analyze each option to achieve an optimum balance between function, performance, quality, safety, and cost. This process identifies and removes uncompetitive options and unnecessary equipment to reduce costs, thereby increasing the value of the project. The option that has the proper balance results in the maximum value, and is likely to be selected for the project.

It should be noted that the optimum development option may not necessarily be the option that has the lowest residual risk from a safety perspective. Nevertheless, all significant process safety and EHS hazards inherent in the proposed development option should have acceptable solutions or solutions are capable of being developed within the timeframe of the project.

4.2   FURTHER DEFINITION OF THE SELECTED OPTION

One of the main benefits of successfully integrating process safety into a project is to reduce residual safety risk. Project teams must balance competing priorities, so frequently the final solution is a compromise. Nevertheless, project teams should seek to drive residual risk to as low as reasonably practicable (ALARP). This goal infers an ISD approach that should place fewer demands on operations personnel, while also limiting potential for major incidents. Further information and guidance on ALARP is available from the following CCPS publication: *Guidelines for Developing Quantitative Safety Risk Criteria,* (CCPS 2009b).

When the project team has selected their preferred development option, based on an understanding of the business case and risks and uncertainties inherent in the development, the option is further developed to produce a preliminary project development/execution plan.

Additional technical and commercial studies are undertaken to produce this plan. The relevant process safety studies and activities are discussed below.

4.2.1   Design Hazard Management Process

A core challenge faced by project teams is how to drive risk to ALARP while keeping the project on schedule and budget. Several questions arise when faced with this challenge:

- What hazards exist?
- How severe can their impact be?
- How frequently do they occur?
- Which hazards pose the greatest threat?
- What are the cost-effective alternatives?

Many companies employ some form of Design Hazard Management (DHM) process to identify and evaluate major hazards, and continuously reduce risk through

---

Functional Safety

Part of the overall safety relating to the process and its control system which depends on the correct functioning of the safety controls, alarms, and interlocks (SCAI) and other protection layers.

(CCPS 2017)

functional safety and other design safety measures. IEC 61511 provides a functional safety lifecycle of activities to ensure the design and integrity of safety instrumented systems (SIS) (IEC, 2016). It is also important to avoid vulnerabilities to the integrity of the control system and its data that could significantly increase risks.

Major accident hazards include loss of containment (LOC) resulting in explosion, fire, and toxic release. A range of initiating events for LOC should be considered, such as corrosion, external impact (dropped loads, vehicle/ship collision, etc.), operator error, and environmental hazards (earthquake, hurricane, flood, etc.). Utilizing a DHM process facilitates incorporating ISD and process safety principles into the project.

Typical key steps in DHM during FEL are illustrated in Table 4.2 below. These typical key steps and other process safety activities in FEL-2 are discussed below.

While the project team may develop a DHM philosophy/strategy to define goals and standards earlier in FEL-1, DHM normally starts in FEL-2 when the development option has been selected, and continues iteratively through the project life cycle. The project team develops a DHM Implementation Plan, which identifies the required risk analysis, environmental, and safety studies that will be required. This plan should be updated during the subsequent FEL-3 stage, so that all of the required studies are known before the start of the project execution phases. The DHM plan may be incorporated within the EHS Plan and/or Process Safety Plan (see below).

Table 4.2. Typical Steps in a Design Hazard Management Process

Project Stage	Step
FEL-1	• Identify major accident hazards in each development option
FEL-2	• HIRA • Optimize layout and design of facilities • Establish design safety concepts/critical design measures (including functional safety)
FEL-3	• Continue ISD optimization • Refine design safety concepts/critical design measures (including functional safety) • Set performance standards • Re-evaluate major accident risk • Finalize important design safety decisions • Finalize Basis of Design (BOD)

In some jurisdictions, owners and/or Operators are responsible for identifying, profiling and managing the major accident risks they create. Design and Operational Safety Case Studies are prepared, which detail the identified risks, document the studies undertaken to evaluate the risks, and describe the measures employed to manage the risks or mitigate the potential consequences the risks represent. During the FEL 2 stage, information needed for the Design Safety Case and studies will need to be gathered. Documentation requirements are discussed in Chapter 12.

Managing hazards in design involves the elimination or minimization of major accident hazards at source (i.e. ISD principles), and preventing those hazards that remain from becoming major accidents (i.e. managing residual risk). The overall goal should be to reduce residual risk to at least a level that meets corporate policy. The management of residual risk during design usually involves a combination of applying ISD principles and adding risk reduction measures, a.k.a. layers of protection or barriers.

These layers of protection may be:

- Hardware equipment, controls (including SIS), vessels, piping, etc.
- Procedural operating procedures, safe work practices, maintenance procedures, etc.

Good practice at the FEL and detailed engineering stages is to avoid, as much as possible, reliance on procedures and the intervention of people to manage the residual risks. Table 4.3 illustrates a hierarchy of effectiveness for various risk reduction measures.

Table 4.3. Hierarchy of Risk Reduction Measures

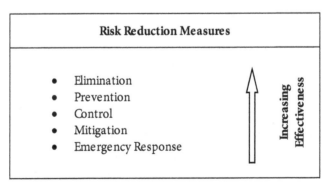

Some examples of risk reduction measures within this hierarchy are:

- Elimination by use of substitution (e.g. use of less hazardous chemical; liquid or solid hypochlorite instead of chlorine gas for water treatment)

- Prevention at source (e.g. use of corrosion resistant alloy to prevent corrosion or designing pressure vessel to withstand worst case upset pressure)
- Control through design features or administrative procedures (e.g. fire/gas detection and emergency shutdown system (ESDS))
- Mitigation to protect personnel (e.g. personal protective equipment (PPE) and blast resistant buildings)
- Emergency Response to prevent escalation (e.g. firewater)

As a general rule, passive measures (e.g. dike, drainage, fireproofing) are more reliable, and therefore preferred, than active measures (e.g. ESDV, SIS). There is a preferred hierarchy regarding the reliability of the measures selected for risk reduction, as follows:

- Passive measures are more reliable than…
- Active measures, which are more reliable than…
- Administrative or procedural measures.

For example, it is more reliable to design a baghouse to withstand a deflagration than it is to install a suppression system than it is to rely on operator monitoring and control.

These principles should be reflected in the DHM process. Nevertheless, the final design is likely to incorporate passive, active, and some procedural safety measures.

Other important aspects of DHM include the provision of:

- Design integrity to control the consequences of a major accident by reducing its severity and potential escalation, and
- Mitigation and protection for personnel and facilities from fire, explosion, and toxic vapors (including smoke and products of combustion).

Appendix D is a checklist that can used to identify and manage design safety issues. Some companies have a design safety 'roadmap,' which outlines the activities recommended during the various project stages.

Further information and guidance on functional safety is available from the following publications: *Guidelines for Safe Automation of Chemical Processes*, 2nd edition (CCPS 2017b); *Functional safety - Safety instrumented systems for the process industry sector - Part 1: Framework, definitions, system, hardware and application programming requirements*, IEC 61511-1 (IEC 2016).

Case Study: Offshore Platform – Piper Bravo

The design of the Piper Alpha platform in the North Sea owed allegiance to designs used in the Gulf of Mexico. It was modified in 1980 to conserve and export gas. The Piper Alpha plot plan was essentially square, and, although modules were originally organized to separate the most hazardous operations from accommodation, the conversion to gas processing ruined this safety concept, and brought together sensitive areas with gas compression, probably the highest risk. In 1988, Piper Alpha suffered the worst offshore oil industry disaster that took the lives of 167 persons and became an industry-changing watershed event.

The subsequent Public Inquiry made a number of recommendations related to inherently safer design (ISD), including studies for fire and explosion hazards, smoke and gas ingress into accommodation, survivability of temporary haven, vulnerability of safety critical equipment, and evacuation, escape and rescue. These studies have been responsible for a fundamental change in platform design.

Future production platform designs invariably have a more exaggerated rectangular plot plan/layout, thereby increasing separation of accommodation from hazardous modules. The layout of Piper Bravo, the replacement for Piper Alpha, set the standard for a new generation of offshore platforms with safety as an inherent feature. In addition to a rectangular plot, other Piper Bravo design features included:

i. Blast walls (as well as firewalls) aid separation of sensitive modules,
ii. Accommodation pressurized and HVAC system equipped with automated fire dampers,
iii. Temporary refuge with at least 2 hours fire/explosion protection,
iv. Control room remote from hazardous modules,
v. Redundant communication systems to avoid common mode failure,
vi. Pipeline emergency shutdown valves (ESDVs) protected from fire and explosion by enclosures,
vii. Pipeline risers remote from accommodation equipped with subsea ESDVs, closing automatically on loss of hydraulic pressure.
viii. Rapid blowdown of process equipment to remove hydrocarbons
ix. Spatial separation of fire pumps with remote start,
x. Shielded escape routes from temporary refuge leading to free-fall lifeboats

Ref: Broadribb, M. P., *What have we really learned? Twenty-five years after Piper Alpha*. Process Safety Progress, Vol. 34: Issue 1: 16–23, March 2015.

4.2.2   Preliminary Inherently Safer Design (ISD)

DHM requires multiple levels of design measures to reduce risk. ISD is one element of the DHM approach and involves the elimination of hazards, where possible, and the optimization of layout and primary structural and system integrity to minimize the impact of those remaining. Therefore an inherently safer design avoids hazards instead of controlling them, e.g. by substituting less hazardous materials, reducing the amount of hazardous material, designing equipment for worst case conditions, and reducing the number of hazardous operations in the facility.

The importance of applying ISD principles during the early stages of a project cannot be over emphasized. The greatest opportunity for achieving a cost-effective, inherently safer design is during FEL, so it is important that a particular effort is made in FEL-2 to identify the key ISD measures for the selected option.

Further guidance on ISD is available from the following CCPS publication: *Inherently Safer Chemical Processes, A Life Cycle Approach, 2nd edition* (CCPS 2009d).

4.2.3   Hazard Identification and Risk Analysis (HIRA)

The HAZID for the selected option may be updated as additional definition of the design becomes available. This updated study is then used as input for risk assessments.

Some companies perform a more detailed preliminary risk analysis of the selected option. This study follows a similar scope as the CRA discussed in Chapter 3, but the availability of more detailed information on the selected option allows the study to be conducted in greater depth than the CRA. For example, the design of the selected option will have evolved such that information on major equipment sizes, design philosophies, preliminary P&IDs, isolation valve placement, and conceptual layout should be available from the BEP. Security risks may be added to the scope of the preliminary risk analysis or alternatively a separate preliminary Security Vulnerability Analysis (SVA, see Section 4.2.13) may be conducted.

The preliminary risk analysis can also be used to evaluate variations in layout, equipment siting/spacing, congestion/confinement, drainage, business interruption due to vulnerable buildings/equipment, and design safety measures such as fire and blast protection. The key objectives are:

- To identify the worst and/or most likely risks so that alternate design plans can be developed to eliminate or reduce the risk or mitigate the consequences of the risk.

- To determine the risk reduction that can be achieved through ISD for use in cost benefit analyses.

- To quantify the risk or qualitatively rank the risk so that any required risk reduction measures can be added to the design to meet the owner's established risk tolerance criteria.

Further information and guidance is available from the following publications:

CCPS, *Guidelines for Hazard Evaluation Procedures, 3rd edition* (CCPS 2008b).

CCPS, *Guidelines for Chemical Process Quantitative Risk Analysis, 2nd edition* (CCPS 2000).

API, *Recommended Practice for Design and Hazards Analysis for Offshore Production Facilities, 2nd edition,* API RP 14J, American Petroleum Institute, 2001 (API 2001a).

ISO, *Petroleum and Natural Gas Industries – Offshore Production Installations – Guidelines on Tools and Techniques for Identification and Assessment of Hazardous Events,* ISO/DIS 17776, International Organization for Standardization, 1999 (ISO 1999).

CMPT, *Guide to Quantitative Risk Assessment for Offshore Installations,* The Centre for Marine and Petroleum Technology, 1999 (CMPT 1999).

4.2.4   Engineering Design Regulations, Codes, and Standards

Before detailed design can start, it is crucial to identify which regulations, codes, and standards will apply to the new project, and therefore should be used in all design studies. National and/or local regulations may specify a range of design requirements, including, but not limited to:

- Competency of design engineers and technicians (e.g. certification, P.E., C.Eng., Chinese design institute, etc.)
- Restrictions on zoning, height of structures, etc.
- Engineering documentation (e.g. OSHA PSM PSI, Safety Case, etc.)
- Design factors (e.g. safety factor for load bearing structure, etc.)

Various global and/or national codes and standards may be applicable. For example:

- Global: ISO, IEC standards
- United States: API, ASME, ANSI, NFPA, ISA, OSHA, or DOT
- European Union: ATEX, PED, EN, BS or DIN
- Other National: Chinese, Japanese, etc.
- Company: Client corporate policy/standards, contractor standards

All selected engineering codes and standards should be updated and 'frozen' before detailed design to avoid later changes and duplication of effort. The project should also establish a formal process for evaluating and approving any departure from selected engineering codes and standards.

In addition, local government may require specific suppliers for raw materials and natural resources, or specific contractors for fabricating equipment. Insurance companies may specify minimum requirements, e.g. spacing/safety measures to reduce escalation/property damage from fires.

4.2.5   Design Philosophies/Strategies

Also before detailed design can proceed, it is essential to develop a number of preliminary design philosophies and strategies, which specify certain approaches and criteria to be followed as the design progresses. Typical examples of design philosophies address, but are not limited to, the following topics:

- Blowdown, pressure relief and flare system
- Control system (e.g. DCS, SIS, alarm system)
- Fire & gas detection
- Fire protection
- Emergency response
- Operations and Maintenance

Based upon past operating experience and loss of containment incidents, the fire and gas detection philosophy, for example, might specify criteria such as hazardous areas requiring detectors, heat/flame/smoke/gas/vapors to be detected, concentration of smoke/gas/vapor to be detected, speed/time of detection, minimum gas/vapor cloud size to be detected, etc. This implies a size of leak or fire that the designer is prepared to accept will be *undetected*. Other aspects, such as the type (e.g. point sensor, optical path or acoustic) and number of detectors, could be added later after evaluation of site specific characteristics. This philosophy then impacts the strategies for shutdown, blowdown, fire protection, etc. and invariably requires an iterative approach to optimize the design.

Other examples of the detail that may be important to include in design philosophies for a specific project include:

i.   The DCS design should allow for routine export of a variety of data and information. This should include but not be limited to: configuration information, HMI, database of tags, historical operating information – both continuous and event driven, alarms and alarm responses. The exported information should be stored in accessible / readable format for quick and full access.

ii. The emergency response plan for extreme events, as well as natural disasters such as flooding, earthquakes etc., should include identification of key process variables to record and retain, for understanding the historical operating conditions of a process such that it can be used in process safety incident investigation and other analysis.

Many larger capital projects also have an EHS philosophy that includes process safety. The EHS philosophy defines the principles and practices that will be applied during the project life cycle, and becomes the basis for developing the EHS and Process Safety Plan (see Section 4.3.1).

4.2.6   Preliminary Facility Siting Study

Siting of permanent and temporary buildings in process areas requires careful consideration of potential effects of explosions, fires, and toxic vapors arising from accidental release of hazardous materials. The American Petroleum Institute (API) has published recommended practices for permanent (API 2009) and portable/temporary (API 2007) buildings, and tents (API 2014). Some national and/or local jurisdictions may have other building codes, such as the International Building Code® (ICC 2018).

A preliminary facility siting study of the selected option should be conducted to assess potential explosion, flammable, and toxic hazards associated with operation of an onshore facility and the impact of these hazards to onsite personnel and buildings. The scope of this study may be combined with the Preliminary Risk Analysis (see Section 4.2.3), and the Preliminary SVA (see Section 4.2.13) if required, with respect to any potential off-site impacts. Information on the preliminary layout/plot plan, hazardous inventories, preliminary PFDs (and preliminary P&IDs if available), preliminary heat and mass balances, and key building locations is required.

Facility siting study results identify hazard vulnerabilities to aid with identification of potential mitigation strategies. Since the layout and equipment may change, this process is designed to be iterative (throughout FEL-3 and Detailed Design stages) to assist the project team with layout decisions from a safety perspective. In particular, it is important to determine the location of temporary project buildings prior to the start of construction.

Further information and guidance is available from the following publications: *Guidelines for Siting and Layout of Facilities, 2nd edition* (CCPS 2018a); *Guidelines for Evaluating Process Plant Buildings for External Explosions, Fires and Toxic Releases, 2nd edition* (CCPS 2012b); *Management of Hazards Associated With Location of Process Plant Permanent Buildings, API RP 752, 3rd edition* (API 2009); *Management of Hazards Associated With Location of Process Plant Portable Buildings, API RP 753, 1st edition* (API 2007); *Management of Hazards Associated with Location of Process Plant Tents, API RP 756, 1st edition* (API 2014).

4.2.7   Preliminary Fire and Explosion Analysis

While preliminary fire and explosion analysis is particularly important for offshore facilities, some onshore facilities may also be subject to separation distance limitations and/or congestion and confinement. Therefore the general principles for offshore analysis may be applied to the layout and equipment of conceptual onshore designs.

In the aftermath of the 1988 Piper Alpha disaster in the North Sea, regulations were introduced for offshore fire and explosion analysis (HM Government 1995). For offshore developments, a preliminary fire and explosion analysis should be conducted using the preliminary platform or rig layout. Toxic consequences, such as H_2S releases, may also be assessed if applicable.

Given the inherent characteristics of offshore facilities with limited spatial separation of personnel from hazards, the focus of the study should be on impacts to personnel, the temporary refuge (usually living quarters), and the platform/rig structure and equipment:

- Fire: Heat, vision obscurity, oxygen depletion, and inhalation of combustion products
- Explosion: Injury/fatality, and equipment/structural damage (due to blast overpressure, heat, projectiles, etc.)
- Toxicity: Inhalation of toxic gases/vapors

As the layout and equipment may change, this process is designed to be iterative to assist the project team with layout decisions from a safety perspective. During FEL-3 and Detailed Design stages, in addition to finalizing the fire and explosion analysis, it may be appropriate to conduct separate studies on smoke and gas ingress into the temporary refuge/living quarters/shelter-in-place; temporary refuge/shelter-in-place impairment; and vulnerability of safety critical equipment.

Further information and guidance is available from the following publications:

(HSE 2016) *Prevention of fire and explosion, and emergency response on offshore installations. Offshore Installations (Prevention of Fire and Explosion, and Emergency Response) Regulations, 1995*. Approved Code and Practice and Guidance, L65, 3rd Edition, Health and Safety Executive, 2016.

UKOOA *Guidelines for Fire and Explosion Hazard Management*, UKOOA, 1995.

4.2.8   Transportation Studies

Many projects involve various means of transportation for delivery of feedstock, catalysts and lubricants, export of products, and disposal of waste, e.g. road, rail, pipeline, and marine. Depending on project-specific factors and local circumstances, one or more studies may be appropriate to evaluate transportation hazards and risks. The scope and level of detail will vary based on these factors and circumstances, but will likely involve consequence analysis and/or risk analysis.

Specific examples of transportation studies are:

Cross-country Pipelines

For cross-country pipeline projects, a high consequence area (HCA) assessment identifies pipeline segments with potential to impact sensitive areas, such as populated areas, drinking water supplies, ecological resources, parks and forests, commercial fishing and recreation water, and other environmentally important areas.

A HCA should be conducted during FEL-2 based on preliminary data for the pipeline route, operating conditions, and pipeline diameter. The HCA should evaluate the consequences of a range of potential hole sizes up to full-bore rupture. The study should be conducted in an iterative manner to assist the project team with pipeline routing decisions from a safety perspective.

Marine

A waterway suitability assessment (WSA) is a requirement for owners or Operators in the USA that intend to build a new waterfront facility handling liquefied natural gas (LNG) and liquefied hazardous gas (LPG and a list of other chemicals). An expansion or modification to marine terminal operations in an existing waterfront facility is also covered. A preliminary WSA must be submitted at least one year prior to operation, and should explain the project (characterization of the port, facility and waterway route (sea to facility)), and address maritime safety/security risk assessment, risk management strategies, and resource needs for maritime safety, security and response in broad terms.

The WSA is based on the waterway route, cargo details, frequency of operation, and it is advisable that the study team includes a member with considerable U.S. Coast Guard experience (e.g. captain level) or equivalent. Further information and guidance is available in *33 CFR 127 Waterfront Facilities Handling Liquefied Natural Gas and Liquefied Hazardous Gas* (subpart 007 Letter of Intent and Waterway Suitability Assessment).

Beyond the USA, International Maritime Organization (IMO) and local national regulations may impact marine vessel design and routing for hazardous cargoes. Similar consequence/risk analysis studies may be appropriate to understand and reduce risks.

Further information and guidance on consequence analysis and risk analysis is available from the following CCPS publication: *Guidelines for Chemical Process Quantitative Risk Analysis, 2nd edition* (CCPS 2000). In addition, the following CCPS references also contain information on various aspects of consequence analysis and risk analysis: (CCPS 1989, 1994a, 1996b, 1998a, 1999, 2002, 2008b).

4.2.9   Preliminary Blowdown and Depressurization Study

A preliminary blowdown and depressurization study ensures that temperatures resulting from auto-refrigeration during depressurization do not lead to a risk of brittle fracture in process equipment and flare systems. The study is typically used to determine minimum design temperatures and material of construction for parts of the facility.

The study should be based upon the blowdown, pressure relief and flare system philosophy, and preliminary information for P&IDs, flare header layout (isometrics if available), and protected equipment.

Further information and guidance is available from the following CCPS publications: *Guidelines for Pressure Relief and Effluent Handling Systems* (CCPS 1998b).

4.2.10   Preliminary Fire & Gas Detection Study

A preliminary fire and gas (F&G) detection study represents a first pass at identifying locations within the facility that require fire and gas detection equipment, such as fire detection, combustible gas detection, toxic gas detection, carbon dioxide detection, or other fire-detection devices and alarms.

The study should be based upon the F&G and control system philosophies, and requires preliminary information and input from plot plans, a facility siting study (or offshore fire and explosion study), and proposed (if any) location and type of preferred F&G detection.

Further information and guidance is available from the following CCPS publications: *Continuous Monitoring for Hazardous Material Releases* (CCPS 2009a); *Guidelines for Fire Protection in Chemical, Petrochemical, and Hydrocarbon Processing Facilities* (CCPS 2003b).

4.2.11   Preliminary Fire Hazard Analysis

A preliminary fire hazard analysis (FHA) develops a cost-effective fire protection strategy involving active and passive (fireproofing, drainage, and containment) fire-protection systems, surface protection, and insulation. The study should be integrated with the ISD work to incorporate strategies such as spacing, layout, hazardous confinement, and material substitution to minimize risks.

The FHA determines the location, size, and duration of potential fires, and is based upon the fire protection philosophy/strategy, preliminary plot plan, process description, and flammable/combustible material inventories.

Further information and guidance is available from the following CCPS publication: *Guidelines for Fire Protection in Chemical, Petrochemical, and Hydrocarbon Processing Facilities* (CCPS 2003b).

4.2.12 Preliminary Firewater Analysis

A preliminary firewater analysis represents an initial evaluation of firewater supply and distribution in terms of supply, distribution, pumps, pump control, fixed fire protection systems (deluge, foam, monitors, hydrants), and portable equipment. Depending on the characteristics of the facility and its location (e.g. Arctic development), a firewater system may not be suitable or required. The preliminary FHA should confirm such requirements.

The firewater analysis should be based upon the fire protection philosophy, FHA, and preliminary plot plan.

Further information and guidance is available from the following CCPS publication: *Guidelines for Fire Protection in Chemical, Petrochemical, and Hydrocarbon Processing Facilities* (CCPS 2003b).

4.2.13 Preliminary Security Vulnerability Analysis

Security vulnerability analysis (SVA) is a methodology for managing the security vulnerability of sites that produce and handle hazardous chemicals. It involves a review of handling, storage, and processing hazardous materials from the perspective of an individual or group intent on causing a major incident with large-scale injury/fatality or supply disruption impacts.

A preliminary SVA identifies and risk ranks possible scenarios by evaluating hazardous material inventories and processes, potential pathways of attack, and proposed security countermeasures. High risk scenarios should be addressed further in FEL-3 as the design develops to assess additional countermeasures to reduce risk.

Further information and guidance is available from the following CCPS publication: *Guidelines for Analyzing and Managing the Security Vulnerabilities of Fixed Chemical Sites* (CCPS 2003a).

4.2.14 Other Engineering Design Considerations

The PMT needs to deliver a project that is both safe to operate and meets the client's EHS policies including risk tolerance criteria. Good process safety performance requires (i) hazards are identified, (ii) risks associated with these hazards are understood, and (iii) risks are managed by *'doing the right thing'*. Managing risks properly to prevent major accidents, or mitigating the consequences of an incident if one occurs, invariably requires going *beyond* any local regulations. Nevertheless, local regulations may require specific deliverables that should be produced.

Different regions have different ways of documenting and communicating their engineering design specifications. It is important to understand what conventions and measurement system are preferred locally for facility instrumentation, drawings, and procedures. It is crucial that the measurement units are clearly understood, including converting them if the original design units are not used at the new location. In particular, the standard units used in the United States will need to be converted to the metric/SI units in other countries.

4.3   OTHER ACTIVITIES

In addition to the various process safety studies needed to further define the selected development option, there are a number of other activities that support FEL and project execution. These activities continue throughout the project life cycle and should be periodically updated.

4.3.1   EHS and Process Safety Plan

The preliminary EHS Plan and a Process Safety Plan developed in FEL-1 should be updated to manage the inherent residual risks associated with the development concept selected. In particular, the plan should identify which process safety, functional safety and additional specialist studies will be performed for the project, their level of detail and timing, and resources required to implement the plan. If functional safety identifies SIS to manage residual risks, then a Functional Safety Plan should be included per IEC 61511. Appendix B illustrates a typical plan content.

4.3.2   Risk Register

The Risk Register developed in FEL-1 should be updated for the selected development option. As the design evolves, any design features (e.g. safety critical equipment) and management processes (e.g. work force competency) that must be maintained to ensure that the risk is adequately managed should be documented. It is essential that both these design features and management processes are clearly understood, as failure to maintain either or both could lead to increased risk. An example of a project risk register is illustrated in Appendix C.

4.3.3   Action Tracking

The project action tracking database or spreadsheet should be updated for the selected development option. Some projects combine an action tracking register with the project risk register (see Section 4.3.2 above), in which case, at the end of the project, it is imperative that any outstanding risks and actions are handed over to the Operator. For example, a noise study may be required after startup.

4.3.4   HIRA Strategy

As part of forward planning, the project team should develop a strategy for future HIRA studies addressing hazard identification, hazard evaluation, consequence analysis, and risk analysis. This strategy may be incorporated into the Process Safety Plan.

The strategy should address the following elements:

- Choice of HIRA methodologies should be based on characteristics and complexity of the project.
- Preliminary and/or intermediate studies may be appropriate before the final HIRA studies.
- Final HIRA studies should be comprehensive and of high quality, including competency/operational experience of the leader and team, quality of P&IDs & design information, process safety information, and documentation requirements.
- Scope of final HIRA studies to include all aspects of the project, including vendor packages, with significant hazard potential. If appropriate, final HIRA studies should address facility siting and human factors.
- Robust system for resolution of findings, including assignment of responsibility and handling of any recommendations outside of project responsibility.
- Proposed timing of final HIRA studies should allow findings to be incorporated into specifications for early-order items.
- Change management should commence no later than the final hazard identification study, e.g. HAZOP.
- Changes after the final HIRA studies to be subject to hazard review.

For simple MOCs and non-process projects, it may be appropriate to use checklists and/or What If studies as the hazard identification methodology. For larger projects with complex processes, the preferred HIRA methodologies among major chemical and oil & gas companies are a combination of HAZOP, LOPA, and QRA. As indicated above, and in the absence of any regulated approach, the final choice of methodology should be based on the nature of the project and be sufficient to ensure that significant hazards are thoroughly addressed.

4.3.5 Documentation

The compilation of process safety information (PSI) and other documentation on the selected development option needs to commence in FEL-2 and continue throughout FEL-3 and project execution. As the detailed design evolves, the early PSI will need to be revised and/or updated. This is discussed in detail in Chapter 12.

4.3.6 Stage Gate Review

When nearing the completion of FEL-2, a stage gate review should be conducted to ensure that process safety (and EHS) risks are being adequately managed by the project. The stage gate review team may use a protocol and/or checklist, such as the detailed protocol in Appendix G. A typical process safety scope for a FEL-2 stage gate review is illustrated in Table 4.4.

The stage gate review team should be independent of the project, familiar with similar facility/process/technology, and typically comprise an experienced leader, process engineer, operations representative, process safety engineer, other discipline engineers (as appropriate), and EHS specialist. At the conclusion of the review, the review team will make recommendations for any improvements needed, and indicate to the Gate Keeper, based on process safety, whether the project is ready to proceed to the next stage, FEL-3.

Table 4.4. FEL-2 Stage Gate Review Scope

Scope Item
Confirm that Process Safety and EHS hazards inherent in the proposed development warranting special attention, or uncertainties that need further investigation, have been identified
Confirm that acceptable solutions for hazards and uncertainties are available or are capable of being developed within the timeframe and organization of the project
Confirm that all Process Safety and EHS concerns relating to the characteristics of the full life cycle of the project, novel technology, and the nature of the location have been identified
Confirm all applicable regulations, standards, and relevant company expectations have been identified
Confirm an adequate Process Safety and EHS plan has been established, communicated to the project team, and endorsed by management for subsequent stages
Confirm an adequate Process Safety and EHS risk management strategy, including future HIRA studies, has been established

4.4 SUMMARY

As previously described, the best opportunity to make a positive impact on the life-cycle of a major capital project is during the early conceptual stages. FEL-2 continues evaluation of the hazards, risks and uncertainties of each development option started in FEL-1 to the point that a preferred option can be selected. This option is then further developed applying ISD principles to prepare a Basic Engineering Package (BEP) containing preliminary process safety information (PSI) on materials, technology and equipment. This BEP represents the foundation for further development in FEL-3 and the preparation of a Front End Engineering Design (FEED) package that can be given to an engineering contractor to complete the detailed engineering.

5 FRONT END LOADING 3

Once a single, commercially viable development option has been selected in FEL-2, the project moves to the next phase of Front End Loading (FEL-3), sometimes known as Define, Definition, Detailed Scope or Front End Engineering Design (FEED), which involves improving the technical definition and project execution plan, such that there is confidence in the design, cost estimate and schedule for the option selected in FEL-2. Figure 5.1 illustrates the position of FEL-3 in the project life cycle.

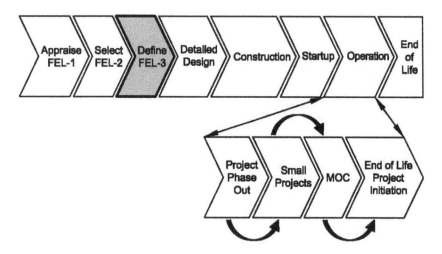

Figure 5.1. Front End Loading 3

Project Management Team

The ultimate goal of the PMT is to confirm the business case to the client and achieve financial approval for project execution. Typical deliverables at the end of FEL-3 include:

- Commercial agreements required for authorization of the project,
- Plan established to manage partner and regulatory approvals,
- Technology options resolved,
- Basis of Design (BOD) finalized,
- FEED Package,
- Cost estimate (typically +25%/-15% or better) and project schedule,
- Contracting and procurement strategy/plan finalized, including contracts/firm prices for main equipment,

- Commissioning/start-up plan and operations/maintenance strategy finalized,
- Training plan,
- Stakeholder outreach plan,
- Overall project risk and uncertainty demonstrated as acceptable, and
- Change management process implemented.

From an engineering perspective, the focus of the PMT is on completing a FEED package that includes all the necessary information required to perform final engineering of the project. This information includes, for example, preliminary details of major equipment, materials of construction, piping/tie-ins, electrical/control system tie-ins, structural steelwork, wiring, buildings, etc. Activities in FEL-2 (Chapter 4) are updated and finalized, and preliminary drawings (e.g. general arrangement/3D model, P&IDs, cause & effect) and datasheets (e.g. relief scenarios, relief valves) prepared.

Environment, Health and Safety

From an EHS perspective, the project team needs to update all EHS risks addressing the full life cycle of the project. Recommendations from EHS studies (including specialist reviews) should be followed-up and satisfactorily resolved. The project EHS Plan needs to be updated to ensure EHS preparedness for commencement of construction. Other FEL-3 activities include confirmation that EHS-related aspects of the engineering design meet or exceed regulatory and corporate requirements and that satisfactory project codes and standards are identified, and design philosophies established. Finally the team needs to ensure that EHS documentation requirements are addressed.

Process Safety

The process safety objectives in FEL-3 generally build on those previously established in FEL-1 and FEL-2, namely:

- Finalize the process safety plan (possibly included in the project EHS Plan),
- Ensure the design meets applicable process safety regulations, standards, and relevant corporate expectations, and
- Further implement the design hazard management (DHM) process to:
  - o Evaluate major hazards through various HIRA studies in line with the established risk management strategy,
  - o Continuously reduce risk through optimization of inherently safer design (ISD),
  - o Refine functional safety, safety critical equipment (SCE) and other design safety measures started in FEL-2 (see Section 5.2.4), and
  - o Set performance standards.

As the level of project definition increases, the evaluation of major hazards involves more detailed, quantitative HIRA studies than was possible in FEL-2. This in turn allows residual risk to be optimized by applying ISD principles, and managed through a diverse range of passive and active safety design measures. Additional process safety activities include developing a resourcing and training strategy, and ensuring integrity management/engineering assurance processes are in place. These process safety studies and activities are discussed below for:

- Evaluating development options and selecting a single option if not completed in FEL-2 (Section 5.1),
- Further definition of the selected preferred option (Section 5.2), and
- Other process safety activities (Section 5.3).

5.1   EVALUATION OF DEVELOPMENT OPTIONS

The evaluation of development options and selection of a single preferred option should have been completed in FEL-2. There may still be minor choices within the preferred option, such as reciprocating vs. centrifugal compressors. However, on occasion more than one development option may be carried forward to FEL-3, especially if one option involves unproven new technology. This is not ideal and implies that insufficient work was performed in FEL-2.

In these circumstances, the process necessary to select a single development option mirrors that described in Chapter 4, Section 4.1. However, the process safety studies performed for each development option may be possible in greater detail as the definition of each option is likely to have increased. For example, instead of high level HAZID studies, preliminary HAZOP studies may be possible if preliminary P&IDs are available. Quantitative risk assessments (QRAs) may even be possible to comparatively assess which option has the lowest residual risk.

5.2   FURTHER DEFINITION OF THE SELECTED OPTION

In FEL-2 (Chapter 4), when the PMT selected their preferred development option, that option was further developed to produce a preliminary project development/execution plan. In FEL-3, the design of the option is progressed to the point where detailed engineering can proceed in the Execute stage of the project. To achieve this, additional technical and commercial studies are undertaken to produce a design package, sometimes known as the FEED package, typically comprising the deliverables detailed in Table 5.1.

Table 5.1. Typical Deliverables in a FEED Package

Deliverables
Cost Estimate
Project Schedule
Final Basis of Design (BOD) - feedstock, product yields and specifications
Process Design
Detailed Process Description
Process Flow Diagrams (PFDs) - Approved for Design
Heat And Material Balance
Operating Philosophy
Utility Design Basis, Philosophy, Loads, Supply Conditions & Detailed Description
Utility Flow Diagrams (UFDs) – Approved for Design
Equipment List (including tag nos.)
Equipment & Instrument Datasheets (service, size, operating & design conditions)
Preliminary P&IDs
Preliminary Cause & Effect Diagrams
Piping Specifications
Final Philosophies (relief, ESD, F&G detection, sparing, effluent disposal, etc.)
Layout / Plot Plans
Preliminary Hazardous Area Classification (HAC)
Preliminary Electrical One Line Drawings
Cable and Pipe Routing Drawings
Civil Work Drawings (foundations, buildings)
Long Lead Equipment Items and Specifications
Long Lead Permits
Register of Safety Critical Equipment
Safety Requirement Specification for SIS

Note: This table is not meant to be all inclusive.

Certain pre-requisites must be determined before the relevant process safety and technical studies and activities can be undertaken to produce the FEED package. Some of the fundamental decisions and data that are necessary may have already been finalized in FEL-2, and include, but are not limited to:

- Finalized statement of requirements (SOR),
- Preliminary project strategy,
- Preliminary cost estimate,
- Financial approval for FEL-3,
- Design codes and standards, which may be national, industry or corporate,
- Design philosophies, e.g. relief, blowdown and flare; fire & gas detection; control system; safety instrumented systems (SIS), emergency shutdown, operations and maintenance; data protection from extreme events, etc.,
- HIRA strategy,
- Meteorological and topographical site data,
- Third-party requirements for project approval, e.g. planning authority, environmental regulator, JV partners,
- Language and units, especially if an overseas project.

With the exception of greenfield developments, new projects are likely to require information about existing facilities and infrastructure. The FEED study will require as-built drawings and up to date equipment records, as well as access to operating and inspection records. Depending on the confidence in available records, a site survey may be required that should address both above and underground equipment, piping and cables.

The relevant process safety studies and activities that contribute to the FEED package are discussed below. It should be noted that, although the studies and activities are listed below sequentially, this is an iterative process that requires some studies and activities to be updated periodically or even repeated under certain circumstances. The checklist in Appendix D can also be used to identify and manage design safety issues.

5.2.1 Design Hazard Management Process

Design Hazard Management (DHM) is a process to identify and evaluate major accident hazards (MAH), and continuously reduce risk through design safety measures. The primary aim is to eliminate or minimize MAH at the source, and prevent remaining hazards from becoming major hazards. This process typically starts in FEL-2 when the development option has been selected with the development of a DHM implementation plan and preliminary HIRA studies.

FEL-3 builds on the earlier work with a strong focus on continuous risk reduction through ISD and more detailed HIRA studies. Other objectives, in the event of a major incident, are to ensure:

- Adequate design integrity to control consequence severity (e.g. fire, explosion, toxic release, etc.) and potential escalation, and
- Mitigation and protection of people, the environment and property.

The DHM implementation plan should be updated to identify all HIRA and other process safety studies that should be completed prior to the project execution stages. The overall goal is a reduction in residual risk to a level that, as a minimum, meets corporate policy by a combination of applying ISD principles and adding risk reduction measures, i.e. design safety measures. Refer to Chapter 4 Section 4.2.1 for additional information and examples of risk reduction measures.

Typical DHM steps in during FEL-3 are:

- Continue ISD optimization
- Refine design safety measures, including functional safety
- Set performance standards
- Re-evaluate major accident risk
- Finalize important safety decisions
- Finalize the BOD

These key steps and other process safety activities in FEL-3 are discussed below.

5.2.2   Inherently Safer Design Optimization

The most effective means of reducing project residual risk is by applying a robust and thorough HIRA study during the FEL-2 and early FEL-3 stages, in which cost effective ISD options are defined and maintained throughout project execution. Therefore, the ISD work started in FEL-2 (see Chapter 4, Section 4.2.2) continues in FEL-3 as the level of project definition increases, until the optimum balance between risk reduction, operability and cost is achieved. By the end of FEL-3, the project should aim to have identified *all* the key ISD measures in the BOD in preparation for project execution. These measures should be recorded in the project risk register (see Section 5.4.2), especially if different resources (e.g. client engineering department and/or contractors) are responsible for FEL and detailed engineering.

If an ISD review was previously conducted on the selected option, it should be updated in FEL-3 using the available information as the definition of the project evolves. The review should consider opportunities for eliminating hazards, optimizing the site layout, and optimizing structural and system integrity, to control the impact of remaining hazards. A combination of What If analysis, checklist and/or brainstorming may be used. The following hierarchical approach to risk reduction (CCPS 2009d) should be used:

- Elimination and minimization of hazards by design,
- Prevention (reduction of likelihood),
- Detection (transmission of information to control point),
- Control (limitation of scale, intensity and duration),
- Mitigation of consequences (protection from effects), and
- Emergency response.

Further information and guidance on ISD reviews is available from the following CCPS publications: *Inherently Safer Chemical Processes, A Life Cycle Approach, 2nd edition* (CCPS 2009d), and *Guidelines for Hazard Evaluation Procedures, 3rd edition* (CCPS 2008b).

5.2.3   Facility Siting and Layout

Site layout and spacing can influence material and construction costs, major accident risk, and the safety of future operations. As a rule, optimizing ISD during layout design reduces cost, complexity and risk. To assist ISD optimization, a facility siting study (FSS) should be conducted to address both off-site and on-site impacts from potential fire, explosion, and toxic hazards. This allows the layout to be adjusted to reduce risks. FSS is discussed in Section 5.2.6.2.

Various aspects of siting and layout are discussed below.

5.2.3.1   Major Accident Risks

Adequate layout and spacing is necessary to prevent a fire or explosion impacting adjacent people, property and equipment, and can minimize the risk of fire or explosion by separating sources of fuel from potential sources of ignition. Proximity to local communities, other industry, offices, shops, public roads and other receptors should be evaluated to reduce risks. For example, there may be several layout options for locating an LPG sphere on a site, but careful evaluation of topography, prevailing wind direction, and location of neighboring buildings that are difficult to evacuate in an emergency (e.g. schools, hospitals, residential, etc.) is necessary to reduce risk. Where possible, consideration should be given to 'green belt' or buffer zones to increase separation.

Toxic releases can travel downwind for long distances before dispersing to concentrations that no longer present acute health impacts. Nevertheless, separation of toxic inventories from local communities can reduce potential impacts, especially if the layout considers the prevailing wind direction. Proximity to sensitive environmental habitats, rivers, and groundwater sources should also be considered.

5.2.3.2   New Layout Considerations

For a greenfield development, there is likely to be greater freedom to optimize the layout than for a brownfield development, where some compromises may be necessary due to limited area, and proximity of local community and neighboring facilities. The FSS should seek to optimize:

- Separation / segregation of hazardous facilities from other hazardous and non-hazardous facilities, such as:
  o   Flammable, explosive and toxic hazards from people,
  o   Flammable hazards from ignition sources, e.g. flares and fired heaters,
  o   Spacing of process units/equipment to limit potential escalation in the event of an incident,
  o   Reactive chemicals from one another.
- Location of process vents.
- Location and protection of access/egress/evacuation routes and emergency response facilities.
- Location and protection of occupied buildings, e.g. offices, control rooms, workshops and living quarters.
- Location and protection of shelter-in-place facilities including HVAC intakes.

Further information and guidance is available from the following CCPS publications: *Guidelines for Siting and Layout of Facilities, 2nd edition* (CCPS 2018a), and *Guidelines for Evaluating Process Plant Buildings for External Explosions, Fires and Toxic Releases, 2nd edition* (CCPS 2012b).

5.2.3.3   Replication of Existing Layout

Some brownfield developments may involve an expansion by the addition of one or more process units that produce the same product(s) as an existing process unit. In this case there is likely to be an incentive to reduce costs by selecting the same technology and design as the existing unit. However, the project should verify whether the existing layout and design meets current industry codes and standards, which may have changed since the original unit was commissioned.

5.2.3.4   General Siting Concerns

Irrespective of whether a project is a greenfield or brownfield development, there are a number of general siting issues that should be addressed:

Spacing

Proper spacing of equipment is one of the most important design considerations for limiting impact from hazards, operability and maintainability as well as facilitating emergency egress and emergency response.   Spacing tables are available for specifying the minimum distance between process equipment.   However, most industry spacing tables for process units and equipment are based on insurance companies' experience of fire consequences (AIA 1968; IRI 1991; NFPA 2015a). These spacing tables are not applicable for process equipment enclosed inside buildings, where vapor dissipation, ventilation, and firefighting accessibility are likely to be impaired vs. open-air process facilities.

If explosion and toxic hazards exist greater spacing distances may be required, and projects should conduct specific consequence analysis or QRA studies to evaluate the required spacing to limit blast damage and acute exposure to toxic vapors.   Even spacing for fire hazards may require variations from the spacing tables based on site-specific circumstances.   In particular, the layout and spacing of multiple aboveground storage tanks (ASTs) containing flammable hydrocarbons requires care to minimize the potential for escalation in the event of a tank fire.

The layout of brownfield sites is sometimes a compromise due to limited spacing flexibility.   However, the provision of passive fire and explosion protection or other design safety measures may justify reduced distances.   Whereas, increased spacing and drainage control may be appropriate for facilities with limited internal emergency response capability.   Local regulations may specify minimum spacing requirements.

Further information and guidance, including spacing tables, is available from the following CCPS publications: *Guidelines for Siting and Layout of Facilities, 2nd edition* (CCPS, 2018a); *Guidelines for Engineering Design for Process Safety, 2nd edition,* (CCPS 2012a).

Drainage and Containment

Well-designed drainage, containment, and sewers ensure that the potential impact of hazardous material leaks and spills is minimized.

Grading and drainage systems should be designed to:

- Carry spills of flammable materials away from equipment and potential sources of ignition, e.g. to prevent pool fires under pressure vessels

- Carry spills of hazardous materials away from occupied buildings and egress/evacuation routes
- Separate clean and oily run-off, discharges and effluents
- Separate incompatible materials, e.g. reactive chemicals
- Remove firewater (at full application rate) from potential incident scenes to prevent hydrocarbons floating on water from spreading to adjacent process areas

Secondary containment systems, such as berms, dikes and curbing, should be provided to prevent spills spreading from major inventories in bulk storage tanks, process vessels and piping, the following parameters should be considered when determining site-specific containment design: area geography, inventory sizes, presence of site personnel, and storm water / firewater drainage requirements. Common practice is to provide, as a minimum, containment for 110% of the largest inventory in the diked area. Alternatively, double-walled tanks may be appropriate for some above ground storage tanks (ASTs) and underground storage tanks (USTs).

Catchment should also be provided for any tank truck loading and unloading operations, and portable container (e.g. drum, ISO tank) storage areas. Catchment generally has less strict requirements, as it is assumed that any spill will be quickly cleaned up since personnel are on-site and would immediately notice a spill. Local regulations and codes may specify the required capacity of the catchment system (e.g. NFPA, ICC, EPA). In the absence of regulation, as a minimum, catchment should hold 100% of the largest compartment of tank truck or the largest container, or 10% of the total volume of all the containers in the area, whichever is larger.

Storage of Bulk Materials

Bulk materials include a wide variety of dry solid chemical feedstocks, products, catalysts and filter media, such as polymers, salts, acids, phosphates, refinery coke, sulfur, diatomaceous earth, and carbon black. They range in particle size from pellets to prills to powders, and require silos, hoppers, bins, and bag storage facilities. These storage facilities invariably require conveyors, pneumatic and fluidization transfer, dust filtration, and bulk bag handling systems.

Further information and guidance is available from the following CCPS publication: *Guidelines for Safe Handling of Powders and Bulk Solids* (CCPS 2004).

Confinement and Congestion

In the event of ignition, confinement of a flammable vapor cloud or combustible dust cloud can result in rapid increases in explosion overpressure. Unconfined clouds usually do not generate sufficient flame speeds to result in overpressure effects. However, even where there is little confinement, high pressures may be generated by turbulence caused by congestion. The factors that dominate the development of overpressure from a vapor cloud explosion (VCE) or dust explosion are:

- Presence of obstacles that cause turbulence,
- Degree of confinement,
- Reactivity of the vapor or dust, and
- Other properties affect the combustion of dusts, including moisture content, particle size, particle size distribution, etc.

Understanding the importance of this mechanism when determining the layout of the project allows an inherently safer design (ISD) with lightly congested plant to reduce overpressures from potential explosions. Chemical and oil refinery plants typically have large amounts of pipework, process vessels and other obstructions that create congestion. Projects should aim to achieve designs where (i) obstacles block less than 40% of the flame path of an explosion, and (ii) avoid closely repeated rows of obstacles that also increase turbulence. Projects should also aim to minimize the volume of confined regions (without increasing flame path congestion), as the explosion is only generated by the confined volume of vapor or dust. The project should also consider methods of handling and processing solids in order to minimize dust formation.

Further information and guidance is available from the following CCPS publications: *Understanding Explosions* (CCPS 2003d), *Guidelines for Vapor Cloud Explosion, Pressure Vessel Burst, BLEVE and Flash Fire Hazards 2nd edition* (CCPS 2010b), and *Guidelines for Safe Handling of Powders and Bulk Solids* (CCPS 2004).

Blast Resistant and Shelter-in-Place Buildings

On-site occupied buildings (e.g. offices, laboratory, workshop, medical center, emergency response facility, security guard house) should be located remote from flammable, explosive and toxic hazards. Where this is not possible, for example in the case of essential personnel who need to be close to process units, adequate protection for the occupants should be provided from the relevant hazards. Siting of permanent and temporary buildings in process areas requires careful consideration of potential effects of explosions, fires and toxic vapors arising from accidental release of hazardous materials.

Protection may take the form of:

- Fire-rated structure to survive potential fires (e.g. jet, pool, etc.),
- Passive/active fire protection to allow occupants to safely egress / evacuate within a reasonable timeframe,
- Blast resistant structure to survive potential explosion overpressures,
- Positive internal pressure, air locks, and tightly closing doors and windows (if any) to exclude flammable/toxic vapors and products of combustion (smoke, CO/CO_2),
- HVAC system equipped with flammable/toxic gas detection, emergency shutdown, and automatic closure of louvres,
- Alternative egress / evacuation routes.

In some cases, occupants may be required to remain in the building for an extended period of time during an incident, e.g. operators controlling safe shutdown of facilities, personnel awaiting evacuation from offshore platforms, and personnel required to shelter-in-place in accordance with the site emergency response plan. In these circumstances, the building structure must maintain its integrity for the required timeframe with an internal environment that does not impair the health and safety of the occupants. Some jurisdictions have regulations that specify minimum requirements, e.g. onshore control rooms and buildings (ICC 2018); frequency of impairment of living quarters/temporary refuge on offshore platforms in the North Sea (HM Government 1995).

Further information and guidance is available from the following CCPS publication: *Guidelines for Evaluating Process Plant Buildings for External Explosions, Fires and Toxic Releases, 2nd edition* (CCPS 2012b). API also offers guidance: *Management of Hazards Associated with Location of Process Plant Portable Buildings, RP 753*, (API 2007), and *Management of Hazards Associated With Location of Process Plant Buildings, 3rd edition, RP 752*, (API 2009).

Utility Routing and Locations

Most projects require new, or modification of existing, utility systems. These systems include, but are not limited to, electrical power (various voltages), steam (various pressures) and condensate, water (cooling, boiler feed, process, potable, firewater), air (plant, instrument), nitrogen, natural gas, and oil (fuel, heating, lube).

Particular care is required in planning the routing and location of these utility systems, which may be above and/or underground. Many of the utilities, with the exception of electrical power, are normally arranged in pipetracks as an integral part of a process unit (located in the center of the unit) or as an arterial part connecting several services to/from other process units.

While economics demand the shortest possible routes for power lines and pipetracks, consideration needs to be given to terrain, topography and access for construction and maintenance of process equipment, and to allow for future expansion. Terrain and topography factors include soil type (e.g. dry land, marsh, mountainous, desert, shallow water), and gradients on piping (e.g. requirement for condensate drainage, avoidance of low points). Access requirements may include:

- Utility headers supplying whole process unit,
- Adequate overhead height for cranes and other tall vehicles,
- Adequate depth of cover for underground services,
- Road, railroad, river/canal, ditch crossings,
- Space for meter runs, pig launchers/receivers, sub-stations, and other equipment,
- Valve access and maintenance platforms.

From a process safety perspective, it may be appropriate to route one or more utilities required in an emergency by separating from major hazardous inventories in vessels and pipelines. HIRA studies and layout reviews should be used to identify any critical utility systems that are vulnerable from major accident hazards.

Future Expansion

Design reviews and HIRA studies should consider layout options for potential future expansion. Future expansion, such as the addition of extra reactors, process units and/or hazardous inventories, could compromise ISD due to area constraints and the proximity of major hazards to on-site and off-site populations.

Unit Accessibility

Access should be provided to all process units and areas of the site from at least two directions to facilitate emergency response in the event of an incident. Design reviews should also consider access within process units for equipment requiring:

- Frequent inspection, testing and preventive maintenance (ITPM),
- Space requirement for repair, e.g. area for pulling heat exchanger bundles,
- Space requirement for eventual replacement of equipment with short life cycle, e.g. wear-out of glass lined vessels,
- Space requirement for loading/unloading catalyst and/or packing in reactors and columns.

Constructability reviews should consider layout requirements for access to install major items of equipment, such as tall fractionation towers, and large compressors in a compressor house.

Other Layout Considerations

Other layout concerns may include a variety of project specific issues depending upon the hazards and design safety measures, such as:

- High thermal radiation exclusion zone around foot of elevated flare stack(s),
- Blast barrier(s) around high pressure vessels and equipment with explosion potential,
- Space for explosion venting,
- Vent vapors to safe location,
- Separation between frequently opened or maintained equipment and people, high temperature vessels / piping, etc.,
- Structures or buildings with lifting devices, such as a compressor house with vertical clearance for a bridge crane,
- Other structures, such as cooling towers, electrical sub-stations, satellite instrument houses, metering stations,
- Low occupancy buildings, such as warehouses with shipping / receiving facilities and road access,
- Effluent treatment facilities
- Coordination of vehicle and pedestrian traffic,
- Storage location of emergency response equipment with respect to hazards,
- Transportation facilities, e.g. road and rail loading/unloading rack, marine jetty, helipad.

5.2.4   Refine Design Safety Measures

When ISD optimization has exhausted opportunities to eliminate hazards, the remaining major accident hazards (MAH) require additional risk reduction through design safety measures, including functional safety, that typically cover a diverse range of passive and active measures. These design safety measures reduce risk by MAH prevention, control, mitigation and/or emergency response.

In reality the design hazard management process is iterative, and Sections 5.2.4 through 5.2.7 should be read together as the refinement of design safety measures may require several iterations before finalization. Various HIRA (see Section 5.2.6) and technical safety assessments (see Section 5.2.7) should be used to identify and evaluate the design safety measures and their performance standards (see Section 5.2.5). The aim should be a reduction in risk that meets or exceeds corporate and/or jurisdictional risk tolerance criteria. Some of the design safety measures are likely to be safety critical equipment/elements (SCE) including SIS (see Section 5.2.7.3).

All MAH (with their causes and consequences) and design safety measures should be recorded in the project risk register (see Section 5.4.2) for eventual handover to the future Operator.

Further information and guidance on functional safety is available from the following publications: *Guidelines for Safe Automation of Chemical Processes*, 2nd edition (CCPS 2017b); *Functional safety - Safety instrumented systems for the process industry sector - Part 1: Framework, definitions, system, hardware and application programming requirements*, IEC 61511-1 (IEC 2016).

5.2.5   Set Performance Standards

Each design safety measure has been selected for its risk reduction, and should have a specified performance standard that is required throughout the life cycle of the facility. The performance standard should specify its required function (including process fluid, parameters), reliability, availability, and survivability to perform during and after a major accident. Examples of performance standards are:

- An emergency diesel generator must start within X seconds so that safety critical equipment are available.

- The structure of a temporary refuge (TR) on an offshore platform must survive a major fire/explosion, and maintain a comfortable safe indoor atmosphere (not excessive temperature, free of smoke/flammable/toxic gas, etc.), for X hours.

- The control system may be required to protect recorded and historical data in extreme events (such as explosions, flooding, earthquakes, etc.) so that it is not compromised or lost completely in the event. Requirements may involve locating servers a minimum distance from hazardous areas, or physically protecting servers from explosion effects, or routine automated off-site archiving and backup of key process data.

Design safety measures should be tested at regular intervals to ensure reliability and availability.

Verification of performance in practice will be demonstrated later through a factory acceptance test (FAT) and mechanical completion and commissioning tests (see Chapters 6, 7, and 8).

Case Study: Emergency Shutdown Valve Performance Standard Inadequate

An oilfield reinjected gas to maintain reservoir pressure. The high pressure injection system tripped, which resulted in the pressure safety valve (PSV) on the gas compressor inlet scrubber lifting to relieve pressure to the flare system.

An investigation found that all safety related devices including the safety instrumented system (SIS) performed according to design. However, the emergency shutdown valve (ESDV) installed at the inlet of the scrubber was not capable of closing fast enough to prevent a buildup of pressure when the downstream ESDV closed. This caused the PSV to lift.

The closing time of an ESDV can be an important performance standard that needs to be established during design.

The investigation recommended establishing a design process to evaluate process risks and scenarios more thoroughly, and consequently specify performance standards for all safety critical equipment (SCE) to be applied during procurement. A second recommendation advocated re-assessing scenarios and performance standards as a result of commissioning experience. These actions were intended as essential elements of the project quality management program.

5.2.6   Hazard Identification and Risk Analysis (HIRA)

The HIRA strategy developed in FEL-2 (see Chapter 4 Section 4.3.4) may need to be finalized, and the HIRA studies and their timing documented in the EHS / Process Safety Plans may require updating for the evolving design in FEL-3. In line with the strategy, increasingly detailed HIRA studies should be conducted to identify intrinsic and extrinsic hazards, and evaluate the risks associated with the hazards. As a general rule, the study facilitator should be independent of the project.

The selection, scope, and methodology of these studies varies based upon the specific project, company preference, and local jurisdiction. For larger projects with complex processes, the preferred HIRA methodologies among major chemical and oil & gas companies are a combination of HAZOP, LOPA and QRA. For simple MOCs and non-process projects it may be appropriate to use checklists and/or What If studies. However, some or all of the following studies may be appropriate:

5.2.6.1   Hazard Identification

Building on the HAZID study(s) performed in FEL-1 and/or FEL-2, a preliminary HAZOP may be conducted as soon as preliminary P&IDs are available. While the HAZID focus is on major accident hazards, the purpose of the HAZOP is to systematically review a process unit to determine whether process deviations lead to undesirable consequences.   It also identifies operability problems that could compromise the unit's ability to achieve design intent and productivity.

For HAZOP studies to be effective, a process safety engineer, or other competent facilitator, should identify nodes (sections of each process unit) and lead a multi-disciplinary team, knowledgeable in the process, identifying consequences of any potential deviations from the design intent of each node.   Then existing safeguards are evaluated against these intrinsic hazards to determine their adequacy.

The preliminary HAZOP should be repeated during the detailed design stage of the project using final P&IDs when change management is initiated to control any late design changes.

As an alternative to HAZOP, some projects may elect to use a checklist, What If or other equivalent methodology to identify hazards.   As the design evolves in FEL-3, some projects may also update the extrinsic (e.g. transportation hazards during construction) hazard findings from the earlier HAZID of the selected option as input to a QRA (see below).   Irrespective of methodology, the findings from all hazard identification studies should be added to the Hazard Register.

Further information and guidance is available from the following publications: *Guidelines for Hazard Evaluation Procedures, 3rd edition* (CCPS 2008b); *API Recommended Practice for Design and Hazards Analysis for Offshore Production Facilities, 2nd edition, API RP 14J,* American Petroleum Institute (API 2001a); and *Petroleum and Natural Gas Industries – Offshore Production Installations – Guidelines on Tools and Techniques for Identification and Assessment of Hazardous Events,* ISO/DIS 17776, International Organization for Standardization (ISO 1999).

5.2.6.2   Consequence Analysis

A number of studies use a consequence analysis approach to quantify the impact of undesired events on people, property and the environment. These studies model the consequences of releases of hazardous materials in terms of vapor dispersion, thermal radiation, blast overpressure, and toxicity.   If the consequences are unacceptable to local regulators and/or the client, a risk analysis approach may be necessary (see Section 5.2.6.3).

Examples of consequence analysis studies are listed in the following paragraphs.

Facility Siting Study

If a preliminary facility siting study (FSS) was conducted in FEL-2, it should be updated as the definition of the project evolves.   This iterative approach usually continues through the detailed design stage of the project.   The FSS is used to determine the off-site and on-site impacts from potential fire, explosion, and toxic hazards, and is therefore integral to ISD optimization and layout development.

Information on the layout/plot plan, hazardous inventories, PFDs and preliminary P&IDs, heat and mass balances, locations of off-site receptors (residential, commercial, industrial, hospital, etc.), and key occupied building locations within the site is required.

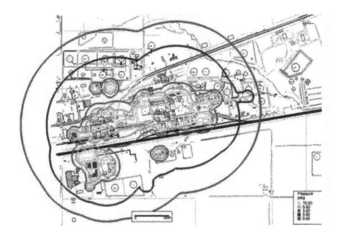

Figure 5.2. Example of Overpressure Contour Plot

Typically, the focus of a FSS is on consequence analysis to determine the location and magnitude of potential fires, explosions and toxic concentrations from a range of release sizes. The results of the FSS may be presented in a range of meaningful forms, such as spreadsheets and contour plots (see example in Figure 5.2). Blast impacts may be presented in terms of building damage level (BDL), and flammable and/or toxic impacts shown in terms of concentration categories predicted to reach each building. Composite contour plots of overpressure, flammability, and toxicity define key endpoints and can provide an overview of areas vulnerable to impacts from the assessed hazards. These vulnerabilities aid in the identification of potential mitigation strategies, such as occupancy reduction, upgrading a building's blast resistance, building relocation to areas less susceptible to damage, and installing flammable or toxic gas detection with automatic or manual ventilation shutdown.

Serious off-site impacts are sometimes difficult to eliminate or reduce, and some companies may evaluate risks within the FSS to see if they meet jurisdictional and/or company risk tolerance criteria. For example, a QRA may be conducted to evaluate the risk to a local community from a certain size of release of a toxic chemical. If the risks exceed the jurisdictional and/or company risk tolerance criteria, the project may need to be relocated or cancelled unless the risk can be reduced. A number of risk reduction options may be possible, including, but not limited to, eliminating or reducing inventory of hazardous material(s), changing the layout to provide greater separation distance, additional design safety measures, etc.

The scope of the FSS may be broadened to combine with one or more of the HIRA studies below.

Further information and guidance on facility siting is available from the following CCPS publications: *Guidelines for Siting and Layout of Facilities, 2nd edition* (CCPS 2018a); *Guidelines for Evaluating Process Plant Buildings for External Explosions, Fires and Toxic Releases, 2nd edition* (CCPS 2012b).

Additional guidance on on-site facility siting is available from the following API publications: *Management of Hazards Associated with Location of Process Plant Buildings*, 3rd edition, RP 752 (API 2009); *Management of Hazards Associated with Location of Process Plant Portable Buildings*, RP 753 (API 2007); *Management of Hazards Associated with Location of Process Plant Tents*, RP 756, (API 2014).

Further information and guidance on consequence analysis is available from the following CCPS publications: *Guidelines for Chemical Process Quantitative Risk Assessment, 2nd edition* (CCPS 2000); *Guidelines for Use of Vapor Cloud Dispersion Models, 2nd edition* (CCPS 1996b); *Estimating Flammable Mass of a Vapor Cloud* (CCPS 1998a); *Wind Flow and Vapor Cloud Dispersion at Industrial and Urban Sites* (CCPS 2002); *Guidelines for Vapor Cloud Explosion, Pressure Vessel Burst, BLEVE and Flash Fire Hazards, 2nd edition* (CCPS 2010b).

Fire Hazard Analysis

The preliminary fire hazard analysis (FHA) conducted in FEL-2 (see Chapter 4 Section 4.2.12) to determine the location, type (e.g. jet, pool), size (thermal radiation, flame impingement) and duration of potential fires should be updated to reflect the latest layout and design. The FHA should be based on a range of release sizes from the flammable inventories, and also address any combustible materials. It should be integrated with the ISD optimization work to evaluate requirements for spacing, drainage, containment, active and passive fire-protection, insulation (from radiant heat), and other design safety measures in line with the project fire protection philosophy. It should identify locations and design/performance criteria for fixed fire protection systems (fire hydrants, monitors, deluge systems, foam systems, etc.).

The FHA requires the following information: fire protection philosophy, plot plan, facility siting study (or offshore fire and explosion study), and HIRA studies.

Further information and guidance is available from the following CCPS publications: *Guidelines for Fire Protection in Chemical, Petrochemical, and Hydrocarbon Processing Facilities* (CCPS 2003b); *Guidelines for Consequence Analysis of Chemical Releases* (CCPS 1999); and *Guidelines for Chemical Process Quantitative Risk Assessment*, 2nd edition (CCPS 2000).

Offshore Fire and Explosion Analysis

For offshore developments, the preliminary fire and explosion analysis conducted in FEL-2 (see Chapter 4 Section 4.2.7) on the preliminary platform or rig layout should be updated to reflect the latest layout and equipment design. This consequence analysis will provide data on the location, type, size and duration of potential fires, and the magnitude of explosion overpressures at various locations. These data will be used by other studies, such as structural integrity, safety critical equipment/element (SCE) vulnerability, and temporary refuge impairment. Toxic consequences, such as H_2S releases may also be assessed, if applicable.

Further information and guidance is available from the following publications: HSE, *Prevention of fire and explosion, and emergency response on offshore installations. Offshore Installations (Prevention of Fire and Explosion, and Emergency Response) Regulations, 1995. Approved Code and Practice and Guidance*, L65, 3rd Edition, Health and Safety Executive, 2016; and *UKOOA Guidelines for Fire and Explosion Hazard Management*, UKOOA, 1995.

Smoke & Gas Ingress Analysis

For an onshore facility with emergency response procedures that include shelter-in-place (SIP), a study should be conducted of the impact of fire, and releases of flammable and toxic materials on each SIP building.

In particular, the study should evaluate:

- Thermal radiation,
- Building porosity/tightness (i.e. closure of doors, windows, HVAC louvres, penetrations, holes),
- Requirement for positive pressure,
- Detection of products of combustion (particulates, carbon monoxide, carbon dioxide, etc.) and flammable/toxic vapors in HVAC inlet, and
- Isolation of HVAC and ventilation systems.

This study may be combined with the facility siting study to ensure that the integrity of the SIP building is not compromised by potential explosions.

In respect of offshore platforms/facilities, a similar smoke and gas ingress analysis (SGIA) should be performed to ensure that in an emergency situation a temporary refuge (TR) can provide life support for a period of time until complete evacuation can occur. The study should evaluate whether the temporary refuge is designed for the relevant accident scenarios and the levels of explosion overpressure, thermal radiation, smoke, and toxic gas to which it could be exposed to. It also ensures that breathable air is maintained in the TR by limiting ingress of smoke, gases, and other combustion productions resulting from external fires, and ensures that smoke does not hinder full and safe evacuation of the installation. Some jurisdictions (HM Government 1995, 2005) specify criteria, such as the duration that the TR must survive without impairment of life support.

Both onshore and offshore studies typically require data, such as: plot plans/3D models, major accident hazards, facility siting/fire and explosion study, and SIP/TR building design.

Further information and guidance is available from the following publications: NIST, *Airtightness Evaluation of Shelter-in-Place Spaces for Protection Against Airborne Chemical and Biological Releases*, NISTIR 7546, National Institute of Standards & Technology, U.S. Dept. of Commerce, Gaithersburg, MD, 2009; MFB, *A Best Practice Approach to Shelter-in-Place for Victoria*, Metropolitan Fire & Emergency Services Board, Victoria, Australia, 2011; HSE 2013, *Modelling Smoke and Gas Ingress into Offshore Temporary Refuges*, Research Report RR997, Health & Safety Laboratory, UK, 2013.

5.2.6.3   Risk Analysis

Quantitative Risk Analysis

Some companies commence a quantitative risk analysis (QRA) of the selected option in FEL-3 that is finalized during the detailed design phase. The QRA builds on the preliminary risk analysis performed in FEL-2, using more detailed information. For example, the design may have evolved such that detailed information on plot plan/layout, PFDs, preliminary P&IDs, heat and mass balances, major equipment, design philosophies, isolation valve placement, and population/occupancy data should be available.

QRAs are typically performed if other HIRA studies indicate that potential consequences of major accident hazards are significant despite ISD and DHM measures at this stage of the project. Depending upon scope, the QRA provides a numerical estimate of onsite and/or offsite risk exposures to people, property, the environment, or other areas of interest. This allows risk levels to be compared with corporate and/or jurisdiction risk tolerance criteria, and provides input on decisions regarding strategies to mitigate risk, such as potential issues with plant layout, building locations, structural blast resistance, etc., which need to be resolved prior to the detailed design stage of the project.

Further information and guidance is available from the following publications: *Guidelines for Chemical Process Quantitative Risk Analysis*, 2nd edition (CCPS 2000); *Guidelines for Developing Quantitative Safety Risk Criteria*, (CCPS 2009b); *Guide to Quantitative Risk Assessment for Offshore Installations*, The Centre for Marine and Petroleum Technology (CMPT 1999).

Transportation Studies

Preliminary studies conducted in FEL-2 for road, rail, pipeline, marine and/or air transportation of hazardous materials should be updated to reflect the evolving design of the facilities. These studies will require the latest information on cargo details, routes, frequency of operation, and facility (road tanker/truck, railcar, pipeline, vessel, terminal) inventories and design.

The aim of these studies is to better understand the following items, so that the PMT can make more informed decisions on transportation:

- Features of various modes of transport
- Types of incidents that might occur
- Characteristics of alternative routes
- Design and complexity of various distribution systems
- Depth and rigor of management systems (types/frequency of inspections/maintenance, shipment tracking, stewardship, etc.)
- Safety performance of carriers and other logistics service providers
- Possible options to reduce risks

Further information and guidance is available from the following CCPS publications: *Guidelines for Chemical Transportation Risk Assessment*, (CCPS 1995a); *Guidelines for Chemical Transportation Safety, Security, and Risk Management*, (CCPS 2008a).

Pipeline safety is regulated by many jurisdictions, e.g. USA (49 CFR 190-199), Canada (CSA Z662), and UK (The Pipelines Safety Regulations 1996). Although these regulations may differ by jurisdiction, they generally address the design, construction and safe operation to safeguard pipeline integrity. Some jurisdictions also cover land use planning to create separation of pipelines carrying hazardous fluids from sensitive receptors, e.g. local communities and environment. Irrespective whether local regulations exist, it is good process safety practice to apply similar risk analysis and design principles to pipeline routes as those for siting of process units.

In respect of U.S. marine traffic, a final waterway suitability assessment (WSA) must be submitted no later than when an application is filed with the relevant authorities. Based on the preliminary WSA, it should identify credible security threats and safety hazards to liquefied natural gas (LNG) and liquefied hazardous gas (LHG, i.e. LPG and other listed chemicals) marine traffic in the port and along the vessel transit route. Additionally, it should identify appropriate risk management strategies, mitigation measures and resources needed to carry out those measures, and address comments from the Coast Guard and other authorities on the preliminary WSA.

Further information and guidance is available in 33 CFR 127 *Waterfront Facilities Handling Liquefied Natural Gas and Liquefied Hazardous Gas* (subpart 007 Letter of Intent and Waterway Suitability Assessment).

Layer of Protection Analysis

A layer of protection analysis (LOPA) is a semi-quantitative risk analysis tool used to determine the risk of individual hazard scenarios. It should be used selectively following a hazard evaluation, e.g. HAZOP study, to assist judgments on the sufficiency of safeguards for certain major accident hazard scenarios, such as, depending on company/project, those potentially resulting in serious injuries or one or more fatalities. In particular, LOPA is typically used for determining if a safety instrumented system (SIS) is appropriate, if an additional safeguard/risk reduction is necessary. If a SIS is chosen as the risk reduction measure, LOPA is also the preferred industry methodology for determining its required reliability, i.e. safety integrity level (SIL) for each safety instrumented function (SIF). LOPA is also used by some companies as an alternative to QRA; for example, combined with cost benefit analysis to determine economic benefits of different risk reduction measures.

The technique involves identifying safeguards that meet specific criteria as independent protection layers (IPLs). IPLs are assigned a performance factor that reflects the reliability of the protection layer along with mitigating factors or conditional modifiers to provide a rough estimate of the likelihood of the scenario that is then compared to risk criteria. The LOPA is typically updated/finalized during the detailed design phase of the project.

Further information and guidance is available from the following CCPS publications: *Layer of Protection Analysis: Simplified Process Risk Assessment*, (CCPS 2001a); *Guidelines for Initiating Events and Independent Protection Layers in Layer of Protection Analysis*, (CCPS 2015b); *Guidelines for Enabling Conditions and Conditional Modifiers in Layer of Protection Analysis*, (CCPS 2013a).

5.2.7   Safety Assessments

There are a variety of technical safety studies that are commonly performed in the FEL-3 stage of a capital project in order to mitigate the risks identified by HIRA studies. For example, the facility siting study findings may identify hazard vulnerabilities that can lead to potential mitigation strategies. While these technical safety studies typically commence in FEL-3, they are unlikely to be finalized until the detailed design stage of the project.

Not every technical safety study described below will be appropriate for every project, especially MOC and other small projects. It is therefore important to be clear about the objective, scope and methodology to be used in the following safety studies:

5.2.7.1   Hazardous Area Classification

If electrical equipment is used in or around a location where flammable gases/vapors/liquids or combustible dusts may be present, there is potential for fire or explosion. Locations where flammable/explosive atmospheres may occur under normal (e.g. sampling, venting) or unplanned (e.g. pump seal leak) operations are referred to as hazardous (or classified) areas. A hazardous area classification (HAC) assessment should be conducted in FEL-3 to determine classified areas. The assessment identifies areas within the facility where electrical equipment may need to be appropriately classified to prevent ignition, and can be used to optimize plot plan layout in terms of potential classified equipment cost. HAC does not apply to catastrophic failures, e.g. vessel or piping rupture.

The HAC assessment should be based on national regulations and/or industry codes. There are two classification systems: the class/division system used predominately in North America (e.g. API RP 500, NFPA 497); and the zone system used in the rest of the world (e.g. EI 15, IEC 60079-10-1, IGEM SR25, CENELEC 60 079, DSEAR[3], Gost R 51330-X-99, API RP 505; IEC 61241-10 / IEC 60079-10-2 apply to combustible dust or fiber hazards).

Once the hazardous areas have been identified, the technical specification of equipment with the potential to cause ignition must meet applicable national regulations and/or industry codes (e.g. NEC 500/505, ISA-12.04.01, ATEX Directive, IEC 60079 multiple parts) for the appropriate area classification.

Further information and guidance is available from the following publications: *Recommended Practice for Classification of Locations for Electrical Installations at Petroleum Facilities Classified as Class I, Division I and Division 2*, 3rd Edition, API RP 500; *Model Code of Safe Practice Part 15: Area Classification Code for Installations Handling Flammable Fluids*, 4th edition, EI 15 (formerly IP 15); *Explosive Atmospheres - Part 10-1: Classification of Areas - Explosive Gas Atmospheres*, 2nd edition, IEC 60079-10-1.

5.2.7.2   Safety Instrumented System Assessment and Safety Integrity Level Determination/Verification

If a particular hazard cannot be eliminated or sufficiently mitigated through ISD principles or other IPLs, such as alarms, pressure relief, control loops, etc., it may be necessary to design a safety instrumented system (SIS) to reduce the risk. For example, risk reduction measures like spacing and segregation should be exhausted before determining any requirement for SIS. LOPA (see 5.2.4 above) is normally used to identify if additional protection layers are necessary, and if they are to be provided using safety instrumented functions (SIFs).

---

[3] HM Government, *The Dangerous Substances and Explosive Atmospheres Regulations*, Statutory Instruments 2002 No. 2776, Health & Safety, UK, 2002.

In order for a SIF to mitigate a major accident hazard, it must typically respond on demand and therefore should have a high reliability. LOPA is most commonly used (although other methods include risk graph or QRA) to determine how reliable each SIF needs to be, i.e. its safety integrity level (SIL). The required SIL rating should then be used to design each SIF, in line with functional safety standards, such as IEC 61511 and ANSI/ISA 84.00.01. Consideration should also be given to functional performance (c.g. speed of response) and survivability (e.g. ability to meet design intent in event of fire/explosion) of each SIF. Finally, a SIL verification (reliability analysis) is performed to show that the selected SIF sensors, logic and final elements will achieve the required reliability performance (SIL target) for the selected testing regime.

Although the capital cost of a SIS may be attractive to the PMT compared to the cost of alternative risk reduction options, the life cycle cost may be significant due to the frequency of ITPM and the demand on operational discipline to meet the SIL rating over the life of the facility (Broadribb & Currie, 2010). The final decision on SIS should be a joint agreement between the PMT and the future Operator, and application of the As Low As Reasonably Practicable (ALARP) concept may assist the decision.

Further information and guidance is available from the following CCPS publications: *Guidelines for Safe Automation of Chemical Processes, 2nd edition,* (CCPS 2017b); *Guidelines for Safe and Reliable Instrumented Protective Systems,* (CCPS 2007c).

5.2.7.3   Safety Critical Equipment

The prevention and mitigation of major accidents relies upon appropriate layers of protection or barriers working on demand. Criticality ranking for process safety purposes is being increasingly applied to identify the subset of equipment that is critical to the management of major accident hazards, and therefore requires a high reliability. These items of equipment are known as *Safety Critical Equipment or Elements* (SCE).

Safety Critical Equipment / Element

Equipment, the malfunction or failure of which is likely to cause or contribute to a major accident, or the purpose of which is to prevent a major accident or mitigate its effects.

(CCPS 2017)

Some jurisdictions define SCE using a consequence approach, rather than the risk-based definition above. Such approaches result in a significant proportion of the facility master equipment list being designated as SCE, which can present challenges in managing ITPM (Broadribb 2016).

It should be noted that SCE can appear on both sides of a typical bow-tie model (CCPS 2018c) as a tool for communicating how barriers may cause, prevent, control and mitigate major accident hazards (MAHs). For example, some categories of SCE are illustrated in Table 5.2.

A study should be performed during FEL-3 to identify SCE and to determine its required function and reliability, which should be documented in performance standards. This is then used as a basis for detailed design of individual items and components of equipment. The two main methods for determining SCE are (i) logic trees (CCPS 2017a, Broadribb 2016), and (ii) identifying safeguards in HIRA studies. The first step involves identification of MAHs, followed by identifying equipment, systems, structures, etc. that can cause, contribute to, prevent, mitigate, or help recover from a MAH. These SCE should be recorded in the risk register and for eventual handover to the future Operator. The following information is required to determine SCE: HIRA study reports, master equipment list (MEL), design intent/function of equipment, equipment data sheets, and SIS assessment reports.

Table 5.2. Typical Examples of Safety Critical Equipment / Elements

Prevention	Detection	Control	Mitigation
Hydrocarbon Containment (vessels, piping, tankage)	Fire Detection (flame, smoke)	ESD System	Firewater Systems (pumps, deluge, monitors, foam, hydrants, etc.)
Ignition Prevention (intrinsically safe electrical equipment)	Gas Detection (flammable, toxic)	Relief, Flare and Blowdown System	Passive Fire Protection
Navigation Aids		Safety Instrumented System (SIS)	Blast Walls
Structural Integrity		Uninterruptible Power Supply	Communication Systems (alarm, public address)
Buoyancy Integrity (ballast system)		Excess Flow Valves	Shelter-in-Place / Temporary Refuge

Further information and guidance is available from the following publications: *Guidelines for Asset Integrity Management* (CCPS 2017a); *Guidelines for the Management of Safety Critical Elements, 2nd edition* (EI 2007).

5.2.7.4   Vulnerability Analysis of Safety Critical Equipment

In addition to being highly reliable, SCE needs to survive major accident hazards, such as fires and explosions, if it is to meet its design function of mitigating the same major accident hazards and/or preventing and minimizing escalation.   A vulnerability analysis should be conducted to systematically review each SCE (including the control system) to determine its vulnerability to major accident hazards that could potentially stop it from functioning.   If an unacceptable vulnerability is identified, the technical specification of the SCE should be modified or suitable protection provided.

The results from other studies (e.g. HIRA, fire hazard analysis, offshore fire and explosion analysis) are used to identify potential impacts (e.g. thermal radiation, blast overpressure) from major accident hazards at the location of the SCE.   The preliminary fire and gas (F&G) detection, and preliminary emergency shutdown (ESD) studies may also provide input.

Further information and guidance is available from the following publications: *Recommended Practice for the Design of Offshore Facilities Against Fire and Blast Loading* (API 2006).

5.2.7.5   Reliability, Availability and Maintainability Study

A reliability, availability, and maintainability (RAM) study should be performed to identify possible causes of production losses.   This high level analysis simulates the configuration, operation, failure, repair, and maintenance of equipment to determine average production levels over the facility life.   By detecting failures early in the design process, decisions regarding alternative process options, such as duplicating process trains and/or adding spare equipment, can be made to optimize efficiency. Other benefits of a RAM study include identification of production bottlenecks, maintenance priorities, and essential equipment spare parts.

RAM studies typically use simulation or analytical models, based on fault tree, block diagrams, Markov, or Petri net methods, and require the following information: P&IDs, electrical schematics, equipment/component configuration and functional specification, expected modes of operation, and maintenance philosophy.

Further information and guidance is available from the following publications: *Product Assurance, Reliability, Availability, and Maintainability*, Army Regulation 702–19, (U.S. Army 2015); *Handbook of Reliability, Availability, Maintainability and Safety in Engineering Design* (Stapelberg R.F., 2009).

5.2.7.6   Temporary Refuge Impairment Assessment

Some offshore jurisdictions require all major accident hazards (MAH) to be identified and their potential for impairment of the installation's temporary refuge (TR) assessed.   The design of the fabric, systems and supporting structure that make up the TR needs to ensure that impairment risk is sufficiently low within the duration required for its survival, i.e. muster and evacuation. A TR impairment assessment should therefore be conducted.

The assessment should use a risk analysis approach. QRA is typically used, although qualitative and semi-quantitative methods may be used where risks are low enough that the impairment risk is not expected to be intolerable. Input from the following process safety and technical studies described above is used in the assessment:

- HIRA (especially MAH identified in the HAZID),
- Offshore fire and explosion analysis,
- Smoke and gas ingress analysis,
- Safety critical equipment/elements (SCE),
- Vulnerability analysis of SCE.

The TR impairment frequency is the sum of all impairment event probabilities, and should be compared to jurisdiction (and company, if appropriate) risk criteria. A similar approach may be applied to onshore SIP buildings.

Further information and guidance is available from the following publications: *Guidance on Risk Assessment for Offshore Installations*, Offshore Information Sheet No. 3/2006, HSE.

5.2.7.7   Evacuation, Escape, and Rescue Analysis

An evacuation, escape and rescue (EER) study should be conducted to evaluate the performance of the emergency response facilities and procedures for an offshore installation.   The EER study addresses the following emergency response equipment:

- Escape routes (including bridge links to other installations, if appropriate),
- Muster area(s) and facilities in the temporary refuge,
- Evacuation equipment (including helicopter and helideck operation, lifeboats, life rafts, and escape chutes),
- Rescue arrangements, such as stand-by boats, SAR helicopters, and non-specific marine craft in the locality.

The EER study is typically undertaken in conjunction with a QRA, and consists of a structured review of the performance of the escape, evacuation and rescue facilities and procedures under representative scenarios. The following information is required: preliminary emergency response plan, FSS/FEA, and QRA.

Further information and guidance is available from the following publication: HSE, *Evacuation, Escape and Rescue (EER) Topic Guidance*, Offshore Division, August 2015.

5.2.7.8   Dropped Object Study

A dropped object study involves a qualitative or quantitative risk assessment of impacts caused by accidentally dropped object loads (or dragging anchors) within the safety zone of an offshore installation. The goal of the study is to ensure that the risks to subsea wellheads and pipelines, and topsides equipment and structures by dropped objects during vessel, lifting and overside operations are understood. Where pipelines and facilities contain hydrocarbons, any loss of containment (LOC) could have potentially catastrophic consequences. The study highlights areas of concern (i.e. risks that exceed jurisdiction or corporate tolerance criteria), and assists decision-making on the most efficient risk reduction measures. A dropped object study may also be performed for an onshore facility.

Factors such as the object's mass and shape, water depth, and sea currents influence the energy of a dropped object when it strikes the seabed. As a general rule, impact energies of greater than 50kJ have the potential to cause significant damage to subsea equipment resulting in likely LOC. Even energies in the range 30 to 50kJ can cause damage and LOC, although the integrity of subsea trees should not be impaired. The following information is required: design and materials of construction details for subsea equipment, pipelines, structures, and topsides equipment; consequence models; load movement details; marine activity; and emergency response plan.

Further information and guidance is available from the following publication: Alexander, C., *Assessing the Effects of Dropped Objects on Subsea Pipelines and Structures*, Paper No. IOPF2007-110, Proceedings of ASME International Offshore Pipeline Forum, October 2007, Houston, Texas.

5.2.7.9   Security Vulnerability Analysis

A security vulnerability analysis (SVA) is a review of handling, storing, and processing of hazardous materials at the facility (including offshore installations) from the perspective of an individual or group intent on causing sabotage/terrorism by deliberately causing a major accident with large-scale injury/fatality or supply disruption impacts. It considers potential scenarios by analyzing inventories and the production process involving hazardous materials, potential pathways of attack, and existing security countermeasure or ring of protection.

While a QRA approach may be applied to the SVA, it is resource intensive and not warranted in many cases. A tiered approach should be used in line with industry guidance (ACC 2001; CCPS 2003a; API 2003a). These methodologies comprise the following steps:

1. Security vulnerability screening using tools, such as the *CCPS Security Vulnerability Enterprise Screening Tool*[4], to produce a list of prioritized facilities.

---

[4] available for download on the CCPS website: www.aiche.org/ccps/security-vulnerability-analysis

2. Identify and characterize credible threats against those facilities.
3. Evaluate the facilities in terms of target attractiveness to each adversary and consequences if they are damaged.
4. Identify potential security vulnerabilities that threaten the facilities' service or integrity.
5. Determine risks by determining likelihood and consequences of each scenario if successful.
6. Rank risks of each scenario occurring and if high propose risk reduction measures.
7. Evaluate risk reduction options, including measures that impact layout, using cost benefit analysis.
8. Re-assess risks to ensure adequate countermeasures are being applied.

The following information is required for the SVA: overall project summary, plot plans, inventories of hazardous materials, HIRA study results, proposed security fences/barriers, and security procedures. The SVA can be facilitated by the process safety engineer(s) working closely with the security experts. In some jurisdictions, the SVA report must be submitted to the appropriate authority for acceptance/approval.

The SVA may draw attention to the potential for cyber security issues. If so, a separate cyber security assessment of control systems and safety systems should be undertaken to identify any vulnerabilities.

Further information and guidance is available from the following publications: *Guidelines for Analyzing and Managing the Security Vulnerabilities of Fixed Chemical Sites* (CCPS 2003a); *Site Security Guidelines for the U.S. Chemical Industry* (ACC et al, 2001); *Security Vulnerability Assessment Methodology for the Petroleum and Petrochemical Industries* (API et. al, 2003a); *Security for Offshore Oil and Natural Gas Operations, 1st Edition, RP 70* (API 2003b); and *Security for Worldwide Offshore Oil and Natural Gas Operations, 1st Edition, RP 70I* (API 2004).

5.2.7.10 Preliminary Simultaneous Operations Study

A preliminary simultaneous operations (SIMOPS) study should be performed during FEL-3 to evaluate potential conflicts if two or more activities are likely to occur in proximity to one another at the same time. Typical activities that could occur simultaneously include construction, drilling, commissioning, maintenance, and production. It is particularly relevant to brownfield developments. The purpose of the study is to ensure that potential conflicts, hazards, and risks are identified and assessed to enable plans to be adjusted to eliminate SIMOPS or apply appropriate safety measures.

A SIMOPS study typically uses a HAZID, What If and/or checklist approach, and requires the following information: plot plans, project schedule/plans, production plans, work orders, and procedures. This preliminary SIMOPS study should be updated and finalized in project execution prior to performing any concurrent activities at or near the same location.

Further information and guidance is available from the following CCPS publication: *Guidelines for Hazard Evaluation Procedures*, 3rd edition (CCPS 2008b).

5.2.7.11 Human Factors Analysis

A human factors analysis (HFA) should be conducted to review risks, issues, and opportunities associated with human factors. In particular, the HFA should analyze the design in respect of physical ergonomics, potential for human error, and issues such as alarm prioritization, labeling/signage, noise and lighting. However, not all aspects of human factors can be assessed at the design stage of the project, especially factors related to organizational or cognitive ergonomics. For example, the culture/working environment of the facility is best addressed during operation, although efforts to ensure the quality of procedures/work practices (particularly for a greenfield development) should be a project objective.

Some jurisdictions mandate consideration of human factors; for example, regulations for Safety Report[5]/Case[6] in the UK, and OSHA PSM[7] in the USA.

As a minimum in FEL-3, the HFA should focus on project plans for the following:

- Operation of key equipment (e.g. isolation valves) – readily accessible, avoidance of hazard zones
- Maintenance access to key equipment (e.g. blinds for LOTO, orientation, spacing, etc.)
- Control system interfaces (including alarms) – provision of effective information, avoidance of alarm flood
- Marking/labeling of equipment and piping
- Emergency exit/evacuation routes – avoidance of hazard zones, visibility/clarity of intended signs
- Communication system – audible/clarity in high noise areas, alarm signals distinguishable
- Lighting
- Emergency response time – duration of required operator calculations or tasks (e.g. close valves, shutdown HVAC) vs. escalation

---

[5] The Control of Major Accident Hazards (COMAH) Regulations, 1999 (and 2005 amendments) No.743, UK.

[6] The Offshore Installations (Safety Case) Regulations 2005, No.3117, UK; The Offshore Installations (Offshore Safety Directive) (Safety Case etc.) Regulations 2015, No.398, UK.

[7] Process Safety Management of Highly Hazardous Chemicals, 29 CFR 1910.119.

• Safety device bypass – ITPM duration when device unavailable, process to put back in-service

A human factors/ergonomics expert may facilitate the HFA with a multi-disciplinary team comprising representatives from operations, maintenance, EHS, process safety, and the project. The following information is required for HFA: design documentation, HIRA results, proposed procedures/work practices (if available). The HFA should identify any additional requirements necessary to support safe and effective performance of critical tasks.

Further information and guidance is available from the following publications: *Guidelines for Preventing Human Error in Process Safety* (CCPS 1994b); *Human Factors Methods for Improving Performance in the Process Industries* (CCPS 2007d); *Human Factors ... a means of improving HSE performance* (IOGP 2006); *Reducing Error and Influencing Behaviour, 2nd Edition,* HSG48 (HSE 1999): *Human Factors & COMAH, A Gap Analysis Tool* (HSE 2010a); *A Manager's Guide to Reducing Human Errors, Improving Human Performance in the Process Industries,* Publication 770 (API 2001b).

5.2.7.12 Fire & Gas Detection Study

A fire and gas (F&G) detection study should be conducted based upon an updated version of the preliminary study (produced in FEL-2) to reflect locations within the latest design requiring F&G detection. The study should ensure that an unplanned release event of a critical size will be rapidly detected and operators alerted by a system of detectors and alarms for combustible and toxic gas, fire/smoke, carbon monoxide/dioxide, and other detection devices.

The study should be based upon the F&G philosophy, control system philosophy, preliminary F&G study, plot plan, HIRA studies, facility siting study (or offshore fire and explosion study). It is increasingly common to set SIL targets on F&G systems (ISA TR84.00.07).

Further information and guidance is available from the following publications: *Continuous Monitoring for Hazardous Material Releases* (CCPS 2009a); *Guidelines for Fire Protection in Chemical, Petrochemical, and Hydrocarbon Processing Facilities* (CCPS 2003b); *Offshore Gas Detector Siting Criterion Investigation of Detector Spacing* (HSE 1993); *Guidance on the Evaluation of Fire and Gas System Effectiveness,* TR84.00.07 (ISA 2010); *Performance-Based Fire and Gas Systems Engineering Handbook,* (ISA 2015).

5.2.7.13 Firewater Analysis

The preliminary firewater analysis produced in FEL-2 should be updated to reflect the latest design. It should estimate firewater demand for the various scenarios addressed in the FHA. Based on these demands, a firewater distribution system should be developed, including water supply sources, pumps, piping to all fixed fire protection equipment, and related equipment (e.g. controls, tankage, ponds, mobile fire protection equipment, foam supplies).

The firewater analysis should be based upon the fire protection philosophy, preliminary firewater analysis, FHA, and plot plan.

Further information and guidance is available from the following CCPS publication: *Guidelines for Fire Protection in Chemical, Petrochemical, and Hydrocarbon Processing Facilities* (CCPS 2003b).

5.2.7.14 Relief, Blowdown and Flare Study

The preliminary blowdown and depressurization study should be updated to reflect the latest design. Calculations, specifications and documentation should be developed for each pressure relief device for all credible overpressure scenarios. Headers and piping should be sized for credible simultaneous relief from multiple devices. Flare stacks (and any vent stacks) should be sized (height, diameter, tip configurations, etc.) based on regulatory compliance and good industry practice.

The study should be based upon the relief, blowdown, and flare philosophy, HIRA studies, P&IDs, flare header layout (isometrics if available), protected equipment, and relief device data. Note that the relief study may affect the LOPA and vice versa.

Further information and guidance is available from the following CCPS publication: *Guidelines for Pressure Relief and Effluent Handling Systems* (CCPS 1998b).

5.2.7.15 Decommissioning

It is prudent to address how the facility would be eventually decommissioned, and to include any necessary provisions in the design. This is particularly relevant for offshore installations that have to be completely removed, but may also be appropriate for some onshore developments. The challenges of decommissioning are discussed in Chapter 11, and consideration during FEL can lessen these challenges, ease deconstruction and/or demolition, and reduce costs.

5.2.7.16 Emergency Response Study

A preliminary emergency response study should be performed to identify strategies and equipment necessary to address worst case major accidents and smaller, more likely incidents. This should include, but not limited to, fire, explosion, toxic and flammable releases, extreme weather, vehicle/railcar/ship collision, personnel rescue, etc. Performance standards for any identified equipment should be drafted. Crisis management arrangements should be identified for greenfield developments.

The study should be based on the results of HIRA, other safety assessments, and the emergency response philosophy. Further information and guidance is available from the following CCPS publication: *Guidelines for Technical Planning for On-Site Emergencies* (CCPS 1995c); *Guidelines for Risk Based Process Safety, Chapter 18*, (CCPS 2007b).

5.2.8   Re-Evaluate Major Accident Risk

Following the DHM reviews, ISD optimization, and specification of design safety measures, the residual major accident risk should be re-evaluated to determine whether it meets corporate and/or jurisdiction risk tolerance criteria. Depending upon the criterion, this evaluation may be qualitative, semi-quantitative or quantitative using one or more of the HIRA methodologies (see Section 5.2.7).

If the appropriate criterion has not been met, the project should return to the DHM process to identify other risk reduction opportunities. If this is not possible, project management should notify the client to discuss options on the way forward. This may entail a variance against the risk tolerance criterion, return to FEL-2 to select a different development option, or terminate the project. In certain circumstances, the regulator may also need to be notified.

5.2.9   Finalize Important Safety Decisions

By the end of FEL-3, the project should have finalized all of the important safety decisions. Most of these decisions will involve the output from the DHM, ISD, HIRA, and design safety measure processes. However, other decisions may be related to other issues, including, but not limited to, plans for:

- EHS and process safety management system,
- Technology options,
- Construction safety,
- Training (including use of process simulator, if appropriate),
- Emergency response,
- Stakeholder outreach, and
- Other safety activities in the project execution stages.

A Decision Register should be compiled for eventual handover to the Operator.

5.2.10   Finalize Basis of Design

FEL should be complete and well documented with technical definition sufficient for detailed design to commence. Sites for all facilities, and pipeline routes and rights-of-way, should be confirmed and other permits secured. This allows basic engineering to be completed and compiled in the Basis of Design (BOD), which should be frozen by the end of FEL-3.

5.3   OTHER ENGINEERING CONSIDERATIONS

5.3.1   Asset Integrity Management

Many of the larger oil and gas, and chemical companies have asset integrity management (AIM) policies and standards that cover some or all of the following the goals:

*   Preventing failure of equipment and infrastructure contributing to major accident risk,

*   Improving operational performance and productivity, and minimizing life cycle costs,

*   Improving EHS and process safety performance and reducing liabilities,

*   Increasing lifecycle value of facilities, and

*   Sustaining company reputation and 'license to operate'.

The life expectancy of the facility should have been determined when the SOR was finalized.   This life expectancy should have influenced the selection of engineering codes, standards, and materials of construction in order to reduce the number and severity of uncontrolled releases of hazardous materials.   For example, if the project involves handling hydrofluoric acid (HF), nickel-containing alloys are the only materials adequately resistant to attack for useful long-term service.   In this regard, it may be beneficial to accept higher capital expenditure to reduce future operating expenditure, e.g. reduce ITPM tasks for the life of the facility.

The project team should evaluate the design in FEL-3 to ensure that sufficient emphasis has been placed on the facility's lifecycle, and its ability to meet the client's AIM goals.   This is an important stage at which the facility's integrity is designed, may significantly impact future reliability and availability, and ultimately the value of the operating facility.

While the process safety studies and activities described in this chapter support AIM, there are a number of technical issues that should also be thoroughly evaluated, including, but not limited to:

*   Integrity of new or extrapolated technology (may require proving trials to demonstrate reliability),

*   Inclusion of integrity within equipment performance standards,

*   Materials selection and corrosion management strategy,

*   Deviations from recognized engineering codes and standards using a formal process for management review and approval,

*   Structural strength required in event of fire, explosion and environmental loads (wind, wave, ice, etc.)

- Structural strength to tolerate accidental loads (e.g. dropped load, vehicle/ship collision),
- Confirmation of strategy for ITPM (e.g. RBI, RCM).

Further information and guidance is available from the following publications: *Guidelines for Asset Integrity Management* (CCPS 2017a).

5.3.2 Quality Management

A quality management (QM) plan should be prepared for the execution stages (detailed design, procurement, fabrication, construction, installation, commissioning) of the project, if not already developed. This will permit the procurement of any long-lead items to meet the project's schedule and quality requirements (in addition to technical specifications and performance standards). The QM plan should address roles and responsibilities for quality assurance (QA) and quality control (QC) activities between the client, contractors, and suppliers.

Quality management is discussed in detail in Chapter 8.

5.3.3 Contractor Selection

A contracting strategy for the execution stages of the project should be developed, if not already prepared. A description of project implementation strategies, and guidance on how various contracting strategies may impact process safety is covered in Chapter 2, Section 2.6.

The finalization of the contracting strategy should permit the timely appointment of the detailed design contractor, if the work is not performed in-house. The strategy should reflect how much work the PMT wants to take on versus sub-contract, e.g. HIRA studies in-house or given to specialist consultant/contractor. The PMT should also be mindful of the number of interfaces to be managed when finalizing the strategy. A detailed scope of work and deliverables should be prepared for the contract(s). This should include the activities in the updated DHM implementation plan.

The contractor selection should be based on a combination of technical competency, EHS and process safety ability, and cost. Cost should not be the sole factor in determining contract award, as familiarity with the process technology is important. The selection process should rank all criteria, including an in-depth competency assessment of the engineering design organization that potential contractors propose. If it is necessary to select an inexperienced contractor, the client should be prepared to take an active management role.

By the end of FEL-3, the project should have contract management and administration practices and procedures in place, including change orders and partner approvals. Major commitments in the execution stages should be finalized and ready for contract award.

5.3.4   Brownfield Developments

If the project is a brownfield development, it is likely that shutdown requirements to tie-in process and utility systems will need to be coordinated with the site management responsible for existing operations. Recognizing the impact on production, any shutdown requirements should be defined in FEL-3, the timing agreed, and added to the project plan.

5.4   OTHER ACTIVITIES

In addition to the various process safety and technical studies needed to develop the project design, there are a number of other activities that support FEL and project execution. These activities continue throughout the project life cycle and should be periodically updated.

5.4.1   EHS and Process Safety Plans

The EHS Plan and a Process Safety Plan should be updated to reflect the latest design and any additional EHS and process safety requirements, such as specialist studies in the project execution stages or changes to required approvals, licenses and permits (Appendix B).

5.4.2   Risk Register

The Risk Register should be updated for the latest design and any new hazards/risks identified in HIRA studies (Appendix C). As the design evolves, any safety design measures (e.g. safety critical equipment) and management processes (e.g. work force competency) that must be maintained to ensure that the risk is adequately managed should be documented. It is essential that both these design measures and management processes are clearly understood and handed over to operations, as failure to maintain either or both could lead to increased risk.

5.4.3   Action Tracking

The project action tracking database or spreadsheet should be updated to include all activity relating to, but not limited to, any legally binding, regulatory or contractual requirements/commitments, specialist studies, peer reviews and other assurance processes.

5.4.4   Change Management

A process for controlling project changes should be in place for all project disciplines, who may work independently of one another. This is particularly important for changes that may affect the DHM process and design safety measures. P&IDs may not be frozen until the final HIRA (e.g. final HAZOP) that may not be performed until the detailed design stage of project execution.

Further information and guidance is available from the following CCPS publication: *Guidelines for Management of Change for Process Safety*, 2008.

5.4.5   Documentation

The compilation of process safety information (PSI) and other documentation, including calculations and design assumptions, should continue throughout FEL-3 and project execution. As the design evolves, the early information will likely need to be revised and/or updated.

Project documentation is discussed in detail in Chapter 12.

5.4.6   Preparation for Project Execution

One of the objectives of FEL-3 is to improve project execution planning to give confidence in the design, cost estimate and schedule. To this end the following plans should be prepared:

5.4.6.1   Detailed Design

In preparation for detailed design, plans may be required to address the following, if appropriate:

- Study program and timing to align with project schedule,
- Verification of performance of design safety measures (e.g. design reviews, etc.),
- Interface management between contractors,
- Interface management with Brownfield touch points, e.g. existing process operations and utility services.

5.4.6.2   Procurement

The procurement strategy and procurement plan (including application of the project quality management plan) should be finalized. Long-lead items of equipment should be ordered. By the end of FEL-3, specifications for all major items of equipment should be prepared, and bids obtained to allow preparation of the project cost estimate for project sanction.

5.4.6.3   Construction

In preparation for the construction stage, plans may be required to address the following, if appropriate:

- Temporary offices, canteen and/or housing for the construction workforce (may involve extension to the facility siting study),
- Security of the construction site(s),
- Safety orientation and procedures for construction workforce,

- Emergency response during construction, including rescue, firefighting (possible requirement for temporary firewater supply), and medical services,
- Health issues during construction, such as excessive heat/cold, malaria, etc.
- Environmental issues during construction, such as protection of adjacent wetlands or protected species, stormwater runoff, etc.
- Temporary utility supplies,
- Unexploded ordinance (major problem in parts of the world),
- Materials management including equipment preservation,
- Verification of performance of design safety measures (e.g. FAT, etc.),
- Pre-commissioning,
- Mechanical completion and handover.

5.4.6.4   Commissioning and Startup

In preparation for commissioning and startup, plans may be required to address the following, if appropriate:

- Hook-up, commissioning and startup plan (including roles and responsibilities, sequence, throughput test runs, etc.) defined,
- Integration of future operations personnel into pre-commissioning and commissioning teams,
- Vendor support,
- Spare parts and maintenance,
- Punch lists and operational readiness review(s).

Further information and guidance on operational readiness reviews and commissioning/sequencing is available respectively from the following publications: *Guidelines for Risk Based Process Safety*, (CCPS 2007b); *Chemical and Process Plant Commissioning Handbook: A Practical Guide to Plant System and Equipment Installation and Commissioning, 1st edition* (IChemE 2011).

5.4.6.5   Operation

The Operations and Maintenance Philosophy should be aligned with the BOD and frozen.

5.5 CASE FOR SAFETY

In some jurisdictions, owners and/or Operators of offshore installations are required to prepare a Design Safety Case to describe the identified major accident hazards (MAH), the studies undertaken to evaluate their risks, and the measures employed to manage the risks or mitigate their potential consequences. Similarly, in some jurisdictions, some onshore facilities are required to prepare and submit a Pre-Construction Safety Report that contains similar data.

Even where there are no regulatory requirements, some companies have elected to prepare a 'Case for Safety' for onshore and/or offshore facilities in order to provide the future operations team with a summary of the MAH and DHM, including specifications and performance standards of all the design safety measures, and administrative / procedural measures that operations should implement. Failure to understand and maintain these measures throughout the facility life cycle would increase risks that might result in a major accident.

Preparation of a Case for Safety during FEL (with updating during project execution) requires all the design elements from each contractor and other sources to be compiled into a single integrated user-friendly document to provide operations with the diverse range of design safety information. This has the following benefits:

- Improved understanding of hazards and risks,
- Enhanced knowledge of technical and administrative / procedural risk reduction measures, and
- Likely reduction in major accidents or their consequences.

5.6 STAGE GATE REVIEW

When nearing the completion of FEL-3, a stage gate review should be conducted to ensure that process safety (and EHS) risks are being adequately managed by the project. The stage gate review team may use a protocol and/or checklist, such as the detailed protocol in Appendix G. A typical process safety scope for a FEL-3 stage gate review is illustrated in Table 5.3.

The stage gate review team should be independent of the project, familiar with similar facility/process/technology, and typically comprise an experienced leader, process engineer, operations representative, process safety engineer, other discipline engineers (as appropriate), and EHS specialist. At the conclusion of the review, the review team will make recommendations for any improvements needed, and indicate to the Gate Keeper, based on process safety, whether the project is ready to proceed to the next stage, Detailed Design.

Table 5.3. FEL-3 Stage Gate Review Scope

Scope Item
Confirm that Process Safety and EHS studies, including specialist reviews, are being satisfactorily addressed and followed up
Confirm that Process Safety and EHS related aspects of the engineering designs meet or exceed regulatory requirements, and that satisfactory project codes and standards have been identified, and design philosophies have been established
Confirm that all Process Safety and EHS concerns relating to the characteristics of the full life cycle of the project, novel technology, and the nature of the location have been identified
Confirm that integrity management / engineering assurance processes are in place
Confirm that Change Management procedures are in place
Confirm that documentation requirements have been addressed
Confirm that a resourcing and training strategy is established
Confirm that project plans ensure Process Safety and EHS preparedness for commencement of construction
Confirm that a risk register has been established for the project and that the risks associated with Process Safety are followed up and formally reviewed by competent personnel

5.7   SUMMARY

As previously described, once a single development option has been selected FEL-3 improves the technical definition and project execution plan. This involves the preparation of a Front End Engineering Design (FEED) package that can be given to in-house engineers or an engineering contractor to complete the detailed engineering. This package includes refined design hazard management (DHM) with reduced risk achieved through understanding major hazards, and optimizing inherently safer design, functional safety, and other safety measures. The ultimate goal is to have confidence in the design, cost and schedule, thereby confirming the business case and receiving financial sanction from the client.

6 DETAILED DESIGN STAGE

Once the project has been sanctioned (i.e. approved by the client), it moves into the first stage of execution, Detailed Design, sometimes known as Detailed Engineering or Design, which involves completion of detailed engineering of the defined scope (FEED package) from the front end loading (FEL) process, management of any scope changes, and procurement of materials and equipment. Figure 6.1 illustrates the position of detailed design in the project life cycle.

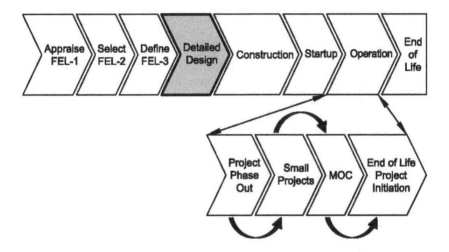

Figure 6.1. Detailed Design

The detailed design stage represents significant financial commitment to the project with authorization to spend in line with the approved financial memorandum for the project. Although the BOD may be frozen, engineering (including inherently safer design and process safety issues) may not have progressed to the point that a proper cost and schedule has been prepared. Therefore the project could still be canceled by the client. Company practices differ, but the Final Investment Decision (FID) is typically made at some point during Detailed Design based on the client's specified % completion of engineering details. Some details may require 100% completion, while others can be completed after FID.

Project Management Team

Many of the activities commenced in FEL require refining and updating to achieve completion prior to procurement and/or construction. The PMT's primary focus should be on implementing the project execution plan, and ensuring delivery of the agreed scope to schedule and within budget. This requires good interface management and regular performance reporting between all parties involved (i.e. contractors, consultants, vendors, suppliers, etc.). All remaining project risks and uncertainties require close management in order to stay on schedule and facilitate efficient commissioning and handover.

Typical project objectives for detailed design include:

- Execute the detailed engineering design against the agreed FEED package/frozen BOD,

- Finalize constructability work (such as maximizing off-site fabrication, use of pre-assembled units, plans for tie-ins to existing facilities, etc.), and ensure construction readiness.

- Oversee all engineering contractor activities to ensure production of relevant procurement, construction, and commissioning deliverables,

- Perform design reviews to verify compliance with regulations, codes/standards, DHM and SCE, performance standards, and operability, maintainability and reliability requirements,

- Update and finalize the project hazard/risk register, quality management plan, and commissioning/startup plan (if necessary),

- Award outstanding contracts and place orders for equipment and materials,

- Prepare a plan for handover to operations (including transition engineering support, and documentation),

- Monitor project performance (expenditure, schedule, quality, process safety, EHS, actions, etc.) closely and report to client.

From an engineering perspective, the focus of the project team is on completing a design package that includes all the necessary information required for construction. This information includes, for example, procurement details for equipment, systems, buildings, structures, etc., and construction drawings, such as isometrics, P&IDs, electrical one line, cause & effect. Any outstanding information (e.g. quality and performance standard data) necessary for the timely procurement of equipment and materials should be a priority. The design work builds on the FEED package from FEL-3, and may require several iterations of some of the various studies and activities before the final design is complete and ready for construction.

Environment, Health and Safety

From an EHS perspective, the project team needs to update all EHS risks addressing the full life cycle of the project. Recommendations from EHS studies (including specialist reviews) should be followed-up and satisfactorily resolved to ensure that EHS aspects are adequately addressed in the detailed design. The project should implement an EHS management system, and ensure that EHS documentation requirements are addressed. The project EHS Plan may need to be updated to ensure EHS preparedness for construction, pre-commissioning and commissioning, including EHS procedures and contractor orientation/training, and emergency response.

Process Safety

The key process safety objectives in detailed design generally build on those previously established in FEL, namely:

- Refine and complete the DHM process (including functional safety, SCE and performance standards),
- Conduct the final HIRA (e.g. HAZOP) and other process safety specialist studies, and address all study findings to ensure that process safety aspects are adequately addressed in the detailed design,
- Implement the management of change (MOC) process fully to evaluate late design changes,
- Update the process safety plan if necessary, to address preparedness for construction, pre-commissioning, commissioning and startup, including operating procedures and training, maintenance management system,

As more detailed design information becomes available, more detailed HIRA and other process safety and technical studies should be conducted than was possible during FEL. Depending upon circumstances, it may be appropriate to update previous studies or perform a fresh study. These process safety studies and activities are discussed below for:

- Detailed design (Section 6.1),
- Procurement (Section 6.2),
- Integrity management (Section 6.3),
- Other process safety activities (Section 6.4),
- Other project activities (Section 6.5),
- Preparation for construction (Section 6.6),
- Preparation for pre-commissioning, commissioning and startup (Section 6.7),
- Stage gate review (Section 6.8).

6.1   DETAILED DESIGN

At the end of FEL, the FEED package includes all the necessary information required by an in-house engineering department or contractor to perform the final engineering of the project. This represents the transition from 'design' to 'engineering'. For example, the FEED package includes preliminary information on project schedule, general arrangement drawings, design philosophies, major equipment and piping specifications, materials of construction, structural steelwork, wiring, etc. By contrast, the detailed design stage is where the design is refined, and details, such as construction drawings (e.g. P&IDs, isometrics, electrical one line), equipment selection, piping details, structural steel supports, buildings, insulation, etc., are engineered ready for procurement. Engineering software, such as finite element analysis and computer-aided design (CAD) programs, is used to evaluate stresses in, and optimize, equipment, components and piping, and their detailed layout. This stage is where the full cost of the project is identified to much greater accuracy.

The PMT should ensure that, as the details in each engineering discipline (i.e. process, mechanical, electrical, control, and civil) progress, the engineering does not depart from the design philosophies. In particular, it is important that each discipline understands the operating and maintenance philosophy. Equipment vendors often contribute to the completion of the design, but care is required to monitor potential impacts from selecting a specific vendor's product. For example, each process control system vendor has a system design that determines the interfaces with process equipment and plant information systems, which may not necessarily meet the Operator's requirements. Proof testing, calibration and maintenance may necessitate piping design elements, such as tap, isolation and bypass.

If not already done so in FEL, all selected engineering codes and standards should be updated and 'frozen' early in detailed design to avoid later changes and duplication of effort. Though one recognizes that some changes are bound to come during implementation, such changes should be resisted, as even seemingly minor changes can have a disproportionate cost after detailed design commences. Temptations to hold down costs by removing apparently inconsequential components can introduce new hazards and risks, and design change requests should be rigorously evaluated.

6.1.1   Design Hazard Management Process

The Design Hazard Management (DHM) process started in FEL should be refined and completed during detailed design to eliminate or minimize major accident hazards (MAH) at source, and prevent the remaining hazards from becoming major hazards. The overall goal is a reduction in residual risk to a level that, as a minimum, meets corporate policy by a combination of applying ISD principles and adding functional safety risk reduction measures, i.e. design safety measures. Refer to

Chapter 4 Section 4.2.1 for additional information and examples of risk reduction measures.

Many capital projects employ one or more engineering contractors to conduct the detailed design. It is important that the contractor(s) understands the DHM process (including previous FEL work), and the deliverables expected by the PMT. Where multiple contractors and/or engineering disciples are involved in the DHM, it is important that the interfaces are well-defined and that frequent information sharing occurs to ensure consistency. Sub-contractors and vendors of packaged units should also be integrated into the DHM process. If a contractor and/or engineering discipline has little or no competency in DHM, the PMT will need to make arrangements for support through in-house process safety or specialist consultant expertise.

The PMT will need to ensure that each contractor delivers on their DHM commitments, and that change management is fully implemented across all design contracts, so that each proposed design change (e.g. additional design safety measure) is properly evaluated for impact on overall hazard management. The PMT should also implement an action tracking system across all design contracts, and verify that actions are closed with the proper authority. Locating members of the PMT in the contractor's offices can facilitate oversight of these requirements, but it is recommended that a competent process safety engineer is appointed to the PMT to verify proper tracking and integration of all process safety information.

6.1.2   Inherently Safer Design Optimization

The Inherently Safer Design (ISD) optimization started in FEL should be refined and completed for:

- detailed design of equipment and systems identified in the BOD during FEL-3, and
- any ISD measures not identified in the BOD during FEL-3, subject to change management,

to achieve the optimum balance between risk reduction, operability and cost.

It is important that the contractor(s), vendors and equipment suppliers understand ISD (including previous FEL work, the hierarchical approach employed for risk reduction), and the deliverables expected by the PMT. If not, an additional effort should be made to deploy competent resources that provide assurance on contractor deliverables. Any changes from the BOD to ISD measures proposed by the contractor(s) should be properly evaluated for their impact on hazard management. Any design assumptions and uncertainties associated with ISD measures and emergency response measures should be progressively resolved. Information should be shared frequently across all contractor and discipline interfaces to ensure consistency and avoid an impediment to continuous risk reduction. The PMT should also ensure a consistent approach and design standards by all contractors and engineering disciplines.

Further information and guidance on ISD is available from the following publications: *Inherently Safer Chemical Processes, A Life Cycle Approach, 2nd edition* (CCPS 2009d), and *Guidelines for Hazard Evaluation Procedures, 3rd edition* (CCPS 2008b); *Process Plants: A Handbook for Inherently Safer Design, 2nd edition* (Kletz and Amyotte, 2010).

6.1.3   Site Layout

Site layout and spacing of process units, other equipment and buildings should have been finalized in FEL-3. If not, any layout issues need to be urgently resolved before the detailed design can proceed as layout and spacing may impact the specification of design safety measures. Refer to Chapter 5 Section 5.2.3 for guidance on site layout and spacing.

6.1.4   Design Safety Measures

After the ISD optimization has exhausted opportunities to eliminate hazards, a diverse range of functional safety and other passive and active design safety measures should be added to reduce risks further in line with risk tolerance criteria and corporate policy. The design safety measure process should continue through detailed design to:

- Identify any required measures that were not identified in the BOD during FEL-3, and, if approved, complete their detailed design, and

- Complete the detailed design of equipment and systems identified as design safety measures in the BOD.

The detailed HIRA (see Section 5.2.6) and other safety and technical studies (see Section 5.2.7) conducted during detailed design may identify and evaluate design safety measures that were not previously identified in FEL. Any new design safety measures proposed should be subject to rigorous change management to evaluate any impact on hazard management, and, if approved, added to a revised BOD. For example, HIRA studies may be performed with and without a safety instrumented system (SIS), such as a HIPPS, to verify the risk reduction achieved.

As the results of detailed studies and change management decisions are finalized, the PMT should ensure that the design contractor(s) and engineering disciplines progressively update and share information on design safety measures, especially safety critical equipment/elements (SCE, see Chapter 5 Section 5.2.7.3). These data should also be added to the project risk register together with the performance standards of the design safety measures. Any assumptions or uncertainties associated with the design safety measures should be progressively resolved.

As part of the overall management of the DHM process, the PMT should also ensure that design safety measures are consistently designed across all the contractor(s) and engineering discipline interfaces. All outstanding design safety actions should be progressively closed.

6.1.5   Set Performance Standards

As part of the continuing design process for functional safety and other design safety measures, performance standards should be updated or refined, if necessary, for equipment and systems identified as design safety measures in the BOD (see Chapter 5 Section 5.2.5). Some of these performance standards may require an assessment to determine response time requirements. Performance standards should also be developed for any new design safety measures identified during detailed design.

In the case of SCE, it is important that the performance standards address their integrity over the life of the facility, as SCE are key to maintaining process safety risks at a desired level and must be reliable over the life cycle. A register of SCE should be compiled as a detailed design deliverable, and should include (or be linked to) the relevant performance standards to ensure integration and consistency during subsequent project stages. This register represents important process safety information for startup and handover to operations.

The design contractor(s) and engineering disciplines should progressively update and share information on performance standards, as each detailed safety study is completed. The PMT should verify that performance standards are captured in design specifications, and added to the project risk register, for handover to operations.

Inspection and test plans should be developed to demonstrate that the required performance for each design safety measure has been achieved. These plans should address factory acceptance test (FAT), site acceptance test (SAT), mechanical completion, and commissioning.

6.1.6   Hazard Identification and Risk Analysis (HIRA)

In order to continue the risk reduction efforts, the various HIRA studies conducted in FEL-3 should be updated as greater design detail becomes available. If any new hazards are identified or changes are necessary for those already identified, additional studies may be necessary. The various HIRA studies include, but are not limited to, the following (see Chapter 5 Section 5.2.6 for more details):

- Hazard identification (e.g. HAZOP, What If, checklist),
- Consequence analysis (e.g. facility siting, fire hazard, offshore fire & explosion, smoke & gas ingress, transportation),
- Risk analysis (e.g. QRA, LOPA)

Eventually the engineering is sufficiently advanced that the final project HIRA, typically a HAZOP study for capital projects, is conducted on the "approved for HAZOP" P&IDs. It is common practice to fully implement management of change

(MOC) at this point, and any further P&ID and design changes should be rigorously evaluated for their impact and potential introduction of new or changed hazards. It is important that the HIRA addresses not only normal operation, but also startup, shutdown (normal and emergency, and other transient operations, such as catalyst regeneration and molecular sieve drying. All HIRA recommendations should be added to the action tracking system, and, as each recommendation is resolved, any approved changes should be reflected on the "approved for construction" P&IDs and related documentation. If a process simulator is going to be used to train operators and to check out control systems, it should be based on the "approved for construction" P&IDs.

HIRA studies for vendor designed equipment and P&IDs are not always available at the time of the final HIRA, and may be delayed to a later date. Furthermore, some technology providers do not meet satisfactory standards for HIRA studies. If the vendor documentation represents a significant proportion of the project, it may be preferable for technology providers to participate in the client/project HIRA.

Some companies also conduct a hazard analysis of the first draft of the operating procedures when they are prepared. The analysis typically focuses on any unusual tasks and requirements, and may use HAZID, What If, checklist or a type of HAZOP methodology.

The PMT should nominate one or more competent coordinators, such as a process engineer or a process safety engineer, to independently verify that all actions derived from HIRA studies (both FEL and detailed design) are well-defined and closed effectively, particularly where they relate to design issues. If the DHM/ISD process has been diligently implemented, a small percentage of the actions may not address design issues, and it is important that the Operations team acknowledges and assumes responsibility for these actions, e.g. specific steps that should be included in operating procedures.

6.1.7   Safety Assessments

In order to continue the risk reduction efforts, the various technical safety assessments conducted in FEL-3 should be updated as greater design detail becomes available. Many of these studies are necessary to mitigate the risks identified by HIRA studies. Where possible, recommendations arising from these studies should be resolved, and the decisions/actions taken documented. Depending on the project scope, some studies may not be appropriate, e.g. fire protection may not be applicable for some utilities or water treatment system. The various safety assessments include, but are not limited to, the following (see Chapter 5 Section 5.2.7 for more details):

- Hazardous area classification,
- Safety instrumented system (SIS) assessment and safety integrity level (SIL) determination/verification,

- Functional safety assessment (FSA) prior to designing SIFs,
- Safety critical equipment (SCE) and their performance standards,
- Vulnerability analysis of SCE (including control system and loss of data),
- Reliability, availability and maintainability (RAM) study,
- Temporary refuge/shelter-in-place impairment assessment,
- Evacuation, escape, and rescue analysis,
- Dropped object study,
- Security vulnerability analysis (SVA),
- Simultaneous operations (SIMOPS) study,
- Human factors analysis,
- Fire & gas detection and suppression study,
- Emergency shutdown study,
- Firewater Analysis,
- Relief, blowdown and flare study,
- Decommissioning study,
- Emergency response study.

6.1.8   Re-Evaluate Major Accident Risk

When the DHM process, ISD optimization, and specification of design safety measures are complete, the residual major accident risk should be re-evaluated to confirm that it meets corporate and/or jurisdiction risk tolerance criteria. Depending upon the criterion, this evaluation may be qualitative, semi-quantitative or quantitative using one or more of the HIRA methodologies (see Section 6.1.6).

Further information and guidance is available from the following CCPS publication: *Guidelines for Developing Quantitative Safety Risk Criteria* (CCPS 2009b).

6.1.9   Other Design Reviews

Eventually the detailed design engineering will have progressed sufficiently that various design reviews should be conducted. In addition to the final HAZOP study of the P&IDs, project related documents, 3D model, and vendor drawings, especially for rotating machinery and instrumentation, should be reviewed as it is likely to be the last opportunity to upgrade technical and safety related issues. In particular, the control system should be checked that it meets all the requirements in design philosophies and performance standards. The checklist in Appendix D can also be used to identify and manage design safety issues, especially for small modifications and MOC projects. If not previously conducted in FEL 2 or 3, a constructability review should be undertaken to ensure that the design does not present any unacceptable construction difficulties and risks (see Section 6.5.6 for more detail).

6.2 PROCUREMENT

The procurement plan including application of the project quality management plan should have been finalized in FEL-3, and long-lead items of equipment ordered to meet the project schedule. Fabrication of these long-lead items and engineering packages will be based on preliminary studies from FEL-3 (e.g. HIRA, LOPA/SIL, RAM, etc.) unless these studies receive priority review and update at the commencement of detailed design. Nevertheless, these studies should be integrated with other detailed design studies to ensure that essential barriers (i.e. layers of protection) identified during design remain healthy to control and mitigate hazards.

The remaining equipment, materials and services will be procured during either the detailed design or construction stages, as appropriate, to meet the project schedule. For example, contracts involving activities at the front end of construction (i.e. demolition/site clearance, grading, access roads, foundations, temporary buildings/camp, etc.) need to be awarded during detailed design to avoid schedule delays.

If the quality management plan requires any updates to deliver the project's quality objectives, this should be completed prior to further procurement activity. The quality management plan should then be implemented. While quality control (QC) inspections and checks may be a contractor's responsibility, quality assurance (QA) audits of design contractors and fabrication of any long-lead items should be undertaken by the PMT or specialist third-party on their behalf (see Chapter 8).

The PMT should oversee the detailed design contractor(s) to ensure that the detailed engineering design is correctly executed against the FEED package/frozen BOD and the agreed codes and standards. In preparation for procurement activities, performance standards for all design safety measures including SCE should be incorporated into equipment technical specifications. Other equipment and systems may have been given a criticality ranking for production, product quality or environmental reasons. Thereafter, the PMT should ensure that the design intent of all equipment and systems is rigorously controlled through procurement, as there may be a requirement to source equipment and materials locally and/or through low cost suppliers. Ideally, equipment and materials should only be procured from reliable suppliers, i.e. an approved list of vendors that should have been agreed early in the project. Sometimes the product quality from low cost suppliers may not meet the technical specification required by the project.

Case Study: Counterfeit Valves

In November 2007, the U.S. Nuclear Regulatory Commission (NRC) became aware that a nuclear power plant had discovered a counterfeit 5-inch 150# Ladish stop check valve on the stator cooling water pump discharge, and another in its warehouse. The installed valve had been in service for 8 months at the time of discovery. Upon discovering the counterfeit valve, the performance of the valve was closely monitored, and was replaced during the next refueling outage in the spring of 2009. The installed valve was being used in a non-safety related system.

Many counterfeit items are not built to the same technical specifications (metallurgy, tolerances, etc.) as OEM equipment, and may fail prematurely in service.

Reference: NRC Information Notice 2008-04

6.3   ASSET INTEGRITY MANAGEMENT

Technical integrity in the context of a project is concerned with the soundness of the design, and assurance of the quality of complex equipment and systems. Engineering codes and standards play a central role in delivering integrity, but are not the sole answer, as competent professional engineers are necessary to interpret them to arrive at the correct engineering solution. The project's processes and procedures should ensure that the design intent is maintained during fabrication and construction.

A number of engineering reviews may have commenced in FEL-3, and should be finalized during the detailed design stage to ensure that the integrity of the facilities will meet the client's policy/expectations. If not previously started, consideration should be given to initiating these reviews and activities, including, but not limited to, the following:

- P&ID reviews to identify errors, omissions, etc.,
- 3D model reviews at key milestones to check detailed design accurately reflects project scope and design input requirements,
- Engineering standards review, including any deviations from recognized codes and standards,
- Corrosion / erosion management,
- Electrical systems protection,
- Reliability, availability and maintainability (RAM) study, including:
  - o Equipment reliability (and required test intervals), particularly SCE,
  - o Design review to minimize impacts related to maintenance,
  - o Inspection, testing and maintenance programs and planning,

- Pipeline integrity monitoring,

- Structural strength, if any fire, explosion, environmental (wind, wave, ice, earthquake, etc.) or accidental (dropped object, vehicle/ship collision, etc.) loads have changed since FEL-3,

- Weight control for offshore structures, towing and major lifts,

- Verification that design safety measures meet their performance standards, including SIS verification reviews,

- Inspection and test schemes for factory acceptance tests (FATs) and site acceptance tests (SATs),

- Inter-discipline reviews (a.k.a. squad checks) to ensure good communication across engineering interfaces,

- Value engineering reviews at key milestones to reduce cost without compromising safety and quality,

- Operability review to improve operation and minimize life cycle costs (including logistics and support issues),

- Peer reviews on specific technical issues (e.g. critical items, new technology, new use, etc.),

- Stage gate review (see Section 6.8).

Following these reviews, integrity requirements should be included within the final performance standards for equipment, and plans/programs developed to deliver these integrity requirements and compile the appropriate engineering and integrity documentation for eventual handover to the Operator. Process safety engineers should be able to provide a significant contribution to these engineering reviews, and the development of related plans and programs.

6.4   OTHER PROCESS SAFETY ACTIVITIES

6.4.1   Case For Safety

6.4.1.1   Safety Case/Pre-Construction Safety Report

If a Safety Case/Pre-Construction Safety Report was prepared in FEL for a local jurisdiction (see Chapter 5, Section 5.5), it should be updated and finalized during the detailed design stage of the project by compiling inputs from the contractor(s), engineering disciplines and other sources of process safety information, such as HIRA studies. When the design is finalized, the future operations team will need to provide input on the administrative / procedural measures that they intend to implement to manage the residual risks. These measures will encompass elements of the proposed management system as well as certain HIRA recommendations that are the operations team's responsibility. The document should be submitted to the relevant competent authority.

6.4.1.2 Operations Case for Safety

If a voluntary Design 'Case for Safety' was prepared in FEL (see Chapter 5 Section 5.5), it should be updated and finalized during the detailed design stage of the project by compiling inputs from the contractor(s), engineering disciplines and other sources of process safety information, such as HIRA studies.

The Design Case for Safety document should be used as the starting point for the development of an Operations Case for Safety. The completed Design Case for Safety should be shared with the future operations team to provide information on the major accident hazards (MAH), and how they are managed through ISD and DHM, including design safety measures and associated performance standards in the final design. A dossier of all safety study work undertaken should be compiled and transferred to operations. Any design limitations for safe operation should also be brought to the operations team notice.

An Operations Case for Safety can then be prepared by adding details of how residual risks are managed by:

- Facility's management system (e.g. operating procedures, employee training, maintenance practices, management of change, etc.),
- Specific administrative / procedural measures that operations intend to implement (including resolution of some recommendations in the final HIRA study(s)),
- Emergency response strategy and provisions.

The CCPS guidance on risk-based process safety provides information on good management practices that may be appropriate for inclusion as part of the facility's management system (CCPS 2007b).

6.5 OTHER PROJECT ACTIVITIES

In addition to the various process safety and technical studies needed to develop the detailed design, there are a number of other activities that support project execution. These activities continue throughout the project life cycle and should be periodically updated. This requires good interface management between the PMT and all the contractors, vendors and suppliers.

6.5.1 EHS and Process Safety Plans

The EHS Plan and the Process Safety Plan should be updated to reflect the detailed design and any additional EHS and process safety requirements, such as specialist studies in the project execution stages or changes to required approvals, licenses and permits (Appendix B).

6.5.2   Risk Register

The Project Hazard and Risk Register should be updated for any new hazards/risks identified during detailed design (Appendix C). As the detailed design progresses, any safety design measures (including their performance standards) and management processes (e.g. administrative/procedural measures) that must be maintained to ensure that risks are adequately managed should be documented. It is essential that both these design measures and management processes are handed over and clearly understood by operations, as failure to maintain them is likely to result in increased risk.

6.5.3   Action Tracking

The project action tracking database or spreadsheet should be updated to include all actions relating to, but not limited to, any legally binding, regulatory or contractual requirements/commitments, technical work, specialist studies, design reviews, peer reviews and other assurance processes. The PMT should also seek to capture actions generated by their contractor(s), and ensure that actions from all sources are progressively resolved, closed and documented. It may be appropriate to appoint an independent competent engineer to coordinate resolution and close-out of design actions.

It is particularly important that the operations team is aware of, and takes responsibility for, all actions that are identified as 'operations actions'. Care should be exercised that recommendations from HIRA and specialist studies are not automatically assigned to operations. Design safety measures are considered more reliable than administrative and procedural measures. Recommendations should be objectively evaluated, and design measures incorporated where feasible.

6.5.4   Change Management

As the design evolved through the FEL stages, change was inevitable, because many decisions were made on incomplete information, assumptions and the project engineers' personal experiences. Although change controls should have been partially implemented in FEL-3 to protect decisions on design safety measures, change management should be fully implemented in the detailed design stage.

Changes at this stage of the project can interrupt workflow, introduce rework, impact safety, and cause delays and schedule slippage, which inevitably escalate costs. For this reason, managing change effectively is vital to the success of the project. Scope change during project execution must be tightly controlled, and some project managers have a philosophy of 'no change' in this regard. A few changes may be entertained where there is good justification, although with the availability of more complete information and the ongoing resolution of assumptions / uncertainties, the number of changes should be limited.

Examples of a change that should be considered are if:

- It responds to a regulatory issue,
- It will significantly reduce risk,
- It will reduce project cost, or
- It will shorten the project schedule.

Up to and including the detailed design stage, change management is primarily concerned with document control and having the latest up-to-date information stored and readily retrievable by project and contractor personnel who need access to perform their tasks. However, when engineering has reached an appropriate level of completion during detailed design, management of change (MOC) should also be implemented to determine if subsequent changes have unintended consequences for process safety and EHS. If not before, MOC should be fully implemented at the time of the final HIRA (e.g. final HAZOP) when the "approved for HAZOP" P&IDs are normally frozen. Any approved changes resulting from HIRA recommendations should be incorporated into the "approved for construction" P&IDs and related documentation.

A critical facet of change management in a capital project is *communication* across the diverse range of interfaces between the client, PMT, contractors, engineering disciplines, vendors, suppliers, and other stakeholders. It is important that all parties, particularly contractors and discipline engineers, understand the change management process, and are aware of changes proposed by others. Any proposed design change should be thoroughly evaluated for the introduction of new or modified hazards, and its potential impact on the DHM process and design safety measures. Approved changes should be documented and communicated to all relevant stakeholders. Typically an individual or small group is assigned responsibility for interface management to communicate between project personnel, EPC contractors, equipment manufacturers, vendors, suppliers, and other stakeholders. Further information and guidance is available from the following CCPS publication: *Guidelines for Management of Change for Process Safety*, (CCPS 2008c).

6.5.5   Documentation

The compilation of process safety information (PSI) and other documentation, including calculations and design decisions, should continue throughout detailed design. The PMT should oversee all contractor, vendor and supplier activities to ensure timely production of relevant documentation. The project should define when documentation is required in order to determine specification requirements before procurement, e.g. SIS components. Whereas the vendor may want to deliver manuals when the equipment is shipped.

It is particularly important to capture information for the operations team on, but not limited to, the following:

- Hazards,
- Hazard management and design safety measures,
- Design limitations for safe operation, and
- Design assumptions on how the facilities will be operated,
- Emergency response strategy and provisions.

The project risk register (see Section 6.5.2) and Operations Case for Safety (see Section 6.4.1.2) may fulfill some of these requirements. Project documentation is discussed in detail in Chapter 12.

6.5.6   Constructability

The concept of constructability was introduced by the Construction Industry Institute (CII) to smooth construction execution and avoid problems that may arise during construction, such as errors, omissions, ambiguities and conflicts. CII promotes a range of tools to facilitate a comprehensive approach to constructability implementation (CII 2006).

Constructability

Optimum use of construction knowledge and experience in planning, design, procurement, and field operations to achieve overall project objective.

(CII 2006)

A constructability review should be conducted to highlight any construction approach issues in the design, and identify potential problem areas and conflicts, thereby avoiding unnecessary requests for information (RFIs), field orders, and change orders. This can also help minimize delays and the risk of disputes, claims and litigation from contractors, vendors and suppliers. Other likely benefits embrace higher quality construction documents, higher quality bids, reduced administrative costs for issuing addenda, and a better understanding of project goals by contractors.

The scope of a constructability review may include, but not necessarily be limited to, consideration of the following aspects:

- Review key elements of design and deliverables,
- Identify conflicts between documents, drawings, and specifications,
- Identify significant construction challenges
  e.g. remote location, limited local labor and/or skills, restrictions on use of expatriate skilled labor, language barriers, limited or poor quality local

equipment and materials, seasonal weather extremes, security issues, limited local support services/housing, etc.,

- Identify and validate philosophy for online vs. turnaround construction including tie-ins for brownfield projects,
- Verify contracting strategy and plans for construction in light of the construction challenges,
- Identify enhancements to design and construction planning, which improve construction sequence, quality, safety, costs, and schedule.

6.5.7 Contractor Selection

A contracting strategy for construction, pre-commissioning and commissioning of the project should be completed, if not already finalized. Refer to Chapter 2 for a description of project implementation strategies and guidance on how various contracting strategies may impact process safety. The project contract management and administration practices and procedures should have been established in FEL-3.

The finalization of the contracting strategy should permit the timely appointment of the construction contractor, if not already appointed. Alternatively, the work may be performed in-house.

A detailed scope of work and deliverables should be prepared for the contract(s). This should include submission of various safety related documents and information, such as:

- Contractor's EHS and process safety performance statistics,
- Contractor's EHS and process safety management system, including safe work practices,
- Contractor's EHS and process safety plan,
- Contractor's competency for jobs requiring special skills,
- Disclosure of any sub-contractors,
- Contractor's infrastructure and equipment, such as cranes, heavy trucks, barges, excavators, etc.

The contractor selection should be based on a combination of technical competency, EHS and process safety ability, and cost. Cost should not be the sole factor in determining contract award, as familiarity with construction practices, construction safety, quality control (QC), and materials management & control are important. Many construction and commissioning contractors have limited process safety capability and processes, and the PMT should consider the requirement for a process safety assessment to identify any weaknesses. Where necessary, the PMT should provide support for key elements, such as process safety information, SCE and performance standards, asset integrity requirements, MOC and risk registers.

6.6   PREPARATION FOR CONSTRUCTION

A number of pre-mobilization activities should be finalized during detailed design to prepare for site construction. Rarely are two projects identical, so factors such as location, contract strategy, construction agreements, site supervision, security, etc. will require different solutions.

However, the following plans, if required, should be developed and ready for site mobilization at the appropriate time:

- Construction planning (task sequence, manpower requirements, required construction equipment, transition from area to system, SIMOPS, etc.),
- Pre-mobilization meetings with all contractors (EHS expectations, hazards and risks, procedures, bridging documents between client and contractor management systems, etc.)
- Site organization (temporary offices, housing, telecommunications, utilities, waste disposal, catering, cleaning services, parking, lighting, fuel, laydown areas, warehousing, etc.),
- Access roads (suitable for transporting heavy equipment),
- Access and foundation for heavy lifts,
- Route planning for equipment and materials,
- Site drainage,
- Security services,
- Emergency response services (first aid, firefighting, rescue, procedures, access/egress),
- EHS services for construction (procedures, orientation training, safety oversight, auditing, incident investigation, EHS performance measurement, etc.),
- Construction equipment (cranes, forklifts, manlifts, trucks, scaffolding, etc.),
- Contractor oversight,
- Administration (control of contracts, contractor personnel, certification of craft skills, insurance, office equipment and consumables, etc.),
- Community liaison,
- Design information and documentation control (receipt, storage, retrieval, updating),
- Engineering design support,
- Engineering queries / design change notice system,
- Process equipment and materials (receipt, certification, storage, preservation, issue, procurement system for shortfalls),
- Project control (planning, progress measurement, reporting).

While the above list is not exhaustive and will vary by project, some of the pre-mobilization activities have process safety implications. For example, receipt of process equipment and materials that do not meet specification could create additional hazards leading to a major incident. The quality management plan should address QC, such as inspections and positive material identification (PMI) of equipment and materials received, and QA audits to provide assurance that the QC program is being correctly implemented by contractors.

If the project is a brownfield development, shutdown requirements to tie-in process and utility systems should already have been coordinated with existing operations. If not, any shutdown requirements should be defined as soon as possible, the timing agreed, and added to the project plan. SIMOPS studies should also be completed as soon as possible to enable any impacts on existing operations to be managed efficiently, e.g. heavy lifts over live process units. The project may also be required to implement the safe work practices (hot work, energy isolation, confined space entry, etc.) of the existing operations.

Other construction activities that affect process safety are discussed in detail in Chapter 7.

6.7   PREPARATION FOR PRE-COMMISSIONING, COMMISSIONING, AND STARTUP

A number of activities should be finalized during detailed design to prepare for pre-commissioning, commissioning and startup. Depending on the scope of the project, plans should be developed as necessary including, but not limited to, the following activities:

Pre-commissioning

- Integration of future operations personnel into pre-commissioning and commissioning teams,
- Check design conformity,
- Prepare QA/QC documentation (inspection/test registers),
- Prepare 'as-built' documentation,
- Check mechanical completion of electrical, mechanical and control systems (including certification),
- Punch-list,
- Run-in machinery,
- Hydro/pneumatic testing,
- Flushing/cleaning,
- Drying,
- Leak detection,

- Load catalyst/desiccant/packing, etc.
- SIMOPS,
- Operational readiness review (aka pre-startup safety review (PSSR)).

These and other pre-commissioning activities that affect process safety are discussed in detail in Chapter 7.

Commissioning and Startup
- Commissioning and startup plan (including roles and responsibilities),
- Functional test of electrical, mechanical, control systems and safety systems, including SIS validation,
- Handover (liaison with Operator, documentation, engineering support, vendor support, punch-lists, etc.),
- Commissioning and operating procedures,
- Competency development and training (including operator training, process simulators),
- Test runs to verify performance goals (throughput, quality),
- Maintenance and inspection programs (including procedures, software data, spares, etc.),
- EHS management system including policies, procedures, emergency response plans,
- Management of change system,
- Document management system.

The above list is not exhaustive and will vary by project. Responsibility for the activities may variously rest with the Project, Operator, or possibly a third party. These and other commissioning and startup activities that affect process safety are discussed in detail in Chapter 9.

6.8 STAGE GATE REVIEW

When nearing the completion of Detailed Design, a stage gate review should be conducted to ensure that process safety (and EHS) risks are being adequately managed by the project. The stage gate review team may use a protocol and/or checklist, such as the detailed protocol in Appendix G. A typical process safety scope for a Detailed Design stage gate review is illustrated in Table 6.1.

The stage gate review team should be independent of the project, familiar with similar facility/process/technology, and typically comprise an experienced leader, process engineer, operations representative, process safety engineer, construction safety specialist, QA/QC specialist, other discipline engineers (as appropriate), and EHS specialist. At the conclusion of the review, the review team will make recommendations for any improvements needed, and indicate to the Gate Keeper, based on process safety, whether the project is ready to proceed to the next stage, Construction.

Table 6.1. Detailed Design Stage Gate Review Scope

Scope Item
Confirm that final HIRA (e.g. HAZOP) is complete and its recommendations are being satisfactorily addressed
Confirm that change control procedures are being applied and that appropriate hazard review of changes has been instigated to maintain Process Safety and EHS integrity
Confirm that appropriate specialist reviews have been carried out and their outcomes are being satisfactorily addressed
Confirm that engineering controls and checks are in place
Confirm that a Process Safety and EHS management system including a Process Safety and EHS Plan(s) is being implemented effectively
Confirm that integrity management programs are being satisfactorily addressed
Confirm that Process Safety and EHS aspects have been adequately considered in the products of detailed engineering and that they are appropriate for construction
Confirm that Project's planning for startup includes development of procedures, training, pre-commissioning and commissioning activities
Confirm that the scope of process safety information is defined and that a plan is in place for formal delivery to Operations
Confirm that an emergency response plan(s) has been developed or updated and that it addresses relevant process safety risks associated with startup and the operation.

6.9   SUMMARY

Detailed Design is the first stage of project execution and involves completion of detailed engineering of the defined scope (FEED package) from FEL. Engineering of inherently safer design, functional safety and other process safety and technical issues requires further development before the client is likely to make a final investment decision. Nevertheless, detailed design represents a significant financial commitment.

Other important project activities include management of any scope changes, procurement of materials and equipment, planning for construction and commissioning, and interface management between multiple organizations including contractors, fabricators, vendors and suppliers. The ultimate goal is to resolve remaining project risks and uncertainties and complete a design package that includes all necessary information required for construction in order to stay on schedule and facilitate efficient commissioning and handover.

Additional information can be found in several publications:

API, *Material Verification Program for New and Existing Alloy Piping Systems*, 2nd edition, RP 578, American Petroleum Institute, 2010.

CCPS (Center for Chemical Process Safety), *Guidelines for Engineering Design for Process Safety*, Second Edition, American Institute of Chemical Engineers, New York, NY, 2012.

7 CONSTRUCTION

Following the Detailed Design stage, the project moves into the Construction stage that is the second phase of project execution. The goal of construction is to safely build the facility in line with the risk assessed design, so that it will startup, operate, and shut down safely. In this regard, it is essential that all engineering drawings and specifications are readily accessible to the construction work crews. Figure 7.1 illustrates the position of construction in the project life cycle.

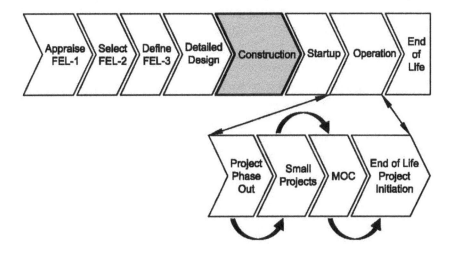

Figure 7.1. Construction

Rarely are two projects identical, so factors such as location, contract strategy, construction agreements, site supervision, security, etc. may require different solutions. Depending upon these and other factors such as the scale of the project and technical skills required, construction may be performed by a contractor(s) or in-house resources (e.g. maintenance team). In some circumstances, the use of a local contractor, local fabrication, local labor and/or local materials rather than a free choice of pre-qualified resources may have been a condition of planning permission for the project. In this case, the PMT may wish to raise the level of oversight of contractors, sub-contractors and suppliers.

The construction stage can potentially involve many parties, for example fabricator(s), contractor(s), sub-contractor(s), consultant(s), vendors, and suppliers. This requires good interface management, and regular performance reporting between all parties involved. In addition, the PMT needs to liaise with all stakeholders, including, but not limited to, the future operations team, local community, national/local government agencies, and NGOs.

Project Management Team

The PMT's primary focus should be on implementing the construction element of the project execution plan within budget and to schedule, while ensuring that the design intent, including process safety aspects, is rigorously controlled and the quality objectives are clearly outlined. A Construction Manager reporting to the Project Manager is normally appointed for major capital projects, and oversees day-to-day fabrication, construction and pre-commissioning to meet the project budget and schedule. Typical project objectives for overall construction management include:

- Planning and continuous progress measurement,
- Site organization and general administration,
- Control of contracts and contractors,
- Receipt, storage, retrieval, and updating of design and technical specification information,
- Receipt, storage, and issue of materials and equipment,
- Site inspection, EHS management, and supervision for control of work activities,
- Quality management and function testing,
- Document management including design, quality, integrity, etc. throughout construction activities,
- Generation of handover documentation, including engineering as-built and integrity baseline data,
- Handover to the Operator.

Any remaining project risks and uncertainities require close management in order to stay on schedule and facilitate efficient handover.

The focus of the PMT from an engineering viewpoint is on closing out any remaining engineering activities not completed in Detailed Design, and providing engineering and technical support for construction, e.g. technical queries raised at the construction site(s).

Environment, Health and Safety

From an EHS perspective, the construction stage is when the PMT has a more direct 'hands on' responsibility for EHS performance. The project needs an effective construction EHS management system, including robust EHS procedures, contractor orientation training, emergency response plan, and auditing of construction site activities. Construction contractors should have been screened for their strong commitment, motivation and abilities with regard to EHS. As the construction representative of the PMT, the Construction Manager liaises directly with contractors to promote good EHS practices and a positive safety culture. HIRA and implementation of safe work practices for construction activities are key to delivering good EHS performance. Finally, the project EHS Plan may need to be updated to ensure EHS preparedness for commissioning.

Process Safety

The key process safety objectives in the construction stage include:

- Manage the process safety risks identified during previous phases that are relevant to fabrication, construction, and installation, including actions from the constructability study,
- Execute asset integrity management (AIM) practices and procedures including QA/QC to deliver integrity and maintain design intent, especially for SCE and other protection systems,
- Assure competence of the construction workforce (all crafts), and deliver induction training,
- Implement the site change management process, and thoroughly evaluate any late design changes,
- Identify and manage key process safety information,
- Update the Process Safety Plan, if necessary, to address preparedness for commissioning and startup, including operating procedures and training, maintenance management system, and emergency response.

These and other process safety activities in fabrication, construction and pre-commissioning are discussed below for:

- Planning (Section 7.1)
- Pre-mobilization (Section 7.2)
- Mobilization (Section 7.3)
- Execution (Section 7.4)
- Other project activities (Section 7.5)
- De-mobilization (Section 7.6)
- Preparation for commissioning and startup (7.7)
- Final evaluation and close-out (Section 7.8)
- Stage gate review (Section 7.9)

Hazard identification and risk management are key factors in delivering good EHS and process safety performance, which also requires active participation from the client, project and contractor(s).

7.1   PLANNING

Initial planning for construction should have started in FEL-3 and continued through Detailed Design. Major capital projects typically have a Project Execution Plan (PEP), Construction Plan, and a Construction EHS Plan. Small projects may only have a simple construction plan, and use company or regulated safe work practices.

The PEP is a high-level plan focused on the main strategies through the execution stages of the project (i.e. detailed design, construction, and startup) up to full production. The PEP also identifies key milestones within project execution at which determinations to proceed or pause (i.e. go/no go decisions) are appropriate. For example, readiness to mobilize resources to the construction site. Contractors' own execution plans for procurement and construction should be consistent with the PEP, and subject to rigorous review by the PMT. Some companies adopt a philosophy of 'no change' to the PEP after it has been peer reviewed and approved by senior executives. In this case, the PEP is only revised in the event that external events impact the project or if a major risk to the business or the public is identified.

The main construction plan comprises a detailed sequence of tasks to be performed together with their timing (start/finish), duration, inter-dependencies, and resources (manpower, skills, equipment) necessary for each task. A capital project may involve thousands of tasks, which are typically documented in a Gantt chart or logic network, e.g. Program Evaluation Review Technique (PERT) chart. Small projects may only use a task list. The plan normally goes through multiple iterations in order to (i) optimize the critical path that determines the construction completion date, and (ii) avoid interference between work crews and crafts working in the same area simultaneously. While the main construction and installation may be planned on an area basis, pre-commissioning is normally planned on a system basis. This transition requires careful planning to avoid interference between ongoing construction and the pre-commissioning activities. Whatever means of planning is used, the plan should be regularly updated to record progress, and allow the PMT to intervene if slippage or other departures from the plan occur.

The construction EHS plan addresses the EHS risks associated with the project scope, and defines the EHS responsibilities, and standards and procedures to be employed during construction to manage the risks. Some companies include process safety requirements in the EHS plan. Key aspects of the plan should be how simultaneous operations (SIMOPS) and emergency response will be managed.

The plan should also address:

- PMT/client's EHS expectations,
- Any gaps or differences in the contractors' (and sub-contractors') EHS management system,
- PMT/client's oversight/assurance practices to monitor EHS performance.

Depending upon the individual circumstances of the project scope, additional plans may address issues, such as temporary offices, storage and laydown areas, camp/housing for construction workforce, and equipment transportation. These and other issues are discussed below (Sections 7.2, 7.3, and 7.4).

7.2   PRE-MOBILIZATION

In addition to planning, a number of other pre-mobilization activities may be necessary to prepare for site construction. Contracts should be in place and their scope reviewed by the PMT for the different construction contractors to ensure that all activities required for adequate completion of the project are included. After mobilization, any additional activities will result in a change order. A key activity is a PMT meeting with all contractors and sub-contractors before mobilizing to the construction site(s) to communicate and ensure understanding of, but not limited to, the following:

- Major process safety and EHS risks during construction, and their management,
- Construction EHS plan, including emergency response and first-aid,
- PMT/client's expectations for process safety and EHS,
- Bridging documents to address gaps or differences in contractors' process safety and EHS management system,
- PMT/client oversight and assurance processes for process safety and EHS, such as walk-through inspections, performance measurement, and audits,
- Site specific issues, such as general site condition, road access, security, waste disposal, etc.
- Material receipt, storage, handling and issue,
- Labor relations.

If the pre-mobilization meeting is conducted as a workshop with participation by all parties, the construction EHS plan may be confirmed as fit for purpose or improvements may be identified. Some companies also conduct a pre-mobilization review to ensure that all process safety and EHS plans are in place, which leads to a 'go/no go' decision by senior leadership on whether to mobilize to the construction site(s).

Other pre-mobilization activities are listed in Table 7.1. This list is not exhaustive, and inevitably will vary depending upon the scope of the project, but covers many of the key activities that require planning before mobilization and commencement of construction.

Table 7.1. Typical Planning Activities at Pre-Mobilization

Pre-Mobilization Activity
Site organization (temporary offices, housing, telecommunications, utilities, waste disposal, catering, cleaning services, parking, lighting, fuel, laydown areas, warehousing, etc.) in accordance with the facility siting study
Access roads - suitable for transporting heavy equipment
Access and foundation for heavy lifts
Route planning for equipment and materials
Site stormwater drainage
Security services and fencing
Emergency response services (first aid, firefighting, rescue, procedures, access/egress)
EHS services for construction (e.g. procedures, induction training, safety oversight, auditing, incident investigation, EHS performance measurement)
Construction equipment (e.g. bulldozers, graders, excavators, cranes, forklifts, manlifts, trucks, scaffolding, hand tools)
Administration (e.g. control of contracts, contractor personnel, certification of craft skills, insurance, office equipment and consumables)
Stakeholder liaison (e.g. local community, regulatory agencies, NGOs)
Design information including equipment datasheets, technical specifications, drawings, underground piping and cables, etc. (receipt, storage, retrieval, updating)
Engineering design support
Engineering queries / design change notice system
Process equipment and materials (receipt, certification, storage, preservation, issue, procurement system for shortfalls)

Pre-Mobilization Activity
Project control (planning, progress measurement, reporting)
Environmental licenses approved
Social / Community agreements (roads use, labor force, local content policies, land ownership, project impacts, etc.
Waste materials management plan

Of particular concern from a process safety perspective is the siting of temporary buildings that may be required for offices, catering, housing for remote locations, etc. If it is intended to locate any temporary buildings in the vicinity of process and storage facilities handling hazardous materials, the guidance provided by CCPS and API should have been followed during the FEL-3 and Detailed Design stages (CCPS 2012b, API 2007). In the case of a brownfield development, it may be possible to locate temporary buildings outside the existing zones deemed hazardous. However, care should be exercised that the construction of a new process unit does not create a new congested volume that extends the hazardous zones such that the temporary buildings are within a hazardous zone. Tents are increasingly being used on construction sites for temporary shelter, catering, and other uses for work crews, and similar care should be exercised in their placement (API 2014).

Further guidance setting safety expectations and subsequent management review is available from the following CCPS publication: *Guidelines for Risk Based Process Safety* (CCPS 2007b).

Further guidance on siting temporary buildings is available from the following publications: *Guidelines for Evaluating Process Plant Buildings for External Explosions, Fires and Toxic Releases, 2nd edition* (CCPS 2012b); *Management of Hazards Associated With Location of Process Plant Portable Buildings, RP 753* (API 2007); *Management of Hazards Associated with Location of Process Plant Tents, RP 756* (API 2014).

Further guidance on pre-mobilization activities is available from the following publication: *HSE Management – Guidelines for Working Together in a Contract Environment, Report No.423* (IOGP 2010).

7.3 MOBILIZATION

Before mobilization can occur it is important that there are sufficient engineering deliverables and materials available to ensure efficient progress. Initial construction activities are likely to include site clearance (and any demolition work required), grading and drainage, access roads, temporary offices, utilities, telecommunications, security fencing, and foundations. In certain circumstances, a camp with temporary housing, catering and related services for the construction workforce may be necessary, especially in remote locations.

One of the key tasks is to ensure that project and contractor personnel mobilize safely to the construction site(s). This entails implementing and communicating the construction EHS plan and PMT/client's process safety and EHS expectations, and then auditing to ensure that each contractor (and sub-contractor) organization sets up operations accordingly. It is important that construction starts with, and maintains, a culture of process safety and EHS owned by project and contractor line management (down to foreman/supervisor level), rather than process safety and EHS advisers (see Section 7.4.3).

Further guidance on safety culture is available from the following CCPS publications: *Essential Practices for Creating, Strengthening, and Sustaining Process Safety Culture* (CCPS, 2018e); *Guidelines for Risk Based Process Safety* (CCPS, 2007); *Building Process Safety Culture: Tools To Enhance Process Safety Performance* (CCPS, 2005).

It should be noted that the size of the initial workforce is likely to be small, but will gradually ramp up to the maximum required to efficiently build the main production and storage facilities. Manpower is then likely to decline until completion of all construction and pre-commissioning activity. Furthermore the balance of skills necessary is also likely to vary as the construction progresses. As a result, the project should retain a capability to identify and provide induction training to all new contract employees when they first access the site(s).

As soon as personnel start mobilizing, the site(s) emergency response plan needs to be in place, and facilities for first-aid, firefighting, and rescue should be operational. A table-top or emergency drill should be conducted as soon as possible to verify the emergency response plan's efficacy and contract employees' compliance with its requirements. Facilities for receipt, certification, storage, preservation, and issue of process equipment and materials will also need to be set up prior to delivery to site.

Any equipment and materials brought on site by contractors should be subject to inspection by a competent person to verify fitness for purpose, and, if appropriate, a safety data sheet (SDS) obtained for site records. Typical equipment could include cranes, forklifts, manlifts, trucks, scaffolding, and hand tools.

7.4   EXECUTION

The key objectives of the Execution phase of construction are to ensure that construction is performed safely, in accordance with the design, construction plan, quality management plan, and performance–managed to ensure safe completion within budget and schedule. Another goal is to manage stakeholder expectations to assure smooth progress. These and other aspects of construction execution are discussed below in more detail, and involve many of the elements of Risk Based Process Safety (RBPS) (CCPS, 2007).

7.4.1   Procurement

While most equipment and materials should have been procured during the Detailed Design stage (see Section 6.2) and long-lead items in FEL-3 (see Section 5.4.6.2), it is normal to retain a limited procurement capability during construction to handle material shortfalls and omissions. The project may also wish to manage cash flow by deliberately phasing purchase orders for some readily available materials and consummables. Nevertheless any procurement during construction should also be subject to the project's quality management (QM) system (see Section 0).

7.4.2   Fabrication

Fabrication of major equipment items, such as fractionation columns and pressure vessels, is likely to continue through the early phase of construction. The construction manager or his designee should maintain oversight of all fabrication work to ensure that engineering standards and technical specifications are being followed, and that manufacturing practices in the fabrication shop do not compromise quality and integrity.

A Quality Management Plan defining the specific quality control checks, hold points, and witnessed tests should have been established and approved by the client as a contractual requirement.

QA inspections/audits at specific hold points in the fabrication process of pressure vessels and high-criticality equipment should be conducted by the project or a third party inspector acting on their behalf. Factory acceptance tests (FAT) of some critical and/or complex equipment should be witnessed at the manufacturer's premises to verify its operability and functionality before delivery. Further information on project QM systems in respect of equipment fabrication is discussed in Chapter 8.

7.4.3   Safety Culture

At the commencement of construction execution, project management with the assistance of contractor leadership should seek to nurture a positive environment where employees (including contractor employees) at all levels are committed to safety. Conduct of operations is closely related to safety culture, and leadership should set expectations for construction tasks to be carried out in a deliberate, careful, and structured manner that follows the EHS and process safety procedures. Managers should set a personal example, ensure that workers perform their tasks in a safe manner, and enforce high standards.

Further guidance on safety culture and conduct of operations is available from the following CCPS publication: *Essential Practices for Creating, Strengthening, and Sustaining Process Safety Culture* (CCPS, 2018e); *Guidelines for Risk Based Process Safety* (CCPS, 2007).

7.4.4   Workforce Involvement

The broad involvement of the workforce in improving construction activities, such as energy isolation (lockout/tagout), confined space entry, and simultaneous operations (SIMOPS), can assist in driving a positive safety culture. Leadership should listen to workforce concerns, and make sure that lessons learned by the people closest to the construction are considered and addressed.

Further guidance on workforce involvement is available from the following CCPS publication: *Guidelines for Risk Based Process Safety* (CCPS, 2007).

7.4.5   Stakeholder Outreach

Before and during construction, the PMT should hold regularly meetings with their key stakeholders to keep them informed, understand concerns, seek alignment, and attain regulatory approval (e.g. permits) in order to smooth construction progress. For example, the local community may have concerns relating to disturbance due to heavy traffic to/from the construction site on a daily basis. Key project stakeholders are likely to include the local community, regulatory agencies, NGOs, emergency services, employees, unions, partners, and contractors.

Further guidance on stakeholder outreach is available from the following CCPS publication: *Guidelines for Risk Based Process Safety* (CCPS, 2007).

7.4.6   Contractor Management

Most projects employ one or more contractors for fabrication, construction and/or pre-commissioning, unless it is a relatively small project that can be handled by the site's engineering and maintenance resources. The contractor(s) should have been selected during the Detailed Design stage of the project when the contracting strategy was finalized.

Following a pre-mobilization meeting with contractor leadership to discuss EHS and process safety expectations, rules and procedures (see 7.2 above), their work crews require orientation training when they first access the construction site (see Section 7.4.13 below). Thereafter, work crews should be briefed daily on the hazards of their work and any hazards adjacent to the job site. This may be accomplished at pre-job toolbox meetings, participation in developing JSAs, or other means. Regular safety meetings should reinforce procedures, and share lessons learned from any incidents that have occurred.

An adequate number of safety specialists and construction supervisors employed by the project and contractors should maintain constant viligance around the construction site(s) to ensure that contract workers perform their jobs safely, and that contracted services do not add to or increase risks. A key aim should be that contractor vehicles and heavy equipment meet project's safety standards, are maintained in safe working order, and are operated by competent operators at all times. There should be a culture of zero tolerance for not following safety policies, rules and procedures.

The construction manager should regularly review contractor(s) performance in meeting the EHS and process safety expectations, rules, and procedures, and rapidly intervene if performance improvement is required. This management review process should also ensure that the contractor(s) is complying with contract conditions for quality, integrity management, client requirements as well as cost and schedule..

Further guidance on contractor management, and management review and continuous improvement is available from the following CCPS publication: *Guidelines for Risk Based Process Safety* (CCPS, 2007).

7.4.7 Transportation

Most projects, except perhaps small MOC works, face a variety of transportation issues. Most equipment and materials are transported by road, but if international fabricators are used, the equipment may be shipped by sea. Even semi-submerisible oil production platforms may be built overseas and then towed or transported on special ships. Some critical or delicate equipment may require special handling to protect its integrity, in which case the project may wish to oversee loading at the fabricator's workshop for transportation to the construction site. Chapter 8 discusses quality aspects of asset integrity management requirements in greater detail.

Some fractionation columns and pressure vessels may be very large and heavy requiring special road permits and very careful route planning to avoid low bridges, tight corners, and urban areas. Even so it may be necessary to disconnect some electricity/telephone cables to allow passage. These very large loads also require an access road into the construction site capable of supporting the weight of the vehicle and equipment. The equipment may need to be stored in a laydown yard and then moved into location at a later date.

Another transportation issue requiring careful planning involves the large number of vehicles that may need to access the site on a daily basis. In addition to the construction workforce's vehicles, there is likely to be a large number of trucks carrying rock/gravel, concrete, equipment, materials, etc. Route planning should attempt to avoid disturbance to the local community, and regular liaison with the community should allow the PMT to intervene if concerns are raised (see 7.4.5 above).

7.4.8 Equipment and Materials Handling

Greenfield developments are likely to require a location for the receipt, certification, storage, and issue of construction equipment and materials. Some brownfield developments may be able to use existing storage facilities, but will need to segregate and identify project materials. Large items will require a laydown area, while smaller components, such as electrical and control equipment, will require covered storage with temperature/humidity control. Some other items may also require special preservation measures while in storage to maintain their integrity and functionality. For example, rotating machinery may periodically require lubrication and hand turning, while some pressure vessels may require an internal positive pressure of nitrogen. Failure to preserve equipment and materials properly could cause, contribute to, or fail to prevent or mitigate, a process safety incident.

All equipment and materials should be inspected on receipt to ensure that they meet specification, and any non-conformances or damaged items should be quarantined. Special steels and alloys may require positive material identification. Further information on project QM systems in respect of receipt, storage, and retrieval of equipment and materials is discussed in Chapter 8.

7.4.9   Hazard Evaluation

Construction and installation activities, such as working at height, heavy lifts, hot work, confined space entry, excavation, and use of multiple vehicles and mobile machinery, involve many hazards. Pre-commissioning adds hazards due to activities like pressure testing, chemical cleaning, air blowing, and live utility systems. Simultaneous activities in close proximity to one another add further complexity. Some offshore projects involving hook-up of anchors, subsea risers and umbilicals to the main installation pose different hazards.

Project's HIRA studies may have already identified some of the construction hazards and recommended safeguards. However it is unlikely that HIRA studies recognized all construction hazards. The project should ensure that safe work practices are rigorously implemented, and each work permit should be supported by a task hazard assessment, such as a job safety analysis (JSA). The JSA (e.g. job hazard analysis (JHA) and task hazard analysis (THA)) should involve the work crew and preferably a safety specialist, identify potential hazards at each step of the permitted job, and determine safeguards to manage the hazards.

Case Study: Onshore Construction Projects

One major operating company reported that incidents involving mobile vehicles and heavy equipment were the greatest source of fatalities and serious injuries during construction.

Root causes were attributed to *inadequate identification of worksite hazards*, inadequate control of work (i.e. safe work practices), inadequate contractor oversight (especially sub-contractors), and lack of required competency to perform the tasks.

Where simultaneous operations, such as two or more of production, drilling, maintenance, construction, and pre-commissioning, occur in close proximity, a SIMOPS study should be conducted to identify and manage potential interactions. This is particularly beneficial for brownfield developments where existing operations continue during construction. Performing the SIMOPS study early can enable impacts, such as shutdowns to tie-in process and utility systems, to be managed efficiently.

All hazards and required safeguards must be communicated to the relevant job crew(s) including any hazards adjacent to the job site. Work crews should also report hazards and unsafe conditions to their supervisor for the attention of project management.

Further guidance on HIRA is available from the following CCPS publications: *Guidelines for Risk Based Process Safety (CCPS, 2007); Guidelines for Hazard Evaluation Procedures, 3rd edition (CCPS, 2008); A Practical Approach to Hazard Identification for Operations and Maintenance Workers (CCPS, 2010).*

Case Study: Loss of Containment during Construction Project

During the brownfield expansion of an operating plant, it was necessary to build a high level piperack over live process equipment. One section of the piperack was being installed close to the High Pressure Separator that was in operation. Three different crews (two at ground level, one above the piperack) were performing welding within 20 meters of the vessel. At the same time a crane was operating in the same area.

Inadvertently, the crane operator moved the hook too close to the vessel and touched the cage of a level controller, cracking the pipe connecting the instrument to the vessel. Immediately, crude oil sprayed out and a vapor cloud formed. Potential ignition sources included the three welding machines and associated sparks from welding rods, plus the crane engine.

The work crews evacuated the area immediately leaving two of the welding machines still on. Fortunately the wind direction was away from the potential ignition sources, and no ignition occurred.

The investigation found that construction personnel were not properly warned of the risks related to simultaneous operations (SIMOPS) work nor the correct emergency procedures in the event of a loss of containment.

7.4.10 Engineering Design

Engineering drawings and specifications are essential to instruct the construction work crews on what to build. Therefore on-site systems for the receipt, storage, retrieval, and updating of these design data must be thorough and efficient. Updates should only be issued after rigorous change management.

Major capital projects employ comprehensive databases (electronic and/or hard copy) to log and record the following design information:

- Receipt of all documents and updates
- Index of latest issues of all documents
- Documents issued to contractors
- Outdated issues retrieved from the field and cancelled or destroyed

Small projects may use a simpler system, but should still address the same aspects. Failure to properly manage engineering documentation could result in incorrect installation leading to expensive rework and/or cause, contribute to, or fail to prevent or mitigate, a process safety incident. Chapter 12 discusses project documentation in greater detail.

Case Study: Construction Documentation

Poorly managed engineering documentation during the construction stage of a project can result in the use of outdated or incorrect documents years after installation, creating confusion and potential errors in the event of new projects, modification, or repair of the original equipment.

For example, a company had assumed for more than 30 years that a tower was fabricated from two different materials, creating the need for low temperature trips. A new project required operation at low temperatures. However, the materials of construction of the the original tower were checked, and found that the tower was fabricated from a single material capable of handling low temperature. The installation, operation and inspection of the trip systemshad been redundant all those years.

Another industry example includes a lack of information on pressure vessel internals (e.g. distributors, baffles, etc.) in documentation handed over to the Operator. This created problems during turnarounds and repair work.

These examples illustrate the importance of properly managing project documentation and handover to the future Operator.

Sufficient engineering resource should be retained during construction to support site activities. This resource may be required to handle queries on the design information, and explain features of the design intent. A system for 'engineering queries' (a.k.a. 'request for information' (RFI)) should be established, and comprise a record with unique number, progress status, and approved answer.

Any field changes should be handled through a 'design change notice' (DCN) system. All DCNs should be logged with a unique number, progress status, and implementation. Most DCNs are likely to involve minor changes, but all should be subject to rigorous change management to evaluate any new or modified hazards.

Further guidance on compliance with engineering codes and standards is available from the following CCPS publications: *Guidelines for Risk Based Process Safety* (CCPS, 2007); *Guidelines for Engineering Design for Process Safety, 2nd edition* (CCPS, 2012).

7.4.11 Safe Work Practices

Rigorous enforcement of safe work practices is critical to safe construction. All construction work crews are responsible for following the approved safe work practices that may be regulated and/or required by the client/project. The client/project may require more stringent practices than local regulations.

The client may also seek to promote adoption of their safe work practices and procedures or an equivalent standard by companies working on their behalf on non-client sites. For example, a project representative may be present at a contractor's fabrication yard to monitor a large process module being constructed using the client's safe work practices.

The safe work practices may cover, but not be limited to:

- Site access control
- Work permitting
- Hot work (welding, grinding, naked flames/sparks, etc.)
- Energy isolation (LOTO)
- Line breaking
- Working at height (scaffolding, man-lifts, fall protection, etc.)
- Excavation and trenching (buried cables/pipes, shoring, sloping, etc.)
- Confined space entry (including excavations, sumps, sewers, etc.)
- Heavy lifts (cranes, lift plans, signalers, forklifts, etc.)
- Electrical systems (high voltage, overhead/buried cables, etc.)
- Vehicles and mobile machinery (bulldozers, graders, trucks, banksman, etc.)
- Pressure testing (hydro, pneumatic) including personnel exclusion zones
- Temporary systems, such as flexible hoses and electrical cables

- Hazard communication (SDS, chemical cleaning, coatings, etc.)
- Radiation (NDT X-ray)
- Bypass/inhibit management of any safety critical equipment
- Inclement weather window (e.g. lifting prohibited due to wind/wave)
- Personal protective equipment (PPE) including lifejackets for work near/over water, gas monitors, respirators, etc.

Every member of the construction workforce will likely require some form of orientation training in the detailed safe work practices and critical safety rules to be employed on the site(s). Thereafter the construction manager should establish daily monitoring and periodic auditing (by a number of project supervisors and safety specialists) to ensure that the safe work practices are being implemented, and, if not, intervene to enforce their implementation. Repeated failure to follow approved practices and procedures should be subject to disciplinary action including dismissal.

Several permit issuing authorities may be necessary to inspect job sites and issue work permits at peak construction activity. It is particularly important that these issuing authorities communicate with each other and consider other permitted work in the area (in three dimensions) when writing permits to avoid interferences between different crews and crafts. Until pre-commissioning activities commence, it may be appropriate to issue 'blanket work permits' in certain areas (e.g. fenced area under control of the construction manager), where the only hazards are construction hazards. These blanket permits may be renewed regularly providing the hazards have not changed. JSAs should support each work permit. Daily toolbox meetings should be held to cover the day's job tasks, hazards, required safeguards, and adjacent activities. Permits, JSAs and meetings should be communicated in the workforce's native language(s).

Project safety specialists and those employed by contractor(s) should maintain high safety standards, including good housekeeping and enforcing exclusion zones behind barriers for work such as heavy lifting, excavation, pressure testing, and NDT work. A competent person should prepare a detailed lifting plan for each heavy lift to manage its hazards and risks.

Further guidance on safe work practices is available from the following CCPS publication: *Guidelines for Risk Based Process Safety* (CCPS, 2007). US OSHA and UK HSE also provide guidance through their websites.

CCPS has also established a website, *Essentials of Safe Work Practices*, to provide guidance on a range of safe work practices. This information may be accessed at https://www.aiche.org/ccps/resources/tools/safe-work-practices.

Each Safe Work Practice (SWP) contains information related to the following eight elements: fundamental intent, need/call to action, potential hazardous consequences, strategies & effective practices to manage and mitigate hazards, possible work flow, common program practices, incidents, and reference materials

7.4.12 Operating, EHS and Process Safety Procedures

The schedule for the development of commissioning and operating procedures should allow the commissioning and operating teams sufficient time for thorough review, comment, revision and familarization prior to the commencement of commissioning. These procedures should cover normal and transient operations, such as startup, shutdown, catalyst regeneration, etc.

Some EHS and process safety procedures may be mandated by local regulations, while the client and/or project may require a higher standard and additional procedures to meet their EHS and process safety expectations. These procedures may cover any of the elements discussed in this Section 7.4 , but the most important that should be required by all construction projects are:

- Hazard evaluation (Section 7.4.9)
- Safe work practices (Section 7.4.11)
- Integrity management (Section 7.4.14)
- Change management (Section 7.4.15)
- Emergency response (Section 7.4.16)

Each project should carefully determine whether any other EHS and process safety procedures are relevant to their construction activities.

Further guidance on operating procedures is available from the following CCPS publications: *Guidelines for Risk Based Process Safety* (CCPS, 2007); *Guidelines for Writing Effective Operating and Maintenance Procedures* (CCPS, 1996).

7.4.13 Training and Competence Assurance

While the contractor and sub-contractor organization(s) would typically be selected on the basis of their competency and capability, sometimes a contractor's resources become over-stretched by successful bids for other clients and/or loss of key personnel. The project should ensure that the contractor(s) have the skills and resources necessary to perform their scope of work. Review of craft skill certifications, audits, and less formal interviews can verify whether the mobilized resources have the necessary skills and experience. Any deficiencies discovered should be addressed with the contractor(s) concerned, and could have contract consequences.

In addition to contractors being responsible for providing trained and competent work crews, the project should ensure that each contract employee (including sub-contractors) receives some form of orientation training appropriate to their job tasks before accessing the construction site(s). This orientation training should cover, but not be limited to:

- Client/project process safety and EHS expectations,
- Site safety rules,
- Site safe work practices (see Section 7.4.11 above),
- Site emergency response plan (see Section 7.4.16 below).

In rare circumstances, a project may decide to provide additional training, especially when it is necessary to employ a less experienced contractor.

Further guidance on training and competence assurance is available from the following CCPS publications: *Guidelines for Risk Based Process Safety* (CCPS, 2007); *Guidelines for Defining Process Safety Competency Requirements* (CCPS, 2015).

7.4.14   Asset Integrity Management

One of the greatest concerns during construction from a process safety perspective is that the installed facilities are in accordance with the technical specifications and design intent, and are fit for purpose. Failure to do so could cause, contribute to, or fail to prevent or mitigate, a major incident.

QM at all stages of procurement, fabrication, equipment and materials handling, construction and pre-commissioning is the primary means of ensuring that facilities meet technical specifications and design intent. QM is discussed in greater detail in Chapter 8. Functional testing of systems during pre-commissioning also contributes to asset integrity (see Section 7.4.21 below).

During the construction stage, all vendors should provide service and maintenance manuals with their equipment, some of which are safety critical. This information needs to be carefully stored for use during commissioning and later by operations personnel to establish maintenance, integrity and inspection tasks. Chapter 12 discusses documentation in greater detail.

Further guidance on compliance with, and implementation of, engineering codes and standards is available from the following CCPS publication: *Guidelines for Risk Based Process Safety* (CCPS, 2007).

Further guidance on asset integrity management is available from the following CCPS publication: *Guidelines for Asset Integrity Management* (CCPS, 2017).

7.4.15 Change Management

Some changes during construction can be expected, but have the potential to impact cost and schedule. These consequences impact not only the original work package but can also impact other work packages. Change orders usually modify the basis upon which contracts were agreed in respect to timing and pricing. The contractor will expect additional compensation for any modification.

Many project managers promote a philosophy of 'no change' within the execution stages in respect of project scope, BOD, design intent, and PEP. If a change is necessary, these project managers set a high hurdle requiring a rigorous change management process to justify and evaluate the change before it is approved. Changes at this stage of the project are likely to be expensive, and priority should be given to those that improve safety, essential for regulatory compliance, or essential for process operation.

Late design changes can occur as a result of a field change (DCN) or an engineering query (RFI). For example, a work crew may find that a pipe spool fabricated in accordance with the isometric drawing cannot be installed because structural steelwork intersects the spool location. Late design changes should be subject to change management to evaluate design options, identify new or modified hazards, and technical review and approval. In the case of the pipe spool above, a seemingly simple change could introduce hazards such as (i) a low point for water to collect, (ii) increased erosion, and (iii) increased pipe stress due to inadequate support, such as a bending moment from a relocated PSV. The last stage of change management requires the engineering documentation to be updated and communicated.

Further guidance on management of change is available from the following CCPS publications: *Guidelines for Risk Based Process Safety* (CCPS, 2007); *Guidelines for the Management of Change for Process Safety* (CCPS, 2008).

Case Study: Sales Gas Plant Expansion Project

The natural gas feedstock for a sales gas plant expansion was supplied from the Phase 2 process trains of the central processing facility. The project Basis of Design (BOD) prohibited feedstock from Phase 1 trains which were to be disconnected, but retained as backup in the event that additional gas was required for injection into the oil reservoir.

A management of change (MOC) should have been initiated by the Operations team to reflect the new operating philosophy and procedures for Phase 1 trains. The Project should have properly disconnected and dismantled Phase 1 train piping that was connected to the main gas header. However, no MOC was issued, and the Phase 1 train piping was not disconnected prior to commissioning and startup of the expansion project in 2010.

That year, a high pressure (HP) valve from the connecting piping was required for use in another project and its removal was assigned as routine work to a non-qualified crew. When the HP valve was disconnected, gas trapped in the isolated pipe was suddenly released, killing one of the crew and injuring another three.

An incident investigation finding identified the lack of updated operating procedures as to why operators did not recognize the risk and depressurize the piping. Another finding was that dismantling of Phase 1 piping was omitted from the scope of work for the engineering contractor.

7.4.16 Emergency Response

The emergency response plan for the construction site(s), and the necessary resources, should have been finalized during the pre-mobilization phase, and a table-top or emergency drill conducted during mobilization or early execution to test its effectiveness. Although the presence of process fluids is likely to be minimal for a greenfield development until commissioning commences, small quantities of hazardous or incompatible materials may be present, e.g. acetylene, nitrogen and other compressed gases, cleaning solvents, etc.

The plan needs to address typical construction issues, such as, but not limited to:

- First aid and medivac,
- Fire and explosion,
- Toxic chemical release,
- Rescue from height/confined space/water,
- Vehicle/mobile machinery accident,
- Electrocution,

- Injury due to slips/trips/falls/struck by/crush, and
- Security incident (trespass, bomb threat, terrorism, etc.).

Brownfield developments will also need to address process fluids from existing process and storage facilities that can cause fires, explosions and toxic releases with potential to impact the construction site.

Two major concerns during construction involve (i) evacuation, and (ii) maintaining access around the site(s). Typically construction may have a greater number of people on site than at any time during subsequent operations. Depending on the nature of the site hazards, it may be practical to evacuate the construction site to a safe location in the event of a major incident. However, if toxic chemicals could be released, shelter-in-place (SIP) may be preferable but the size of the workforce may exceed the capacity of any buildings. In this case, special arrangements for early detection, warning alarm, and evacuation across wind may be appropriate, and will need to be clearly communicated to the workforce.

In respect of the second issue, i.e. access around the site, some road closures within the construction site are inevitable due to activities such as major crane lifts and trenching. A system should be established for closing roads, such that there is always an alternative means of access to any location within the site. This system should be designed to permit the emergency services to respond to any incident without delay.

Periodic emergency drills should be conducted as the number and composition of the workforce changes over the course of construction and pre-commissioning.

Further guidance on emergency management is available from the following CCPS publications: *Guidelines for Risk Based Process Safety* (CCPS, 2007); *Guidelines for Technical Planning for On-Site Emergencies* (CCPS, 1995).

7.4.17 Incident Investigation

Irrespective of what causes an incident on the construction site, if it's a serious incident the media are likely to refer to the client organization. The project should set up a system for reporting all incidents including, but not limited to, injury, illness, fire, chemical spill, and property/vehicle damage occurring within the construction site(s). All contractors and sub-contractors should be required to use this system to immediately report incidents.

The project should also establish a system to investigate all incidents and near-misses to identify root causes, and make recommendations to prevent recurrence. Corrective actions should be tracked to completion, and lessons learned communicated to the workforce. Evaluating incident trends on major capital projects may facilitate further project management intervention to reduce similar incidents.

Further guidance on incident investigation is available from the following CCPS publications: *Guidelines for Risk Based Process Safety* (CCPS, 2007); *Guidelines for Investigating Chemical Process Incidents, 2nd edition* (CCPS, 2003).

7.4.18 Auditing

As discussed above, key project objectives for construction are to ensure that the installed facilities are fit for purpose (i.e. meet technical specifications and design intent), and that contractors meet the client/project expectations for EHS and process safety performance. A number of audits can assist with delivering these objectives.

Project QA audits can help verify that contractors, suppliers and vendors are correctly implementing QC activities during fabrication, equipment and materials receipt and handling, construction, and pre-commissioning. The construction manager should intervene promptly to correct any adverse audit findings. This is discussed in greater detail in Chapter 8.

Project EHS and process safety audits of any of the elements discussed in Section 7.4 above can alert project management to any construction issues that could give rise to poor EHS and process safety performance. Initially project may wish to focus on contractor compliance with safe work practices and EHS rules to help prevent injuries and environmental damage. Other focus areas for auditing could be determined by any incident trends, observations, and employee concerns.

All audit findings, recommendations, and improvement opportunities should be recorded, and corrective actions tracked to closure. Follow-up audits should verify that corrective actions have resolved the original findings.

Further guidance on auditing is available from the following CCPS publications: *Guidelines for Risk Based Process Safety* (CCPS, 2007); *Guidelines for Auditing Process Safety Management Systems* (CCPS, 2011).

7.4.19 Performance Measurement

In keeping with its mission to improve the cost effectiveness of capital projects, CII started a Benchmarking and Metrics (BM&M) program to validate and share the benefit of best practices. The BM&M program measures five aspects of project performance, notably:

- Cost,
- Schedule,
- Safety,
- Change, and
- Field rework.

Most project managers, in accordance with CII guidance, monitor the project's schedule and costs (vs. budget) by regularly measuring construction progress and expenditure. As part of monitoring the project schedule, the project should also require regular updates from suppliers and vendors on delivery dates. Labor productivity and craft utilization are other popular metrics to measure construction efficiency. These data allow intervention and plan changes if slippage occurs, and are also reported to keep key stakeholders informed, e.g. the client and partners.

Many clients also require reporting of EHS and process safety key performance indicators (KPIs) for employees and contractor employees. As a minimum, KPIs for injuries (e.g. first aid, recordable, lost-time), and environmental spills and emissions may have to be reported to the local regulator.

Increasingly, companies have become aware of and implemented leading and lagging indicators of process safety performance, including incident and near-miss rates as well as metrics that show how well key process safety elements are being performed. The application of these metrics to construction projects is less mature. It is not practical to measure every aspect of process safety. Therefore a few metrics should be focused on barriers that are perceived to be weak. For example, process safety metrics are ideally suited to the management of late design changes that, if not rigorously evaluated, can result in increased risk of a major incident. Other important process safety barriers for construction are hazard identification, safe work practices, and QM.

Any metric is only as good as the quality of data collected, analyzed, and actions taken to improve performance.

Further guidance on measurement and metrics is available from the following CCPS publications: *Guidelines for Risk Based Process Safety* (CCPS, 2007); *Guidelines for Process Safety Metrics* (CCPS, 2009).

7.4.20 Operations Case for Safety

If preparation of a voluntary Operations 'Case for Safety' commenced during Detailed Design (see Chapter 6 Section 6.4.1.2), it should be finalized during the construction stage of the project.

This Operations Case for Safety should address how the residual risks (remaining after the final design) should be managed during startup and ongoing operations by:

- Facility's EHS and process safety management system (e.g. operating procedures, employee training, maintenance practices, management of change, etc.),
- Specific administrative / procedural measures that operations intend to implement (including resolution of some recommendations in the final HIRA study(s),
- Emergency response strategy and provisions.

The CCPS guidance on risk-based process safety provides information on good management practices that may be appropriate for inclusion as part of the facility's management system (CCPS, 2007).

The Operations Case for Safety should be communicated to all commissioning and operations personnel to ensure that they have a good understanding of the inherent hazards and necessary safeguards to ensure safe and reliable operations of the new facilities. Thereafter, the Operations Case for Safety should be kept up-to-date and used for training (refresher and new employees).

7.4.21 Pre-Commissioning

Terminology sometimes differs from client to client and project to project. Some define pre-commissioning activities as starting when the plant, or system, achieves mechanical completion. However, for the purposes of this book, pre-commissioning is a phase of construction that should be completed prior to certification of mechanical completion. In reality, construction, pre-commissioning, and mechanical completion often overlap within different parts of a major project.

Pre-commissioning activities are typically performed in small packages (i.e. systems), whereas construction usually proceeds on an area basis. The transition from area construction to system completion is a key project milestone that should be planned. A systems basis allows pre-commissioning activities to commence earlier in the schedule, thereby substantially reducing the peak workload of final pre-commissioning. This ensures a somewhat smoother transition from construction to commissioning, although management of systemized turnover is more complex to co-ordinate due to the parallel activities that have to be performed safely. Frequent SIMOPS studies should be conducted to help manage the risks, and safety specialists should be extra vigilant in their oversight.

System

Section of a facility that can be pre-commissioned independently, but in parallel with other sections of the facility under construction.

Pre-Commissioning ensures that the facilities have been constructed according to the technical specifications and design intent, and that commissioning can proceed safely and effectively. Pre-commissioning therefore involves all the checks that should be completed prior to commissioning (see Table 7.2).

Pre-Commissioning

Verification of functional operability of elements within a system, by subjecting them to simulated operational conditions, to achieve a state of readiness for commissioning.

Pre-commissioning activities typically commence upon substantial completion of construction activities (approximately 70% erection), and proceed in a phased manner - system by system. A detailed pre-commissioning plan should be developed that identifies all of the major activities by system. Color coding of P&IDs can assist understanding of each system in the plan. One of the most labor intensive activities is visual inspection of every item of equipment, including every bolt in every flange, to check the condition of equipment, the quality of the installation, and that it conforms with the design (see Chapter 8 for more detail on quality management). It is important that the installation complies with project drawings and specifications, manufacturer's instructions, safety rules, codes, standards and good engineering practices. The integration of operations personnel into the pre-commissioning team encourages early familiarization with the facilities that they will be operating.

No two projects are exactly the same, but some typical pre-commissioning activities are illustrated in Table 7.2. This list is not exhaustive, and other activities may be required depending on the scope of the project. Some companies and/or projects may perform some of these activities as part of mechanical completion (see Section 7.4.22).

Some of the pre-commissioning activities require utility supplies, such as water, compressed air/nitrogen, and electricity, and some equipment will be energized, e.g. electrical motors to check direction of rotation. This marks a significant change in hazards present within the construction site that require additional safeguards and management. Some of these safeguards include, but are not limited to:

- Communication of changed status to the whole workforce
- Removal of 'blanket work permits' in affected areas
- Introduction of work permits for energy isolation (LOTO)
- Barriers to exclude personnel from specific areas, e.g. hydrostatic pressure testing

Pre-commissioning activities may variously involve construction, commissioning, and/or operations personnel, and clear accountabilities and responsibilities should be established to avoid confusion and potential delays, especially if the sequence of system turnovers does not align with some personnel's expectations.

Table 7.2. Typical Pre-Commissioning Activities

Pre-Commissioning Activity
Pressure testing – hydrostatic & pneumatic
Design conformity checks – visual & review test certificates
Internal inspection
Flushing and chemical cleaning
Dewatering and drying (air, nitrogen, vacuum, methanol/glycol swabbing)
Air blowing
Instrument loop checks & calibration
Control system checks
Safety system functional tests/validation
Electrical continuity and motor rotation checks
Leak testing
Equipment static/de-energized tests
Machinery cold-alignment and guarding
Machinery lubrication
Machinery running-in
Pipeline gauging to identify buckling, dents & other damage
Punch-list – non-conformances & incomplete work
Preservation measures until commissioning
Documentation received from suppliers & vendors

One of the pre-commissioning activities is punch-listing. This is performed to identify, record and correct damaged, incomplete and incorrect fabrication and installation. Identified items are typically categorized into three lists, shown in Table 7.3:

Table 7.3. Typical Punch-List Categories

Category	Description	Correct By	Examples
A	Items of a safety nature or that prevent commissioning	Prior to commissioning	Missing & damaged items Incorrectly fitted, e.g. loose bolts & wires
B	Items that may be completed during commissioning	Prior to handover to operations	Missing signs & labels Long bolts in flanges
C	Items that are cosmetic & do not prevent startup	Schedule agreed with operations	Painting & non-critical insulation

Some examples of non-conformances and poor installation are illustrated in Figure 7.2 through Figure 7.5 below. A system should be established to track outstanding items to completion by system, category and craft.

Figure 7.2. Improperly Installed Electrical Cables

Figure 7.3. Damaged Instrument Cable

Figure 7.4. Improperly Installed   Figure 7.5. Improper Handling
Tubing   of Pressure Safety Valve

7.4.22 Mechanical Completion

Once again terminology sometimes differs between clients and projects, but for the purposes of this book, Mechanical Completion is defined as the point in the project when pre-commissioning is complete, i.e. the facilities have been built per engineering specifications, all equipment, materials, electrical and instrumentation installations have been completed and tested. Category A punch list items should also be complete. Therefore mechanical completion represents a significant project milestone that is the interface between construction and commissioning.

Mechanical Completion

Construction and installation of equipment, piping, cabling, instrumentation, telecommunication, electrical and mechanical components are physically complete, and all inspection, testing and documentation requirements are complete.

The project should have a formal system for mechanical completion with clear responsibilities for approval and documentation requirements. In some jurisdictions, the authority having jurisdiction may also be involved in the approval process.

Mechanical completion (a.k.a. Ready for Commissioning (RFC)) certificates are typically issued for each subsystem and system, when they are declared complete. Many projects compile a dossier of these certificates, and track outstanding systems and subsystems to completion.

7.4.23 Documentation

An important activity during construction is the compilation of all of the process safety information (PSI) and other project documentation required for commissioning and subsequent handover to Operations. This documentation includes, but is not limited to:

- Contracts, purchase orders, correspondence,
- Engineering design drawings and technical specifications (including design intent, codes & standards, etc.),
- Risk register,
- Fabrication QA/QC records (including weld radiographs, certificates, etc.),
- Pre-commissioning QA/QC records (including weld radiographs, checklists, baseline data, certificates, etc.),
- Mechanical completion dossier,
- Operating and maintenance manuals from suppliers and vendors,
- Change management records and DCNs,
- Commissioning and operating procedures,
- Master equipment list and ITPM requirements (including SCE),
- EHS and process safety procedures (including emergency response plan, incident reports, audit reports, training records, etc.),
- Equipment preservation,
- Punch-lists,
- Action tracking from HIRA, operational readiness reviews, stage gate reviews, and construction studies and reviews (e.g. vibration analysis, piping stress analysis, corrosion, etc.),
- Commitments to third parties (e.g. regulator, NGO, community, etc.).

The development of as-built drawings and technical information should commence as soon as possible, and ideally be complete prior to handover to Operations. A copy of red-line drawings should be provided to Operations if the final CAD drawn as-builts are not available.

During FEL and Detailed Design, the project should have set up a document management system for storage, management, updating and retrieval of this information. This system should also track outstanding deliverables, especially any documents required prior to commissioning and startup. Project documentation is discussed in greater detail in Chapter 12.

Further guidance on knowledge management is available from the following CCPS publications: *Guidelines for Risk Based Process Safety* (CCPS, 2007); *Guidelines for Process Safety Documentation* (CCPS, 1995).

7.5   OTHER PROJECT ACTIVITIES

In addition to the various process safety and technical activities needed for construction, there are a number of other activities that support project execution. Some of these activities continue throughout the project life cycle and should be periodically updated. This requires good interface management between the PMT and all the contractors, vendors and suppliers.

7.5.1   EHS and Process Safety Plans

The EHS Plan and the Process Safety Plan should be updated to reflect construction, pre-commissioning, commissioning, handover and startup activities (Appendix B). Contractor and sub-contractor EHS and process safety plans should be reviewed for consistency with the overall project EHS and process safety plans.

7.5.2   Risk Register

The Project Risk Register should be updated for any new or changed hazards/risks identified for construction, pre-commissioning, commissioning, handover and startup (Appendix C). Individuals should be identified as responsible for developing a response plan to manage each item. The PMT should regularly review the register and response plans.

7.5.3   Action Tracking

The project action tracking database or spreadsheet should be updated with particular focus on actions relating to vendor packages that may not have previously received the same attention as the design of the main facilities. The PMT should also capture actions generated by their contractor(s), and ensure that all actions are progressively resolved, closed and documented.

7.5.4   General Construction Management

In addition to measurement of progress and expenditure (see Section 7.4.19), a number of other general management activities should continue throughout construction and pre-commissioning, including, but not limited to, the following:

- Administration of contractor personnel
- Control of contracts
- Regular progress meetings (typically weekly) with contractors
- Regular liaison (telephone, meeting) with suppliers/vendors

The Construction Manager should also keep a daily diary/logbook as a record of construction progress detailing significant areas of activity. This logbook should include details of issue dates of design documents to contractors, workforce numbers and equipment on site, accomplishments, test results, pictures, labor disputes, and any weather or other delays. This information is particularly important in settling or challenging contractor claims at the completion of the contract.

7.6   DE-MOBILIZATION

Towards the end of construction activities, the project and contractor(s) may start to progressively de-mobilize resources (personnel and equipment). Some resources may need to be retained until the facilities are in full operation and have met any production performance targets, and therefore a De-Mobilization Plan should be developed to cover the orderly and effective shut down and removal of all construction resources from the project site. While this plan focuses on the project's resources, it may also address key contractor resources.

The de-mobilization plan should include, but not be limited to, all activities and costs for removal of the following:

- Redeployment of personnel and closure of agency staff contracts,
- Construction equipment,
- Surplus construction materials,
- Temporary facilities (e.g. cleaning and disassembly of offices, buildings and other facilities assembled on the site specifically for the project),
- Disconnection of utilities (telecoms, gas, electricity, water, etc.),
- Leased / rental equipment (copier, fax, desks, chairs, etc.),
- Supplies not required or included in contracts,
- Archiving of project files, documents and records (after handover of documentation required by Operations),
- Site clean up.

De-mobilization is a time when project and contractor employees can potentially lose focus on safety, as their minds may be more concerned with future employment and where the next pay check is coming from. It is essential that the PMT and contractor management ensure that the hazards of de-mobilization are identified and understood by all, and re-enforce EHS and process safety requirements daily.

7.7   PREPARATION FOR COMMISSIONING AND STARTUP

A Pre-Operations Plan should have been developed during Detailed Design to ensure readiness for commissioning and startup. This plan should be updated and the relevant activities progressed during construction to ensure timely completion. The plan should include, but not be limited to, the following activities:

- Preservation of all equipment (installed & awaiting installation),
- HIRA (including any SIMOPS),
- Recruitment – operators, technicians, engineers, EHS, admin (if necessary)
- Commissioning and operating procedures,
- EHS and process safety procedures (e.g. safe work practices, MOC, emergency response, incident investigation, etc.),
- EHS equipment (e.g. ambulance, fire truck, fire extinguishers, etc.),
- Training (including vendor training, use of process simulator),
- Maintenance management system build (baseline data, ITPM tasks, etc.),
- Spare parts, consumables, etc.,
- Chemicals, lubricants, catalysts, etc.,
- Document and data management (e.g. OEM/vendor manuals, as-built drawings, etc.),
- Technical and vendor support,
- Operational readiness review.

Many of the commissioning and startup activities in the pre-operations plan involve different hazards and risks than construction, and are analogous to activities required for normal operations. As such, most commissioning and startup activities fall within the elements of risk-based process safety (CCPS, 2007).

A key activity towards the end of pre-commissioning is an Operational Readiness Review (a.k.a. Pre-Startup Safety Review (PSSR)) to evaluate whether the facilities can be safely started. This review should be very comprehensive as it is the first time that the facilities will startup. In this regard it should be much more thorough than reviews conducted, for example, after a utility failure, trip or precautionary shutdown for inclement weather. It should include a walk-through inspection of all facilities, and, as a minimum, address the adequacy of:

- Construction of all equipment, controls and structures, including SCE and other protective devices, conforms with design, (e.g. IEC stage 3 functional safety assessment (FSA)),
- Resolution of punch-list items (category A & B),
- HIRA studies to meet regulatory and company requirements,

- Action resolution (e.g. HIRA studies and other recommendations),
- Safety, operating, maintenance, and emergency procedures,
- Training of all employees,
- Updated red-line drawings.

Further guidance on operational readiness is available from the following CCPS publications: *Guidelines for Risk Based Process Safety* (CCPS, 2007); *Guidelines for Performing Effective Pre-Startup Safety Reviews* (CCPS, 2007).

7.8 FINAL EVALUATION AND CLOSE-OUT

Final evaluation and close-out involves the process of completing all tasks and all documentation to close out construction contracts. This may be the initial phase of the overall project close-out that, in some instances, may not be fully complete until a year or so after handover and startup, when actual equipment performance can be compared against any contract warranties.

Key objectives are to reimburse all construction contracts for services and materials supplied, capture lessons learned during construction for future projects, and evaluate the performance of the construction contractor(s). This latter item should include the contractor's EHS and process safety performance, which should be documented as a reference for contractor pre-qualification and selection for future projects.

7.9 STAGE GATE REVIEW

A stage gate review should be conducted to ensure that construction process safety (and EHS) risks are being adequately managed by the project. This stage gate review may be conducted in two parts: part one soon after mobilization to evaluate the construction plans, and part two around 50% construction completion to verify implementation of the construction plans including management of field changes. The stage gate review team may use a protocol and/or checklist, such as the detailed protocol in Appendix G. A typical process safety scope for a construction stage gate review is illustrated in Table 7.4.

Table 7.4. Construction Stage Gate Review Scope

Scope Item
Confirm that construction workforce training, competency, and performance assurance arrangements are adequate and being implemented
Confirm that a construction Process Safety and EHS management system is adequate and being implemented
Confirm that owner, contractors and vendors have clarity in regard to their scope and responsibilities for the mechanical completion, and that the construction team have a robust process to manage all interfaces
Confirm that asset integrity management processes including quality management are sufficient to deliver the design intent and facility integrity
Confirm that change management is being applied
Confirm that project plans for pre-commissioning, commissioning, and pre-startup are adequate
Confirm that progress on Operations training and development (or update) of operating procedures is adequate
Confirm that the Operations Team is involved as necessary in preparation for pre-commissioning and commissioning activities.
Confirm that plans for a site Process Safety and EHS management system and procedures are adequate
Confirm that a document management system has been implemented and is performing as expected

The stage gate review team should be independent of the project, familiar with similar facility/process/technology, and typically comprise an experienced leader, operations representative, process safety engineer, construction safety specialist, QA/QC specialist, discipline engineers (as appropriate), and EHS specialist (as appropriate). At the conclusion of the review, the review team will make recommendations for any improvements needed, and indicate to the Gate Keeper, based on process safety, whether the project is ready to proceed to the next stage, Pre-Startup, i.e. Commissioning.

7.10 SUMMARY

Once Detailed Design is complete, the project moves into Construction with the objective of safely building the facility in accordance with the design. This involves multiple interfaces with fabricators, contractors, sub-contractors, suppliers, and vendors that require good management and regular performance reporting. A number of process safety activities are essential to the success of this stage, including risk management, safe work practices, asset integrity management, management of change, and training/competence assurance. Achievement of these and other activities is necessary for a thorough mechanical completion, so that the project may safely proceed to commissioning and startup.

8 QUALITY MANAGEMENT

Quality is a somewhat subjective attribute that one person may perceive differently to another person. It is an inherent feature or property that implies a degree of excellence.

Quality

The degree to which a set of inherent characteristics fulfills requirements.

(PMBOK Glossary (PMI 2013)

In an engineering context, quality may be equated to fitness for purpose, which infers a state of being free from defects, deficiencies and significant variations. Quality is achieved by rigorous commitment to particular standards in order to satisfy specific requirements, i.e. technical specifications. For example, the focus of the ISO 9000 series of international quality management standards (ISO, 2015) is on a quality audit program that verifies that specifications are met. If these specifications are consistently met for (say) a widget, the manufacturer of that widget receives ISO 9000 certification.

However, this is not necessarily a good indicator of excellent quality or even safety. For instance, if the specification is for a cheap, low quality widget but it is consistently manufactured to that specification, it meets the ISO 9000 requirements. In the case of safety, the ISO 9000 audit program does not review failures in design (i.e. incorrect technical specifications) or equipment failures that occur in the field. The ISO 9000 series of standards is described below in Quality Management. Whereas quality is invariably seen as a standard, many companies today view process safety as a core value. In reality there is an interdependency between quality management and process safety management. Customers often include safety as an important component of quality, and quality is an important part of process safety management. In the context of a project for the process industries, the PMT must implement **both** quality and process safety at a high level in order to deliver a safe, reliable, and operable facility to the client.

Quality Management

In the 1980's there was a surge of interest in quality management (QM) with a number of initiatives known as total quality management (TQM), lean principles, six sigma, etc. While much of the focus was on manufacturing, some of the tools are commonly used in projects today, especially during fabrication and construction, such as Plan-Do-Study-Act (Shewhart cycle), 14 Points For Management (Deming, 1982), and just-in-time (JIT). CII has sponsored research into, and published guidance on, the application of quality management in capital projects (CII, 2010a, 2010b, 2010c). For example, lean construction involves waste elimination, meeting or exceeding all client requirements, focusing on the entire value stream, and pursuing perfection in the execution of the constructed project (CII, 2004). Similar principles can also be applied to design and procurement.

QM in the context of a capital project involves the practices and activities that:

- Formulate quality policy,
- Set quality objectives and responsibilities, and
- Execute quality planning, quality control, and quality assurance,

such that the project deliverables meet their design intent and specifications.

While QM ensures that project deliverables meet design requirements, Process Safety, especially risk management, provides tools to 'stress test' the design to ensure that it is fault tolerant, and can safely handle abnormal conditions.

QM also includes activities conducted to improve the efficiency, contract compliance, and cost effectiveness (by reducing waste and rework) throughout the life cycle of the project, including the design, engineering, procurement, construction, commissioning and startup stages.

Quality Management

All the activities that an organization uses to direct, control and coordinate quality.

(CCPS Glossary)

QM systems are necessary because of the numerous human errors that are possible throughout the life cycle of a project. A well-designed and implemented QM system should identify and correct these human errors that can involve many facets of project design, procurement and construction. A few examples of relatively common human errors in projects are listed in Table 8.1. Note that this list is not exhaustive.

Table 8.1. Typical Human Errors That Occur in Projects

Human Error
Calculation error during equipment design
Design or construction based on wrong or out-of-date information
Poor weld due to lack of skill or failure to follow welding procedure
Wrong material of construction supplied by vendor
Wrong component requisitioned from warehouse
Visually identical items installed in wrong locations
Equipment damaged by poor storage and handling
Deficiency not identified due to wrong NDT procedure

If deficiencies and non-conformances are discovered, the causes should be identified and corrective actions developed necessary for achieving the appropriate quality level. This may entail changes to the project quality plan and increasing the required quality levels.

Further information and guidance on human error is available from the following CCPS publications: *Guidelines for Preventing Human Error in Process Safety* (CCPS, 2004); *Human Factors Methods for Improving Performance in the Process Industries* (CCPS, 2007).

Typical activities in a capital project that involve QM are listed in Table 8.2.

Table 8.2. Typical Project Activities Involving Quality Management

Activities
Development of Quality Management strategy for quality assurance/quality control (QA/QC) and documentation
Development of Quality Plan and resources
Design reviews and verification
Procurement and inspection of equipment/materials received
Fabrication oversight and inspection
Vendor inspection and auditing

Activities
Workforce qualification and training (e.g. welders, crane operators, etc.)
Installation/construction/pre-commissioning oversight and inspection
Inspection and testing registers and dossiers
Audit planning and implementation
Project control and reporting, including interface management & communication

Some engineering design and construction contractors have certifications for quality management to ensure their services consistently meet customer's requirements. These certifications are usually related to ISO 9001 within the ISO 9000 family of standards:

- ISO 9000:2015 - covers basic quality concepts and terminology,
- ISO 9001:2015 - sets out requirements of a QM system,
- ISO 9004:2009 - focuses on how to make a QM system more efficient and effective,
- ISO 19011:2011 - sets out guidance on internal and external audits of QM systems,
- ISO/TS 29001:2010 - sets out requirements of a QM system for the petroleum, petrochemical and natural gas industries.

The QM principles in these standards are described in more detail in an ISO publication (ISO, 2015).

Quality assurance (QA) and quality control (QC) are two of the main activities that are required to ensure a quality project. They work together to help ensure that appropriate tools, materials and workmanship combine to provide a project that performs to meet its design intentions. QA and QC are closely related, and are sometimes used interchangeably, but they are different. The terms QC and QA can carry different connotations in different organizations. However, for the purposes of this book, QA and QC are defined as follows:

Quality Assurance (QA)

QA is a set of activities that ensures that development processes (i.e. design, engineering, procurement, construction, etc.) are adequate in order for the project to meet its objectives. In other words, QA can be thought of as a means of preventing quality problems, and detecting quality issues related to work practices.

Quality Assurance

Activities performed to ensure that equipment is designed appropriately and to ensure that the design intent is not compromised, providing confidence throughout that a product or service will continually fulfill a defined need the equipment's entire life cycle.

(CCPS Glossary)

QA activities include audits and reviews to determine if project deliverables meet the scope of work, basis of design, and technical specifications. QA is normally undertaken by the client organization or by a third party inspector on behalf of the client. For example, the PMT may commission an audit to determine by random sampling if the agreed QC program is being properly implemented by the contractor.

Quality Control (QC)

In contrast, QC is a set of activities designed to evaluate the developed project deliverables, so that QC may be thought of as detecting errors in the design, and procured, fabricated and installed equipment and materials.

Quality Control

Execution of a procedure or set of procedures intended to ensure that a design or manufactured product or performed service/activity adheres to a defined set of quality criteria or meets the requirements of the client or customer.

(CCPS Glossary)

QC activities include a variety of checks, measurements, and inspections to reveal any defects or failures in the equipment and materials that make up the project facilities. QC is normally undertaken by the contractor's organization, a sub-contractor inspector or persons performing the work. Some examples of QC include:

- Senior engineer checking design calculations performed by a junior engineer,
- Radiography of a percentage of welds in a piping system to detect flaws,
- Positive material identification for specific metallurgy, e.g. special alloys.

This chapter discusses QM activities for the following life cycle stages of a project:

- Design/engineering (Section 8.1)
- Procurement (Section 0)
- Fabrication (Section 8.3)
- Receipt (Section 8.4)
- Storage and retrieval (Section 8.5)
- Construction and installation (Section 8.6)
- Operation (Section 8.7)

8.1   DESIGN/ENGINEERING

Quality management in design starts with the selection of engineering codes and standards in FEL. This represents the main opportunity to establish and 'build in' quality and safety to the equipment and materials employed in the project. Some companies have their own engineering standards, often based on industry codes and standards, but supplemented by lessons learned from their operations. Once the applicable engineering codes and standards have been determined, technical specifications including performance standards can be finalized in design to establish important process safety attributes, such as reliability, availability, survivability, and other pertinent factors. Thereafter QM activities during design and engineering are generally focused on preserving the design integrity.

The project should appoint, or have access to, experienced personnel knowledgeable in quality practices. For example, many companies bring in subject matter experts (SMEs) external to the PMT to perform technical quality checks. A QM Plan for the project should be developed early in FEL, and certainly before any contracts are awarded. The QM Plan may be part of an overall strategy for procurement and supply chain management. Quality should be planned into the project in order to prevent unnecessary rework, waste, cost, and delays. Any design/engineering errors or quality non-conformances could also result in a process safety incident if not identified and corrected. The plan should be an integral part of the project management system, and define:

- How quality will be managed throughout the life cycle of the project,
- Required quality assurance activities (i.e. practices and procedures),
- Required quality control activities (i.e. practices and procedures),
- Acceptable levels of quality in project deliverables and work processes.

It should be noted that the plan addresses both the quality of the deliverable (i.e. project design) and the quality of the process (i.e. practices and procedures) for achieving the design. The plan should address identified QM risks, including supply chain integrity issues. The plan should also evaluate and check the design at each design stage, and also establish QM activities to be used during procurement, fabrication and construction.

Sometimes the FEL work, and often the detailed design engineering, is outsourced to an engineering contractor. Even if the work is performed in-house, competent and experienced engineers are necessary to correctly interpret the standards, and develop and apply specifications. Potential providers of engineering and HIRA services should be evaluated against quality, competency, experience, capability, and other applicable criteria.

A range of QA and QC activities may be appropriate throughout the design process. The PMT should maintain close oversight of all design activities, and conduct audits at different stages to verify that the design complies with the scope of work, basis of design, design philosophies, agreed codes and standards, local regulations, and any commitments to third parties. The PMT may also commission a number of safety and design reviews, such as P&IDs, 3D model, and technical peer reviews, to assist with ensuring the quality and integrity of the design. Meanwhile the design organization should be self-checking the design for accuracy and integrity. Many engineering contractors have their own checklists and methods, and CCPS has published an extensive hazard evaluation checklist that companies can use for their in-house design work (CCPS, 2008).

A range of typical activities during FEL and detailed design are listed in Table 8.3.

Further information and guidance is available from the following publication: *Guidelines for Asset Integrity Management,* (CCPS, 2017).

Table 8.3. Typical Quality Activities During FEL and Detailed Design

Project Stage	General	Quality Assurance (QA)	Quality Control (QC)
FEL - 1	Establish Project QM Plan and QM System	Include QA strategy	Include QC strategy
FEL - 2	Develop procurement quality program	Include QA program	Include QC program
	Pre-qualify key service providers	Consider competency, experience, etc.	Consider QC plans and capability
	Select development option	Conduct peer review	
FEL - 3	Refine design and specifications	Audit to verify compliance with design and specs.	Review design, check calculations, accuracy, etc.
		Specify QA programs based on criticality	Specify QC programs based on criticality
	Order long lead items	Apply QA requirements	Apply QC requirements
	Select engineering contractor	Consider competency, experience, etc.	Consider contractor's QC program and capability
Detailed Design	Finalize design and specifications	Audit to verify compliance with design and specs.	Review design, check calculations, accuracy, etc.
	Finalize QM Plan for construction	Include QA strategy	Include QC strategy
	Select equipment and materials suppliers	Consider vendor specifications	Consider suppliers' QC program
	Order equipment, materials and services	Apply QA requirements	Apply QC requirements

8.2   PROCUREMENT

QM during procurement focuses on ensuring that purchases adhere to the approved final design (i.e. technical specifications), and that qualified suppliers and vendors are used. Vendors (or a 3rd party) may issue a declaration of conformity that their product complies with a specific standard, but for design safety measures, and other equipment identified as high-criticality (for production, quality, environment, etc.), the project should evaluate whether the methods and information used by the vendor are consistent with **all** requirements for certification. Potential suppliers should be assessed for capability and competency based on previous supplier performance. Pre-qualifying suppliers and vendors, and limiting purchases to them can help to eliminate improper or sub-standard equipment and materials.

It is also important that in-house or contracted procurement services understand the project's change management policy and procedures, especially with regard to the acceptability of substitutions and approval thereof. There may be a requirement to source equipment and materials locally and/or through low cost suppliers, and it is essential that the technical specification of all purchases is rigorously controlled. Less-expensive alternatives may not meet technical specifications. Sometimes a vendor may request a relaxation or minor change to a specification, which should be managed through an amendment/change order that has been formerly assessed to ensure the change is acceptable from both a technical and safety standpoint. Even something as simple as a substitution for an electrical enclosure could be a potential ignition source in a hazardous area if it does not meet the appropriate technical specification for hazardous area classification. The procurement plan including application of the project QM plan should have been finalized in FEL-3. Procurement activities for a capital project typically span several project stages from FEL-3 to construction, as follows:

- FEL-3   long lead items of equipment,
- Detailed Design   most items of equipment and some materials,
- Construction   remaining items of equipment and materials.

Smaller projects and MOC work may condense this timescale, but most projects, irrespective of scale, try to manage the purchase and delivery of equipment and materials to optimize cash flow. This just-in-time (JIT) approach is relatively easy for some services, such as site grading/excavation, and standard equipment, such as small carbon steel gate valves and piping. Other services (e.g. offshore lift vessels for very large loads) and equipment (e.g. complex machinery, exotic alloys) may require extensive research and planning in advance of procurement to ensure deliverables meet the project schedule.

Regardless of timing, the QM requirements should be determined in advance and written into contracts and purchase orders. In particular, contracts/purchase orders for all design safety measures (including SCE) should include the relevant performance standards and required testing that are incorporated into the technical specification of the equipment item. The manufacturer is required to implement its internal procedures effectively so that their product(s) meet all quality requirements. The QM requirements, depending on criticality, may include necessary supplier/vendor inspections (QC), project audits (QA), documentation, and other deliverables, such as quality certificates and radiographs of welds. The project, or a third-party quality inspector acting on behalf of the project, may wish to witness certain QM plan stages during manufacturing/fabrication that are specified in contracts.

If procurement has been out-sourced (e.g. as part of an engineering, procurement and construction (EPC) contract), the project may also wish to audit the procurement activities to provide confidence that engineering standards, technical specifications, quality requirements, and change management are being properly managed by the procurement service provider.

Another aspect of procurement is expediting equipment and material delivery. Experience has demonstrated that without monitoring and forceful expediting, delays can occur, often with a knock-on effect to the project schedule.

8.3 FABRICATION

Quality management of fabricated equipment focuses on verification that engineering standards and technical specifications are being followed, and that manufacturing practices in the fabrication shop do not compromise quality and integrity.

Case Study: Fabrication Specifications Not Followed

A low pressure storage tank containing up to 50,000 gal. of lubricating oil failed catastrophically when it was accidentally overpressured. Compressed air was injected below the liquid level and caused internal mixing within the tank.

The wall-to-floor seam of the tank failed with a massive seam tear instantly draining oil, pulling a vacuum on the tank, and partially collapsing the tank wall. Fortunately there were no injuries and no fire.

The investigation found that the weak weld seam on the wall-to-roof connection that allows the roof to separate safely while liquid is contained within the tank had been compromised. The roof construction method had not followed the fabrication specifications. Some internal brackets (not on fabrication drawings) strengthened the wall-to-roof seam such that it was stronger than the wall-to-floor seam that failed.

Reference: Sanders, R.E., *Chemical Process Safety, Learning from Case Histories*, 4th edition.

The first line of defense to ensure a quality product is the use of a code-approved fabrication shop, which some jurisdictions require for the fabrication of some equipment (e.g. pressure vessels). These shops are subject to periodic inspection and certification by personnel authorized by the jurisdiction, as are some of their craftsmen, such as welders. Nevertheless, the project may wish to pre-qualify a fabrication shop based upon their own inspection and approval process.

The quality of fabricated equipment is the responsibility of the fabricator, and a good shop will have a comprehensive system of procedures, inspections and checks. Their procedures should cover all aspects of fabrication, including measuring/cutting, rolling/forming/shaping, assembly of parts, welding, casting, and post-weld heat treatment of metallic equipment. The shop's QC activities typically include tests and inspections, such as positive material identification, weld radiography, borescopic inspection, hydraulic pressure testing, and other non-destructive testing (NDT). All QC activities should be fully documented.

While QC is the fabricator's responsibility, the project may also wish to conduct QA inspections. Hold points may be identified in the fabrication process, especially for pressure vessels and high-criticality equipment. At these hold points a project quality inspector (or third-party inspector acting on behalf of the project) should inspect the work to date, such as root weld passes on a pressure vessel. In addition to hold point inspections, the QA inspector may also review the calibration of QC testing and inspection equipment, welding procedures, documentation, and the qualifications of personnel responsible for welding, welding inspection, and NDT. The project may also compile baseline information, such as thickness measurements of a finished pressure vessel, for handover to the Operator to assist subsequent in-service inspections.

A study of 364 chemical process incidents identified that 25% of incidents were caused by piping system failures, of which a technical contributor of 3% was due to poor fabrication, i.e. poor heat treatment of welding (Kidam & Hurme, 2013).

Manufacturers of other process equipment, such as pressure safety valves, rotating machinery, piping, structural steelwork, electrical and control equipment, should also have comprehensive fabrication procedures and QC practices. The project should have a program of QA inspections and/or audits at manufacturers' premises before and during manufacture to oversee the QC activities. This QA program is likely to be risk-based with greater oversight of critical equipment. Certain critical and/or complex equipment (e.g. compressor, SIS) may be subject to a factory acceptance test (FAT) at the manufacturer's premises before delivery to verify its operability and functionality. A representative of the project normally witnesses the FAT. The project may also conduct pre-shipment inspections of some critical or delicate equipment requiring special handling and oversee loading for transportation to the construction site.

8.4   RECEIPT

The project should have a formal system for material control to manage the acquisition of, in the case of a large capital project, thousands of different components and equipment required to build the project. Most projects have some form of secure warehouse or covered storage for equipment and materials received at the construction site. Larger items, such as pressure vessels and piping, are normally stored in a laydown yard.

The project's QA activities should be focused on verifying that equipment and materials are received in good condition, and meet the design technical specifications in the contract or purchase order. These inspections vary with the type of equipment and materials involved based upon criticality, and range from simple visual examination and cross-checking the packing list against the PO to material testing and positive material identification for special alloys. Clear procedures and training are necessary for personnel receiving equipment and materials, so that they know when additional QA tests and inspections are required. Any items with damage or nonconformances should be held in a quarantine area until the basis for their rejection has been resolved.

Case Study: Materials of Construction of Valve Components

A major integrated oil and gas company contracted with an engineering and construction firm to engineer, procure, and construct a large-scale gas processing facility. Design specifications required metallurgy in the inlet of the plant to be NACE compliant to 150 ppm H_2S. The engineering contractor provided vessels, piping, and equipment that met the NACE requirements. Valves, however, did not entirely meet the NACE requirements.

The valve bodies were manufactured to meet the NACE requirements for 150 ppm H_2S. However, certain valve components such as springs, valve stems, and sealing materials did not meet NACE requirements. When the owner asked the engineering contractor if the valves met NACE requirements, the contractor replied "these are the specifications for the valves that are installed."

The plant owner and its partners were faced with the dilemma as to a strategy for replacing non-compliant valves. The issue of concern was valve stem ejection if the non-NACE valve stems break. While it was thought that no major releases would occur in the ball and butterfly valves that contain non-NACE components, the potential existed for smaller leaks should a valve stem fail and eject from the valve body.

Further discussions between the company and the engineering contractor were ongoing at the time this book was written.

All documentation and quality certification accompanying the equipment and materials should be reviewed, recorded and filed in line with the project's document management system and the level of traceability required.

8.5   STORAGE AND RETRIEVAL

After thorough checking upon receipt, equipment and materials should be put in their designated area and, if necessary, stored under optimum conditions. Optimum storage conditions are equipment-specific and may require heating, air conditioning, and/or humidity control to preserve quality. These storage requirements may depend on local environmental conditions; e.g. in a tropical climate, many items should not be stored outside.

Other requirements may include electrical parts that require static control, pressure safety valves stored upright, and rotating machinery like pumps and electric motors turned by hand periodically. Equipment stored outside a warehouse may require extra preservation measures, such as wrapping to seal all penetrations and/or maintaining an internal positive pressure of nitrogen. Vendor's storage instructions should be followed. Some project examples of poor preservation are illustrated in Figure 8.1 and Figure 8.2 below.

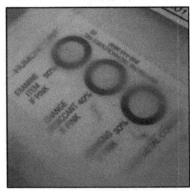

Figure 8.1. Corroded Solenoid

Figure 8.2. Wrapped Equipment with Expired Desiccant

Case Study:

A serious fire occurred during normal operation of an oil refinery Hydrotreater when piping to a heat exchanger failed. A pipe spool failed at a flange, releasing hot high pressure hydrogen that ignited, producing a large jet fire that burned for several hours. Damage was extensive but limited to the flame path, resulting in deformed piping and structural steelwork. Fortunately there were no injuries.

An investigation found that there were three pipe spool 'elbows' of identical dimensions and appearance. However, two spools were constructed of alloy steel and one of carbon steel. A contractor inadvertantly swapped the position of one of the alloy steel spools with the carbon steel spool, exposing it to high-temperature hydrogen for which it was not resistant to degradation.

QC checks for positive material identification of piping components can avoid simple errors with potential for serious consequences.

It is important that all items are properly labeled and segregated to facilitate later retrieval and avoid confusion between visually similar but different specification equipment and parts. Similar errors can occur in the Operations stage during maintenance turnarounds. Human error can be minimized by a well-designed storage and retrieval system using bar coding or similar means. It is also essential that items are handled correctly. For example, control equipment such as control valves require careful handling. Poor storage practices also increase the risk of theft of high value items.

Failure to preserve and handle equipment properly can lead to premature failure that could cause, or fail to mitigate, a process safety incident. The inadvertent retrieval and installation of equipment or components of the wrong specification can also cause, or fail to mitigate, a process safety incident. Many loss of containment incidents have been caused by installing piping and valves constructed from carbon steel instead of exotic alloys or low temperature stainless steels.

The project or their third-party quality inspector should conduct QA audits to ensure that equipment is being correctly stored, handled and retrieved prior to installation. These audits should particularly review the procedures, QC checks, and their implementation being employed by personnel responsible for managing storage and retrieval.

8.6   CONSTRUCTION AND INSTALLATION

If any equipment or component is damaged or does not meet the correct technical specification, the last opportunity to detect this deficiency or nonconformance is during construction and installation. Damage or inadvertent installation of the wrong equipment or component (e.g. wrong gasket type in a piping system) can also occur during construction and installation. Therefore quality management during construction should focus on detecting damaged/deficient equipment and nonconformances with the final design and 'approved for construction' drawings.

The diversity of work being performed during construction and installation requires widely differing criteria for inspection of equipment and components. QM activities are typically based upon criticality assessments. For example, high pressure/high temperature process vessels generally receive more attention than minor civil work.

Most projects also involve a significant amount of material handling, much of which requires deliberate care to maintain quality and ensure the correct items are installed in accordance with the final design. Procedures are required for the management of equipment and materials, which, following inspection on receipt at the construction site, subsequently control their issue and use for construction.

The contractor (or company personnel) responsible for construction and installation should conduct, or have a sub-contractor conduct, a range of QC activities appropriate to the type and criticality of equipment and materials. Typical QC checks during construction include, but are not limited to, the following tasks:

- NDT of a percentage of field welds,
- Torquing bolts on flanges,
- Checking gasket materials,
- Hydro pressure tests (and pneumatic tests in special cases),
- Instrument loop checks and SIS functional tests,
- Alignment of rotating machinery,
- Selective positive material identification.

Some of these QC tasks may impact adjacent construction or, in the case of a brownfield development, operations. For example, exclusion zones are required around radiography and pressure testing for personnel safety. A SIMOPS review should be performed to identify and manage the risks.

While QC is generally the construction contractor's responsibility, the PMT normally requires QA inspections. Hold points may be identified in the construction and installation process, especially for critical equipment. At these hold points a project quality inspector (or third-party inspector acting on behalf of the project) should inspect the work to date. In addition to hold point inspections, the QA inspector may also review the following:

- Welding procedures and consumables,
- Calibration of QC testing and inspection equipment,
- Implementation of other QC activities (by random sampling),
- QC documentation and records, and

- Qualifications of personnel responsible for welding, welding inspection, NDT, heavy-lift cranes, and high-voltage electrical equipment.

An important QM activity is monitoring the installation and testing of SCE in order to verify that it complies with the relevant technical specifications and performance standards. By definition, SCE requires a high reliability/availability in order to work on demand to prevent and mitigate major accident hazards. It is therefore essential that there is particular focus on QA/QC tasks for SCE.

Towards the end of the construction stage, it is common practice to conduct a detailed inspection of the final installation involving representatives of the PMT, construction contractor and future Operator to identify errors, nonconformances and incomplete work. These items are normally added to a 'punch list', and may include a number of quality issues that require resolution prior to commissioning. This inspection and punch list may form part of an operations readiness review (a.k.a. pre-startup safety review (PSSR)), which are discussed in more detail in Chapter 9.

Finally, all QA/QC activities during construction and installation should be documented and full records retained for handover to the Operator.

8.7 OPERATION

Quality issues during the operating stage of the project involve routine and breakdown maintenance, and repairs as a result of equipment deficiencies. For example, QM for repairs that require welding and post-weld heat treatment are addressed in codes and standards. Although on a somewhat smaller scale than most capital projects, many of the QM activities during operation are similar to those discussed above for procurement, fabrication, receipt, storage and retrieval, and construction and installation. Some temporary repairs like pipe clamps need special QM and care during design and installation, and regular QA inspections throughout operation until such time as a permanent repair can be made at the next turnaround.

Some equipment deficiencies may be addressed by re-rating the equipment for operation under less severe operating conditions. For example, the maximum allowable working pressure (MAWP) of a pressure vessel may be reduced to take account of significant corrosion, although the adequacy of the relief system should also be verified. In this case, special QM requirements are necessary to prevent the potential for catastrophic failure, and are covered in the applicable codes and standards, e.g. API Pressure Vessel Inspection Code 510 (API, 2014). Similar QM requirements apply in the case of a debottlenecking project where a pressure vessel is uprated. In both instances, the change of MAWP should also be covered by the plant's MOC procedure and fully documented.

Further information and guidance on quality management is available from the CCPS publication: *Guidelines for Asset Integrity Management*, 2017.

8.8   DOCUMENTATION

Documentation is an important aspect of QM, and typically starts with a Project Quality Plan (PQP) that defines the quality policy, philosophy, QM system, and responsibilities of the various parties (including sub-contractors, suppliers and vendors) involved in each stage of the project. An Inspection and Test Plan (ITP) may form part of the PQP or be a stand alone document(s). The ITP covers the detailed approach to QC of the equipment, materials, components, systems, structures, and software to ensure that they conform to the relevant technical specifications and, if appropriate, performance standards. This will describe activities, such as visual inspection, dimension checks, NDT, function tests, FAT, positive material identification and hydrotest, and where these activities will be performed (i.e. vendor's premises or construction site). Many of these QC tasks are defined in codes and standards, including the required qualifications of persons performing the tasks.

Each of the QC activities should be documented, and supported by signed forms/certificates, radiograph negatives, data from testing equipment, photographs/videos, etc. All results should be reported, including non-conformances, deficiencies and damage, so that appropriate measures may be taken to correct faults.

QA activities conducted by the project (or a third party specialist on their behalf) should also be documented. Generally this is likely to take the form of audit reports assessing the implementation of the QC system, but on occasion certain QC tests may also be performed at random or for cause (e.g. suspected damage or deficiency). QA reports should also be supported by relevant records, such as QC records, copies of qualications for welders, electricians, and QC inspectors, etc.

All QA/QC documentation and records should be retained for handover to the future Operator. All aspects of project documentation, including document retention and control systems, are discussed in greater detail in Chapter 12.

Further information and guidance on documentation is available from the CCPS publication: *Guidelines for Process Safety Documentation*, 1995.

8.9   SUMMARY

Although generally safety is viewed as a 'value' and 'priority', and quality as a 'standard', there is an interdependency between quality management and process safety. For safe, reliable and operable facilities, it is necessary for project teams to implement **both** quality and process safety at a high level. In constructing new plants and equipment, it is important that equipment as it is designed and fabricated is suitable for its process application. Appropriate quality checks and inspections must be performed to ensure that equipment is installed properly and is consistent with design specifications and manufacturer's instructions. Later in the lifecycle, equipment deficiencies outside acceptable process safety limits must be identified and corrected, and maintenance materials and spare parts must be suitable for their process application.

Additional information can be found in several publications:

API, *Material Verification Program for New and Existing Alloy Piping Systems*, 2nd edition, RP 578, American Petroleum Institute, 2010.

API 570, *Piping Inspection Code: Inspection, Repair, Alteration, and Rerating of In-service Piping*, American Petroleum Institute, Washington, DC.

API 610/ISO 13709, *Centrifugal Pumps for Petroleum, Petrochemical and Natural Gas Industries*, American Petroleum Institute, Washington, DC.

API 620, *Design and Construction of Large, Welded, Low-pressure Storage Tanks*, American Petroleum Institute, Washington, DC.

API 650, *Welded Steel Tanks for Oil Storage*, American Petroleum Institute, Washington, DC.

API 653, *Tank Inspection, Repair, Alteration, and Reconstruction*, American Petroleum Institute, Washington, DC.

ASME (American Society of Mechanical Engineers), *International Boiler and Pressure Vessel Code*, New York, NY.

ASME B31.3, *Process Piping*, American Society of Mechanical Engineers, New York, NY.

ASME B73.1, *Specification for Horizontal End Suction Centrifugal Pumps for Chemical Process*, American Society of Mechanical Engineers, New York, NY.

ASME B73.2, *Specifications for Vertical In-line Centrifugal Pumps for Chemical Process*, American Society of Mechanical Engineers, New York, NY.

ASME PCC-2, *Repair of Pressure Equipment and Piping*, American Society of Mechanical Engineers, New York, NY.

ASTM E1476-97, *Standard Guide for Metals Identification, Grade Verification, and Sorting*, ASTM International, West Conshohocken, PA.

IEC 61511, *Functional Safety: Safety Instrumented Systems for the Process Industry Sector - Part 1: Framework, Definitions, System, Hardware and Software Requirements*, International Electrotechnical Commission, Geneva, Switzerland.

NBBPVI, *National Board Inspection Code*, National Board of Boiler and Pressure Vessel Inspectors, Columbus, OH.

NFPA 70, *National Electrical Code*, National Fire Protection Association, Quincy, MA.

Pipe Fabrication Institute, *Standard for Positive Material Identification of Piping Components Using Portable X-Ray Emission Type Equipment*, New York, NY, 2005.

UL 142, *Steel Aboveground Tanks for Flammable and Combustible Liquids*, Underwriters Laboratories Inc., Northbrook, IL.

9 COMMISSIONING AND STARTUP

As construction nears completion, the Project and the Client begin to anticipate handover and commercial operations; but, before the facilities can be put into service, there are other steps that must take place first. So following construction, pre-commissioning and mechanical completion of the facilities, the project moves into the Startup stage that is the final phase of project execution. Figure 9.1 illustrates the position of Startup in the project life cycle.

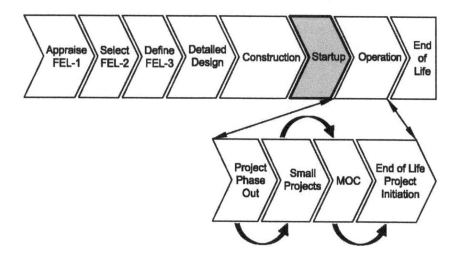

Figure 9.1. Startup

Successful projects that startup efficiently and operate reliably invariably involve the future Operator throughout development, but this is especially important during the Startup stage activities. The integration of the future operations personnel into the pre-commissioning and commissioning teams is essential and ensures that the Operator 'owns' the facilities from an operability and maintainability standpoint. It also brings their learning curve forward enabling earlier achievement of operating competence in the new facilities.

The Startup stage comprises to two main steps: commissioning and startup. Terminology sometimes differs between clients, projects, and countries. Some define commissioning as including pre-commissioning activities and startup of the facility, while others define startup as the transitional phase between construction completion and commercial operations, including all of the activities that bridge those two phases (CII, 1998). In reality, the terms *commissioning* and *startup* are sometimes poorly defined and frequently used ambiguously. However, for the purposes of this book, these terms are characterized as follows:

i. *Pre-commissioning* is a phase of construction that is completed prior to certification of mechanical completion (see Chapter 7 Section 7.4.21),

ii. *Commissioning* is a phase of the startup stage when utility systems are live and process systems are first made operational, typically with low hazard chemicals, such as air or water, to test, calibrate, and prove all systems prior to startup.

iii. *Startup* is when process chemicals are first introduced, and the facility is brought into actual operation.

Based on the above characterization, commissioning and startup are defined as:

Commissioning

The process of assuring that all systems and equipment are tested and operated in a safe environment to verify the facility will operate as intended when process chemicals are introduced.

Startup

The process of introducing process chemicals to the facility to establish operation.

Depending upon the scale of the project, commissioning and startup may be performed by the operations team (small projects) or a separate commissioning team (large greenfield projects) with support from operations and contractor personnel as required. In either circumstance, suppliers and vendors of specialist technology may be required on standby to provide support. As a result, like the construction stage, commissioning and startup can involve many parties. This requires good interface management and communication between all parties involved. In addition the PMT may need to liaise with a number of stakeholders, including, but not limited to, partners, local community, national/local government agencies, and NGOs.

Project Management Team

The PMT's primary focus should be on safely turning a collection of vessels, tanks, pumps, compressors, valves, piping and controls into a fully operational facility meeting the client's requirements, while doing so within cost and schedule.

Typical project objectives for commissioning and startup include:

- Safety and environmental performance during commissioning and startup (e.g. no process safety incidents or lost time injuries),
- Approvals for startup obtained,
- Achievement of reliable operation (without equipment damage),
- Performance testing to verify production throughput, product quality and individual equipment specifications (e.g. on specification product within reasonable timeframe),
- Provision of engineering and technical support to the commissioning and operations team(s),
- Generation of handover documentation,
- Handover to the Operator.

Environment, Health and Safety

From an EHS perspective, the EHS risks of commissioning and startup should be identified, understood, and managed to reduce risk. The project EHS risk register should be updated accordingly. The project EHS plan should also be updated to ensure robust EHS procedures and emergency response plan suitable for commissioning and startup, and to address preparedness for operations, including the requirement for an effective EHS management system.

Process Safety

The key process safety objectives in the startup stage include:

- Operational readiness reviews have been conducted and their recommendations satisfactorily resolved before startup commences,
- Competent commissioning and startup team(s),
- Availability of adequate commissioning and startup procedures,
- Maintenance of asset integrity during commissioning and startup,
- Process Safety Plan updated, if necessary, to address preparedness for operations, including operating procedures and training, maintenance management system, and emergency response.

These and other process safety activities during commissioning and startup are discussed below for:

- Preparation (Section 9.1)
- Operational Readiness (Section 9.2)
- Commissioning (Section 9.3)
- Startup (Section 9.4)
- Common Process Safety Elements (Section 9.5)
- Other Project Activities (Section 9.6)
- Performance Test Runs (Section 9.7)
- Handover (Section 9.8)
- Preparation for Ongoing Operation (Section 9.9)
- Project Close Out (Section 9.10)

Hazard identification and risk management are key factors in delivering good EHS and process safety performance, which requires active participation from both the Project and Operator.

9.1 PREPARATION

9.1.1 Planning

Safe and efficient commissioning and startup of new facilities requires careful and detailed planning. This planning should have commenced during Detailed Design (see Chapter 6, Section 6.7) and been refined throughout the Construction stage (see Chapter 7 Section 7.7). Not all projects plan sufficiently in advance or in enough detail, but its importance cannot be over emphasized.

In reality, for large projects, a Commissioning Manager should be appointed who oversees the development of a Commissioning and Startup Plan in significant detail, and estimates the budget necessary to implement the plan. This plan may start as a philosophy for the sequence of major process units and/or systems to be commissioned that is progressively developed into greater and greater detail. A typical plan for a greenfield capital project should include, but not be limited to, the content illustrated in Table 9.1. Some of this content, such as EHS and process safety procedures, should already exist at brownfield sites.

Table 9.1. Typical Commissioning and Startup Plan

Content
Scope of facilities to be commissioned
Commissioning and startup organization (number of personnel, required competencies, roles & responsibilities, etc.)
Training for commissioning and startup team (e.g. vendor instruction on complex machinery, etc.)
Contracts for third party support (e.g. technology vendors, engineering design, etc.)
Resource requirements (e.g. equipment, radios, PPE, etc.)
Schedule & sequence of systems (at individual task level)
Commissioning and startup procedures (incl. safe operating limits, consequences of deviation, etc.)
ITPM to maintain asset integrity (incl. preservation, procedures, spares, etc.)
EHS and Process Safety management system (incl. policies, procedures, SIMOPS, management of change, emergency response plans, etc.)
Test runs* to verify performance goals (e.g. throughput, product quality, site acceptance test (SAT) for individual equipment item, etc.)
Documentation requirements & management system
Measurement and metrics (e.g. KPIs, progress reporting, etc.)

Occurs after steady state operation achieved

Commissioning may be performed in phases as mechanical completion of the construction progresses. On a large capital project, pre-commissioning (see Chapter 7, Section 7.4.21) and commissioning activities of different systems may overlap. It comprises many different activities that either (i) verify that equipment or a system functions as intended or is ready to operate or (ii) involve actually operating individual items of equipment, systems, or parts of systems. Some systems may be deliberately prioritized to facilitate commissioning of other systems. A number of hold points may be appropriate for a large development as certain tasks must be completed before other tasks may commence. For example, the flare system must be commissioned before process units. Typically, detailed instructions of how each item of equipment or system is to be commissioned are documented in commissioning packages.

Scheduling should take into account the potential impact(s) of SIMOPS, especially for brownfield developments. As utilities, process chemicals or other materials are introduced into a process unit or area, extreme caution must be exercised to ensure that potentially hazardous materials are confined to specific known areas and do not create a hazard to adjacent and simultaneous activities.

Commissioning and startup activities can become hectic, but safety must be paramount. In this respect, scheduling must allow sufficient time to perform each commissioning task safely, in order to avoid undue pressure on the commissioning and startup team that could otherwise be distracted and overlook safety concerns.

Further information and guidance is available from the following publications: *Chemical and Process Plant Commissioning Handbook* (IChemE, 2011); *Achieving Success in the Commissioning and Start-up of Capital Projects* (CII, 2015).

9.1.2   Safety

Safety should always be the first priority during commissioning of a new facility, as commissioning of an unproven facility can be a time of particular risk. For example, even after pre-commissioning leak testing, leaks from flanges, valves and pump seals can develop during commissioning due to no/poor preservation, thermal expansion/contraction, and vibration. In addition, large quantities of utilities are often used, including nitrogen for purging. It is also a time when materials (hazardous and non-hazardous), especially utilities, are moved to locations previously free of hazardous materials. Adjacent simultaneous activities can add further hazards through potential interaction.

Case Study: Refinery Major Expansion Project

A specialist company was employed to store major turbine rotors for a major refinery expansion. These rotors were correctly preserved in a temperature and humidity controlled environment, and operated without problem when commissioned. A contractor was responsible for a large number of hydrocarbon pumps that were stored in the open or under cover for a lengthy period before installation. The Client had no QA or oversight role with respect to pump storage.

A large number of process safety incidents (loss of primary containment) occurred during commissioning, as most of the pumps experienced seal leaks. It was discovered that the seals had deteriorated due to improper preservation. In many cases it was quicker to replace the entire pump due to replacement seals not being readily available or insufficient craft personnel available to properly rebuild the pumps. This resulted in significant cost and schedule delay. The causes were related to:

- Lack of plan to properly preserve the pumps
- Lack of oversight by the client

With the end of the project in sight and project demobilization near, commissioning is also a time when safety can be overlooked. For example, some project personnel may be pre-occupied with other issues, such as whether or not they have another job. Plans should be developed to provide additional occupational and process safety specialists in the field during commissioning and startup activities.

These safety specialists should have several roles and responsibilities to:

- Reassess hazards continuously as activities progress and conditions change,
- Monitor work permits with regard to SIMOPS,
- Rigorously enforce safe work practices, and ensure safeguards are in place,
- Stop work in the event of unsafe acts and unsafe conditions.

Plans should also ensure that all commissioning and startup procedures, practices, and checklists emphasize hazards and required safety precautions.

9.2   OPERATIONAL READINESS

Commissioning and startup activities are the first occasion on which the project facilities will be operated, and it is essential to verify that the facilities are safe to operate. As a minimum, this involves confirming that construction is in accordance with design, all recommendations from HIRA have been satisfactorily resolved, all required procedures are available, and personnel are adequately trained. Depending upon the scope of the project, one or more readiness reviews are appropriate to either (i) affirm the facilities are ready to safely startup or (ii) identify a number of actions necessary to achieve readiness. Planning for these operational readiness reviews should have commenced no later than the Construction stage (and preferably in Detailed Design), and these plans updated during commissioning.

These readiness reviews may involve one or more of the following:

- Pre-Startup Stage Gate Review,
- Operational Readiness Review (ORR) (a.k.a. Pre-Startup Safety Review (PSSR)),
- Startup Efficency Review (SUE).

These operational readiness reviews are described in more detail below.

9.2.1   Pre-Startup Stage Gate Review

A stage gate review should be conducted for larger projects to ensure that process safety (and EHS) risks during commissioning and startup are being adequately managed by the project and that the facilities are safe to startup. The review should focus on the plans and management system to be applied, and not duplicate the 'nuts and bolts' level of detail that an ORR addresses. For example, one task for the stage gate review should be to evaluate the plans for a ORR (team, scope, timing, method, etc.).

This stage gate review should be conducted at least two months prior to the introduction of process chemicals in order to allow time to address any actions necessary without adversely affecting the project schedule.   In certain circumstances, it may be appropriate to conduct the review in two parts.

Two examples are:

i.   Major Capital Project
- Part 1 early enough to address adverse findings,
- Part 2 *immediately* prior to introducing process chemicals to verify corrective actions and other necessary preparations are complete.

ii.   Offshore Installation
- Part 1 when installation/topsides in the construction yard,
- Part 2 after riser hook-up in the field.

The stage gate review team may use a protocol and/or checklist, such as the detailed protocol in Appendix G. A typical process safety scope for a pre-startup stage gate review is illustrated in Table 9.2.

The stage gate review team should be independent of the project, familiar with similar facility/process/technology, and typically comprise an experienced leader, operations representative, process engineer, process safety engineer, discipline engineers (as appropriate), and EHS specialist. At the conclusion of the review, the review team will make recommendations for any improvements needed, and indicate to the Gate Keeper, based on process safety, whether the project is ready to proceed to Startup.

Table 9.2. Pre-Startup Stage Gate Review Scope

Scope Item
Confirm that pre-commissioning has been satisfactorily completed and the facilities are ready for commissioning
Verify that project and/or site is implementing a comprehensive process to confirm preparedness (e.g. PSSR) and obtain approvals for startup
Confirm that integrity of the design has been maintained, and deviations from design have been satisfactorily addressed and will not compromise Process Safety and EHS performance
Confirm that the commissioning, startup and operations teams are adequately trained, equipped, and competent and that all necessary procedures are available
Confirm that the Client / site have made adequate preparations for startup
Confirm that emergency response arrangements and procedures have been established

9.2.2   Operational Readiness Review

An Operational Readiness Review (ORR) (a.k.a. Pre-Startup Safety Review (PSSR)) should be conducted during pre-commissioning and commissioning (which invariably overlap) to evaluate whether the facilities can be safely started. Whereas the pre-startup stage gate review is conducted early and concentrates on plans and management systems, an ORR is focused on the detailed implementation, often using detailed checklists. The purpose of this key activity is to assess whether the facilities can be safely started. As this is the first occasion that the facilities will be started, the review must be very thorough, and far more detailed than reviews conducted after a trip, utility failure, or precautionary shutdown (e.g. hurricane/typhoon).

The review should include a walk-through inspection of all facilities, and, *as a minimum*, address the adequacy of:

- Construction and installation of all equipment, controls, and structures conforms with design specifications, including performance standards for SCE and other protective devices (i.e. validation of SIS and other IPLs),
- Functional safety assessment (FSA),
- Resolution of punch-list items (category A & B),
- Completion of all pre-commissioning activities,
- HIRA studies to meet regulatory and company requirements,

- Action resolution (e.g. HIRA studies and other recommendations),
- Safety, operating, maintenance, and emergency procedures,
- Training and competency assurance of all employees.

Most major operating companies have developed detailed checklists, and CCPS has also provided example checklists (CCPS, 2007) that are risk-based and may exceed the requirements of a local jurisdiction. Table 9.3 represents an example of the typical categories covered by these checklists that may comprise up to several hundred items. These categories may be customized to suit a particular project.

The operational readiness review represents one of the final opportunities before startup to identify discrepancies between the design and installation. Any discrepancies or other QM issues identified must be evaluated, corrected (if necessary) prior to startup, and documented. The evaluation process should be equivalent to a management of change (MOC) review.

If safe chemicals (e.g. water, nitrogen, etc.) are used, the review is preferably conducted towards the end of commissioning just prior to startup with process chemicals, but may take several weeks for large capital projects. In certain circumstances, it may be appropriate to conduct the review prior to commissioning (i.e. after pre-commissioning and mechanical completion), and/or have more than one review team for large projects. The operational readiness review should preferably be led by someone not involved in the design or construction, and the multi-discipline team members may be drawn from Project and Operations.

Operational readiness review checklists typically include details of items to be verified during the facility inspection, including, but not limited to, cleanliness/housekeeping, provision of EHS equipment (e.g. fire extinguishers, SCBA, PPE, first aid boxes, etc.), signage, lighting, and removal of scaffolding, cables and hoses.

In addition to conducting a facility inspection, the review team should review a wide selection of project documentation, including, but not limited to:

- Organization charts for commissioning, startup and operations teams
- Commissioning and startup plans, procedures, checklists, technical support
- Engineering datasheets, drawings, specifications, SCE register, SIS validation, deviations from standards,
- Verifying vessel name plate information vs. design documentation,
- Verifying size of installed PSVs vs. design documentation,
- Construction documentation, field changes, QA/QC records, punch-list items (category A & B),
- HIRA report(s) and action resolution
- Operating procedures, manuals, checklists,

- ITPM program, practices,
- Process safety and EHS procedures (e.g. safe work practices, MOC, incident investigation, etc.), equipment, SDSs, emergency response plans
- Training program, training materials, records

Table 9.3. Typical Operational Readiness Review Checklist Categories

Category
Cleanliness and housekeeping (including cables, hoses)
Scaffolding removal
Provision of EHS equipment (e.g. fire extinguishers, SCBA, PPE, first aid boxes, etc.)
Signage
Startup & Operations plans, organization, and preparedness
Process Safety & EHS organization, procedures, equipment, regulations and permits
Technical support organization
Asset integrity management (including organization, procedures, ITPM program, CMMS, spares, etc.)
Safety equipment and safeguards in service (e.g. PSVs, SCE, locked open/locked closed valves in correct position, inlet valve to 100% spare PSVs closed, etc.)
Other affected organizations (e.g. engineering, purchasing, laboratory, utilities, adjacent operations, security, etc.)
HIRA (incl. information on safeguards, SCE)
Operating procedures & manuals
Commissioning plan, controls, records, technical assurance
Training programs, status, competency verification, and records
QA/QC during Construction
Equipment review and fitness for purpose (including punch-list items, coupling alignment for large rotating equipment, ergonomics, lighting, etc.)
Process safety information and other documentation (see Chapter 12)

Further guidance on operational readiness is available from the following CCPS publications: *Guidelines for Risk Based Process Safety* (CCPS, 2007); *Guidelines for Performing Effective Pre-Startup Safety Reviews* (CCPS, 2007).

9.2.3   Start-Up Efficiency Review

Some companies also conduct a Start-Up Efficiency (SUE) review to *look beyond startup* and assess potential risks to delivery of the facility's first year of operation and beyond. This allows the SUE team to identify potential opportunities to mitigate the risks and improve the first year of operation without compromising long term operational efficiency or integrity.

Although the focus of the SUE review is on commercial and production issues, it also addresses a number of process safety aspects that have the potential to impact the early years of operation. Many of these aspects should have been addressed earlier in the project lifecycle, but the SUE review represents a final chance to identify opportunities for improvement before startup.

A typical scope for a SUE review includes, but is not limited to:

- Project schedule and critical path to first production
- Commercial arrangements, contracts, operating plan
- Commissioning, handover, and startup plans, contingency plans
- Process safety and EHS strategy, plans, management
    - Asset integrity management plans (e.g. condition monitoring, corrosion, erosion, vibration, etc.)
    - Management of design deficiencies, modifications, premature failure of equipment, incidents / process upsets, etc.
- Operations & maintenance strategy (including inspection, equipment sparing), procedures, organization, training/competency, interfaces with other production facilities (e.g. pipelines, communication, etc.)

The SUE review is normally performed by an independent multi-discipline team that works with Project and Operations personnel. However, due to the potential overlap and duplication of effort with the startup stage gate review, the SUE and stage gate reviews may be combined or at least share some common sessions. An integrated capacity model (a.k.a. choke model) is sometimes developed to identify strengths and weaknesses in 1st year operation based on capacity and limiting factors in the production value chain from feedstock supply to point of sale.

A report should record the output of the review is terms of:

*   Facility risks and opportunities to delivery of 1st year operation agreed with Project and Operations, and recorded in an opportunity matrix,
*   Recommendations for high priority risks and opportunities to optimize access to the opportunities and close the gaps.

9.3   COMMISSIONING

Commissioning is a centuries-old process that owes its origins to the shipbuilding industry. A ship that has been commissioned has gone through extensive quality verification processes including sea trials, all of which are intended to identify any operational problems or deficiencies that need attention. All of the ship's various equipment systems will have been inspected and tested with any deficiencies systematically identified, cataloged, and corrected. Additionally, crew members will have been trained to ensure that they know how to properly operate all of the ship's various systems.

Commissioning a process facility follows the same basic approach, whether it is a process unit, support system (e.g. utilities), or individual item of equipment. Historically commissioning has focused on specific tasks rather than a holistic approach, its value was not generally recognized, and as a consequence there was a tendency to under-resource commissioning activities. However, over the last decade or two, the methodology for process plant commissioning has advanced significantly largely due to project cost and schedule pressures.

A well-planned, systematic and rigorous methodology should be instituted to ensure that the integrity of the design has been maintained and that the facilities are operable. In essence, commissioning verifies that:

*   What was specified *was* installed, and
*   It functions properly (i.e. testing of the full functionality of the process including the control and safety systems).

The process should be fully documented and requires a well-managed engineering approach continuing through start-up and handover of the facilities to Operations. This helps facilitate successful handover to the end-user, compliance with local regulations, and safe long-term operation.

Case Study: Multiple Small Projects at Chemical Plant

Multiple small projects at a chemical plant involved replacement of piping. In numerous cases, the material specifications of the piping, flanges, and gaskets did not match the original design, but were treated as 'replacement-in-kind'. Superficial field checks of the completed work did not identify the changes.

In preparation for a major project, several discrepancies relating to material specifications were identified (e.g. ANSI class 600 gaskets in class 900 flanges). Subsequently, a detailed plant review found many similar problems requiring significant effort by additional external resources to rectify. This latent error had the potential for a process safety incident(s) involving loss of containment of hazardous chemicals.

The principal causes were related to:

- Work orders did not specify correct material specifications,
- Poorly understood and implemented MOC process, and
- No system to properly verify the quality of installed equipment.

Before a facility is ready for commissioning, it is important that pre-commissioning and mechanical completion of that facility has been satisfactorily completed, including resolution of any defects identified (see Chapter 7). However, for large capital projects, it is common for pre-commissioning, mechanical completion certification, and commissioning of various process units, support systems and/or individual items of equipment to occur simultaneously. Hence the need for a well-managed approach that is systematic and cognizant of hazards and their risks.

9.3.1   Equipment Testing

Whereas pre-commissioning was conducted 'dry' with no chemicals in the process equipment, commissioning normally involves first introducing water or another relatively safe material. This approach enables the process equipment to be operated in a way that replicates as closely as possible normal operation.

While this approach is certainly safer than using process chemicals and may appear hazard free, it is not without risk. The possibility of process upsets and unexpected incidents cannot be completely eliminated, and attention to detail and strict safety precautions are critical while acquiring operating experience during commissioning.

Utility systems (e.g. electrical power, instrument air, process water, nitrogen, etc.) will be live, and pose their normal hazards. If, for example, a fractionating column and reflux drum are first made operational with water, the equipment design seldom addresses commissioning activities. The specific gravity of water may be considerably heavier than the process chemicals (e.g. light hydrocarbons). Therefore commissioning operations must address, and be conducted within, the design limitations of the system (e.g. foundations, structural steelwork, etc.). Care should be taken to avoid over-pressure, overload, thermal shock and stress of the equipment, which could result in damage and loss of containment. Even failure of equipment (e.g. pressure burst) containing safe chemicals could cause personnel injuries.

Equipment should be set up in a closed loop with safe chemicals (water, air, nitrogen) continuously recycled. In the example of a fractionating column, it is started up by raising temperatures gradually, stabilized at near normal operation for a period, and then shutdown to allow the operators (preferably each shift) to familiarize themselves and gain confidence with the operation before process chemicals are introduced. Operation with a safe chemical also allows leak and flow testing at elevated temperatures, clears dirt/debris (pre-commissioning flushing rarely cleans all debris), and provides an initial indication of how well the control systems work. It also permits any problems to be rectified without having to purge the facilities of hazardous process chemicals. An alternative to commissioning with safe chemicals is the use of training simulators to allow operators to gain experience of startup, normal operation, and shutdown. However this does not provide the other benefits of operation with safe chemicals discussed above.

Commissioning requires a slow and methodical approach with the commissioning team following detailed procedures (see Section 9.3.2 below), and exhibiting a sense of vulnerability and sound operating discipline. All team members must remain vigilant to identify leaks and equipment deficiencies as soon as possible. Design or construction errors not previously identified may become apparent during operation. Some deficiencies (e.g. heat exchanger tube leaks) may not be immediately obvious, and may require sustained operation at elevated temperatures for a period of time before detection.

Some 'hot' testing of equipment may be performed with the assistance of a manufacturer's representative, but should be witnessed by members of the commissioning and/or Operations personnel. A Site Acceptance Test (SAT) may be performed to inspect and dynamically test systems or major equipment items to support the earlier factory acceptance test (FAT) and verify that no damage occurred during shipment and installation. Operator training in the system or equipment may be combined with the checks and functionality verification of the SAT.

Site Acceptance Test

The system or equipment is tested in accordance with client approved test plans and procedures to demonstrate that it is installed properly and interfaces with other systems and equipment in its working environment.

It is particularly important to thoroughly test and evaluate any new technology to determine any operational implications. Depending on findings, it may be necessary to revise commissioning, startup and operating procedures.

After operation with water, air or nitrogen, dynamic testing may progress to simulated operation with one or more safe solvents or proceed immediately to process chemicals. Solvents are sometimes used to operate systems and equipment at or near design conditions prior to introducing more hazardous process chemicals/fluids. However, a thorough review should be conducted to verify that the solvent does not create unacceptable process safety hazards or issues, such as reactivity with materials of construction, catalysts, etc.

If a solvent is used, it should be a 'relatively safe' fluid that has properties close to those of the process chemicals/fluids. It may be necessary to drain, dry and purge the system first. Temporary piping and tanks may be necessary for the solvent supply and subsequent removal, and their design, installation and operation should be subject to thorough hazard and technical review equivalent to the management of change (MOC) for temporary changes. Operation with a solvent allows equipment to be operated close to design throughput and operating limits, instruments to be calibrated, and offers an excellent training opportunity for operators.

9.3.2   Commissioning Procedures

Prior to conducting any equipment testing, detailed commissioning procedures must be available, and their development should have commenced no later than the construction stage of the project (see Chapter 7, Section 7.4.12). These procedures should cover:

- Operation of individual systems, equipment, and/or facilities with water, air or nitrogen,
- Operation of individual systems, equipment, and/or facilities with safe solvent (if appropriate).

Specific operating procedures for utility systems, individual items of equipment, process systems, and facilities operating with safe chemicals should have been developed by the commissioning team. These procedures should cover temporary, normal and transient operations (such as startup and shutdown), provide clear step by step details, and address the hazards of each activity with distinct warnings and cautions where appropriate.

Utility systems are normally commissioned first, including, but not limited to, electrical power, process and cooling water, boiler feed water, steam, process and instrument air, nitrogen, natural gas, oily water and contaminated rain water sewers, effluent treatment, and relief and flare systems. All safety systems should also be operational, e.g. fire and gas detection, and fire protection. Procedures are required to establish steady state operation in order to permit other commissioning activities to proceed. In some instances, sampling procedures may be necessary to confirm satisfactory operation, e.g. oxygen content of nitrogen before use for inerting.

There may also be special procedures for operating some equipment for the first time. For example, fired heaters may require operation with pilot burners and a few main burners for a period of time in order to dry refractory material. Other equipment may require draining water at low points and thorough drying before introducing safe solvents.

Detailed procedures help ensure that equipment condition and function can be verified without the risk of damage, and should address the protection of the newly installed facilities during the commissioning. In this respect, the procedures must ensure that operations remain within design limitations to avoid overpressure, overload, and temperature shock and stress that could result in a process safety incident, costly damage and startup delays. Equipment preservation must also be covered to avoid degradation, such as corrosion.

The equipment manufacturer or vendor should be consulted for guidance on commissioning procedures for certain equipment, such as skid mounted packages. Their attendance at the initial commissioning may be required.

There should be clear roles and responsibilities, especially for brownfield developments and/or where contractors are involved in commissioning, as to which personnel may perform certain tasks, including, but not limited to:

- Open process valves (including existing operating areas),
- Start electrical equipment,
- Manipulate graphic pages on an existing control system.

These responsibilities must be strictly adhered to during commissioning activities to avoid potential process safety incidents.

All procedures should have a thorough review, comment, revision and familiarization by the commissioning team prior to their use. It is important that all members of the commissioning team understand the procedures, local regulations, and any instructions that may vary between sites.

Further guidance on operating procedures is available from the following CCPS publications: *Guidelines for Risk Based Process Safety* (CCPS, 2007); *Guidelines for Writing Effective Operating and Maintenance Procedures* (CCPS, 1996).

Thereafter the facilities are ready for startup and the introduction of process chemicals.

9.4   STARTUP

Startup involves the introduction of process chemicals instead of safe chemicals or safe solvents.   Before a facility is ready for startup, it is important that commissioning activities have been satisfactorily completed, including resolution of any defects identified (see Section 9.3 above).   Some major companies have a formalized 'Go/No Go' practice/ prior to significant operations like startup.

9.4.1   Preparation for Startup

In preparation for startup with process chemicals, it is important that full safety procedures are established and all protective devices are working as if it were a live operating facility.   All startup activities should be planned with individual tasks assigned, and detailed operating procedures developed.

Safe chemicals (e.g. water) or safe solvents may need to be drained from equipment and systems at low points on piping, control valve loops, process vessels and machinery.   Thorough drying may also be necessary before introducing process chemicals to avoid hydrate formation that could result in a process safety incident. Drying can be achieved by blowing air or nitrogen through systems or alternatively by oil circulation followed by repeated draining at low points.

Another common practice prior to introducing process chemicals is to purge and inert the facilities with nitrogen, which involves pressuring (typically 50 to 100 psig) and de-pressuring several times to achieve a low oxygen content (typically less than 3%).   Further guidance on purging, inerting and explosion prevention is available from the following publication: *Standard on Explosion Prevention Systems*, NFPA 69, 2014.

Nitrogen combined with helium as a tracer gas can be used to hold pressure and monitor helium loss for an hour or two in order to facilitate a final check for leaks that may have been caused by the stress of dynamic testing with safe chemicals.   In this manner, leaks as small as 100 scf/year can be detected.   Monitoring pressure decay also confirms that no vents or drains are open or passing.   If the rate of helium and/or pressure loss indicates a leak(s), it must be found and rectified prior to the introduction of process chemicals.

Other startup preparations include charging any catalyst or molecular sieve, and ensuring adequate supplies of raw materials and spare parts.   Finally, formal approval for startup may be required from a local jurisdiction.

9.4.2   Calibration of Instruments and Analyzers

Most instruments will have been factory calibrated and their calibration checked during commissioning using safe chemicals.   Some instruments may need to be calibrated at site before installation using an appropriate calibration fluid.

In certain circumstances after commissioning with safe chemicals, final calibration with process chemicals may be necessary, and may be partially performed on closed loops before startup using process chemicals/fluids or an appropriate calibration fluid. Similarly, process analyzers should be calibrated using an appropriate calibration fluid.

Detailed procedures for calibration of instruments and analyzers should have been provided by the equipment manufacturer or vendor, and form part of the documentation to be ultimately handed over to the Operator.

9.4.3   Startup with Process Chemicals/Fluids

A disproportionate number of process safety incidents occur during transient operations, resulting in loss of primary containment or a process upset outside safe upper and lower limits (Duguid, 2008; Ostrowski & Keim, 2010). Historically approximately 50% of major incidents occurred during operations, such as startup, shutdown, and abnormal/emergency events. The majority of these major incidents involved process unit startups. It is therefore essential that extreme care is taken during startup of a facility for the first time. Typical safeguards include, but are not limited to:

- Exclusion zone around the facility for non-essential personnel and SIMOPS,
- Thorough team briefing on hazards, procedures, etc. before commencing startup,
- Thorough check of valve alignment, energy isolation, feedstock, utilities, etc. before commencing startup,
- Extra operators patrolling the facility alert to abnormal signals,
- Slow step-by-step progression through the startup operating procedure with hold points before high risk steps,
- Regular communication between all parties involved,
- Propensity to halt the startup and make facility safe in event of abnormal situation/uncertainty.

Detailed operating procedures should have been developed for the facility. These procedures should cover normal and transient operations (such as startup and shutdown), provide clear step by step details, and address the hazards of each activity with distinct warnings and cautions where appropriate.

The startup team should include the operations personnel to supplement their training and provide practical experience with operation of the new facility. If appropriate, a representative(s) from the technology licensor should be present to support the startup by providing advice and guidance.

All procedures should have had thorough review, comment, revision and familiarization by the startup team prior to their use. It is important that all members of the startup team understand the procedures, local regulations, and any instructions that may vary between sites.

Further guidance on operating procedures is available from the following CCPS publications: *Guidelines for Risk Based Process Safety* (CCPS, 2007); *Guidelines for Writing Effective Operating and Maintenance Procedures* (CCPS, 1996); *Guidelines for Process Safety During Transient Operations* (CCPS, 2018b).

9.5 COMMON PROCESS SAFETY ELEMENTS

The following process safety elements apply to both commissioning and startup activities.

9.5.1 Hazard Evaluation

Hazard identification and evaluation is central to the prevention of process safety, occupational safety, and environmental incidents. It is common for one or two members of the commissioning team to attend various project HIRA studies. However, while the HIRA studies may have identified some commissioning and startup hazards, it is unlikely that they have identified *all* commissioning and startup hazards. Commissioning and startup procedures should also be reviewed by a HIRA team before commissioning commences. Any late field changes should also be carefully reviewed from both safety and technical bases to ensure that any new hazards are properly addressed.

The commissioning team should ensure that safe work practices are rigorously implemented, and that each work permit is supported by a task hazard assessment, such as a job safety analysis (JSA), to identify hazards and the appropriate safeguards at each step of the permitted job.

Where simultaneous operations, such as two or more of production, drilling, maintenance, pre-commissioning, commissioning, and startup activities, occur in adjacent areas, a SIMOPS study should be conducted to identify and manage potential interactions. This is particularly relevant to brownfield projects where existing operations may continue during commissioning.

Good communication is essential. All hazards (including hazards adjacent to the job site) and required safeguards must be communicated to the job crew(s) and adjacent operations. Work crews should also report hazards and unsafe conditions to their supervisor for the attention of project management.

Further guidance on HIRA is available from the following CCPS publications: *Guidelines for Risk Based Process Safety (CCPS, 2007); Guidelines for Hazard Evaluation Procedures, 3rd edition (CCPS, 2008); A Practical Approach to Hazard Identification for Operations and Maintenance Workers (CCPS, 2010).*

9.5.2   Safe Work Practices

Rigorous enforcement of safe work practices is critical to the safety of commissioning and startup activities. All commissioning personnel are responsible for following the approved safe work practices that may be regulated and/or required by the client/project. The client/project may require more stringent practices than local regulations. Every member of the commissioning team will likely require some form of orientation training in the detailed safe work practices and critical safety rules to be employed on the site(s).

The safe work practices are likely to cover a similar range of activities as those discussed in Chapter 7 Section 7.4.11. However, as commissioning progresses electrical systems will be energized, and hazardous materials moved to new locations. This will necessitate extreme care and emphasis upon the following safe work practices:

- Energy isolation (clear labeling of isolated and energized systems),
- Line breaking (for removal of blinds),
- Barriers (with tags to identify activities in exclusion zones),
- Work permits (including hot, cold/safe, confined space entry, energy isolation/LOTO),
- Management of any startup bypass/inhibit of SCE.

As previously discussed, historically a disproportionate number of process safety incidents involving loss of containment occur during transient operations, such as startup. Many companies create exclsion zones for commissioning and startup activities around process units. Limiting access for non-essential personnel, especially for plants with high hazard potential, is good practice and reduces risk. It is also not unknown for large quantities of liquid from passing PSVs to overwhelm flare knockout drums resulting in 'burning rain' from elevated flares, and, irrespective of any thermal radiation considerations, an exclusion zone should be established until any initial operational problems are resolved.

There should be clearly defined responsibilities for implementation of the work permit system within the commissioning areas. Safety specialists in the field should monitor and enforce safe work practices on a daily basis, and periodically conduct permit audits. Above all else, good communication is essential on a daily basis between all parties to raise awareness of planned activities, changed status of certain areas, and changed safety conditions prior to the activities taking place.

Further guidance on safe work practices is available from the following CCPS publication: *Guidelines for Risk Based Process Safety* (CCPS, 2007). US OSHA and UK HSE also provide guidance through their websites.

9.5.3   Procedures

In addition to commissioning and startup procedures, procedures are also required for:

- EHS and Process Safety, and
- Asset Integrity (including ITPM practices).

9.5.3.1   EHS and Process Safety Procedures

The EHS and process safety procedures (e.g. MOC, incident reporting, incident investigation, etc.) may be the same as those intended for future operation of the facilities after handover to the Operator. If those procedures are still under development, the commissioning team will need to develop adequate procedures in the interim. Additional procedures or site instructions may be required for specific commissioning activities. For example, if water is used to commission some systems, its disposal may be subject to local regulations requiring the water to be re-cycled or routed to a particular sewer.

The commissioning team must observe and actively enforce all site EHS and process safety procedures and instructions, especially safe work practices, restricted areas/exclusion zones, and the use of personnel protective equipment.

9.5.3.2   Asset Integrity Procedures

The asset integrity procedures may be the same as those intended for future operation of the facilities after handover to the Operator. If those procedures are still under development, the commissioning team will need to develop adequate procedures in the interim. Commissioning activities may stretch over several weeks or months, and must ensure that all equipment is appropriately preserved prior to startup (see Chapter 7, Section 7.4.8; Chapter 8, Section 8.5). Manufacturer and equipment vendor instructions should be available for maintenance and preservation.

Further guidance on asset integrity is available from the following CCPS publications: *Guidelines for Asset Integrity Management* (CCPS, 2017); *Guidelines for Risk Based Process Safety* (CCPS, 2007).

9.5.4   Training and Competence Assurance

Training of the Operations team (including managers, supervisors, operators, technicians, engineers, and EHS specialists) should have been completed while the facilities were being constructed. Where appropriate, some of this training should be delivered by the licensor of proprietary technology and the manufacturer / vendor of complex equipment and machinery. A curriculum should have been developed for each discipline, but all personnel should receive training in the EHS and process safety procedures and site instructions.

For the operators and their supervisors their training should have focused on the operating procedures, i.e. how to safely startup with process chemicals, normal operation including product quality and troubleshooting, shutdown, and limits of safe operation. A written test demonstrated that the operators have reviewed and understood the procedures. Where practical, the training should have included visits to the facilities to learn locations of equipment, piping and valves. Control board operators should have received training on simulators or similar plants, and mock operational drills.

Experienced operators are generally brought in from other plants to support, and sometimes take the lead, on commissioning and startup. The commissioning team, which should include many of the operators, should have received training in the commissioning procedures, and also undergone all other relevant training. By participating in the commissioning, the operators gained practical experience to supplement their classroom training. Again, instruction by the technology licensor and manufacturers / vendors, where appropriate, may have been necessary to ensure that the commissioning team can safely discharge their duties.

In the case of a greenfield development, especially in a rural area with little industry, some of the operators recruited locally may have little or no process experience, and may not be part of the formal commissioning team. Their classroom training during construction likely focused on general instruction (i.e. process equipment and systems, such as fractionation columns, heat exchangers, furnaces, pumps, compressors, piping, valves, flares, and control systems) and operating procedures with limited time to walk around the completed facilities before commissioning commenced. Nevertheless, commissioning is ideal for these new operators to observe and learn on-the-job from experienced operators before they are qualified to work without close supervision.

Further guidance on training and competence assurance is available from the following CCPS publications: *Guidelines for Risk Based Process Safety* (CCPS, 2007); *Guidelines for Defining Process Safety Competency Requirements* (CCPS, 2015).

9.5.5   Management of Change

The Pre-Startup Stage Gate Review and Operational Readiness Review should have addressed any changes to the facilities that occurred prior to commissioning and startup (see Section 9.2). While it is unlikely that changes to equipment and facilities will be necessary during commissioning, it is not totally unheard of. If necessary, some project engineering resources that were retained to support operations may be able to handle the design and technical reviews required. Large and complex changes may require referral to the engineering design contractor and/or technology licensor.

It is more likely that a few changes to commissioning and startup procedures will be necessary based upon initial operating experience. All changes, whether equipment, chemicals or procedural, should be subject to a rigorous safety and technical review that evaluates any hazards and their appropriate safeguards. Each change should be documented and approved by senior management.

Further guidance on management of change (MOC) is available from the following CCPS publications: *Guidelines for Risk Based Process Safety* (CCPS, 2007); *Guidelines for the Management of Change for Process Safety* (CCPS, 2008).

9.5.6 Incident Investigation

The EHS and process safety procedures, intended for future operation of the facilities after handover to the Operator, should include instructions for reporting and investigating incidents and near-misses. These procedures should include, but not be limited to, injury, illness, fire, chemical spill, property/vehicle damage, and near-misses.

If the Operator's procedures are still under development, the project will need to develop adequate procedures in the interim for any incidents occurring within the commissioning site(s).

All investigations should aim to identify root causes, and make recommendations to prevent recurrence. Corrective actions should be tracked to completion, and lessons learned communicated to the workforce.

Case Study: Production Facility Expansion Project

During the startup of an expansion project, a gas leak was detected at a pipe flange while pressurizing the process systems. A flammable gas cloud formed, and emergency alarms and procedures were activated. Fortunately the cloud did not ignite and personnel evacuated safely.

The investigation found that the defective flange was procured locally from a non-qualified vendor despite the QA/QC system employed for the commissioning stage.

The other main finding was related to Operational Discipline associated with operating procedures for safe startup. As the project had been delayed, the Operations team decided to perform leak testing using process gas instead of nitrogen/inert gas, which contravened the commissioning and startup plan and also corporate life-saving rules (a.k.a. golden rules).

Some companies also report and investigate asset integrity incidents (e.g. equipment failure) and process upsets, such as a demand on a protective system (e.g. PSV lifting), temperature or pressure excursion, or alarm flood. Some process upsets can be expected during commissioning and startup activities as the commissioning and operations teams familiarize themselves with the new facilities. Reporting and promptly investigating these upset incidents should result in smoother commissioning and startup activities.

Further guidance on incident investigation is available from the following CCPS publications: *Guidelines for Risk Based Process Safety* (CCPS, 2007); *Guidelines for Investigating Chemical Process Incidents*, 2nd edition (CCPS, 2003).

9.5.7   Emergency Response

As previously discussed, commissioning and startup of a new unproven facility can be a time of particular risk. Although water and safe chemicals are used initially, electrical and other utility systems will be live, and large quantities of inert gases may be required. Some leaks should be expected due to thermal expansion / contraction and vibration, and the unintentional opening of the wrong valve can allow materials to move to an area that was previously isolated, free of hazard, and lacking the proper safeguards. Operational incidents (e.g. fires, explosions and toxic releases) on brownfield sites also have the potential to impact the commissioning site.

The emergency response plan, and the necessary resources, for commissioning and startup is likely to be the same as that intended for future operation of the facilities after handover to the Operator. The plan should address issues, such as, but not limited to:

- First aid and medivac,
- Fire and explosion,
- Toxic chemical release,
- Rescue from height/confined space/water,
- Vehicle accident,
- Electrocution,
- Injury due to slips/trips/falls/struck by/crush, and
- Security incident (trespass, bomb threat, terrorism, etc.).

The site emergency services should be on standby throughout commissioning and startup activities to respond to any incident without delay. A table-top or emergency drill based on likely commissioning incidents should be conducted to test the plan's effectiveness.

Further guidance on emergency management is available from the following CCPS publications: *Guidelines for Risk Based Process Safety* (CCPS, 2007); *Guidelines for Technical Planning for On-Site Emergencies* (CCPS, 1995).

9.5.8   Auditing

EHS and process safety audits of any of the elements discussed in Section 9.3 above can alert management to any issues that could give rise to poor EHS and process safety performance. Given the sometimes hectic nature of commissioning activities, the initial focus should be on compliance with safe work practices (especially work permits), and safety rules. Other focus areas for auditing could be housekeeping, or determined by any incidents, observations, and employee concerns.

All audit findings, recommendations, and improvement opportunities should be recorded, and corrective actions tracked to closure. Follow-up audits should verify that corrective actions have resolved the original findings.

Further guidance on auditing is available from the following CCPS publications: *Guidelines for Risk Based Process Safety* (CCPS, 2007); *Guidelines for Auditing Process Safety Management Systems* (CCPS, 2011).

9.5.9   Documentation

In addition to any ongoing effort to prepare 'as-built' documentation for handover to Operations (see Chapter 7 Section 7.4.23) and develop step by step commissioning and startup procedures, the results of running each item of equipment, system test, commissioning operation with safe chemicals, and startup with process chemicals should be fully documented.

The commissioning documentation should contain the results of the tests, and specifically any follow-up action or steps that are required to ensure the equipment is ready for startup. In many cases the commissioning activities will be multi-step, take place over a period of time, and be completed by different personnel. Full documentation minimizes the opportunity for miscommunication or omission of a step in the process. Having a comprehensive file for each system and item of equipment showing the status of each step that has been performed is critical to the success of commissioning. Startup documentation is similar but also includes the data and sample analyses from performance test runs.

Both the commissioning and startup teams should keep operating logs, shift handover notes, and records of implementing each step in the commissioning and startup procedures. These commissioning logs must be detailed as some deviation from norm is only to be expected and a detailed log helps in troubleshooting.

Other examples of documentation are related to temporary operations, but some may also have application to ongoing operations. For example, blind lists should be available to record the location and status of isolations, and temporary strainers on rotating machinery should be recorded for later removal. It is also an opportunity to activate ESD systems at each shutdown for testing to demonstrate reliability, which should be recorded.

Documentation is discussed in greater detail in Chapter 12.

Further guidance on knowledge management is available from the following CCPS publications: *Guidelines for Risk Based Process Safety* (CCPS, 2007); *Guidelines for Process Safety Documentation* (CCPS, 1995).

9.5.10 Performance Measurement

Performance indicators used during commissioning and startup activities vary by the scale of the project and by company. Typical capital project metrics for commissioning and startup are:

- EHS and process safety performance
- Schedule (i.e. commissioning tasks completed)
- Cost vs. budget
- Status of preparations for operation

Projects almost universally measure schedule and cost continuously throughout the project. Schedule for commissioning and startup normally comprises measuring completion of the tasks required to commission each equipment item, system, and facility, including operating with safe chemicals and process chemicals.

Key performance indicators (KPIs) for EHS and process safety may be stipulated by the local regulator, and are likely to include injuries (e.g. first aid, recordable, lost-time), environmental spills / emissions, and process safety incidents (fire, explosion, release of hazardous material). Some companies also measure leading and lagging indicators of process safety performance that indicate the strength of key barriers, such as the adequacy of commissioning procedures. For example, deficiency reports related to unclear / not understood commissioning procedures, and incidents that recommend changes to commissioning procedures are indicators of adequacy.

Further guidance on measurement and metrics is available from the following CCPS publications: *Guidelines for Risk Based Process Safety* (CCPS, 2007); *Guidelines for Process Safety Metrics* (CCPS, 2009).

9.6 OTHER PROJECT ACTIVITIES

In addition to the various process safety and technical activities needed for commissioning and startup, there are a number of other activities that support project execution.

9.6.1 EHS and Process Safety Plans

The EHS Plan and the Process Safety Plan should be updated to reflect commissioning, startup, and handover activities, such as lessons learned from pre-commissioning and commissioning that require changes or additions to future EHS and process safety activities (Appendix B). It is important that all EHS and process safety activities required for a smooth transition from Project to Operator are incorporated.

9.6.2 Risk Register

The Project Risk Register should be updated for any new or changed hazards/risks identified for commissioning and startup (Appendix C). In particular, key risks associated with handover and the transition to the Operator must be identified and managed. Individuals should be identified as responsible for developing a response plan to manage each item. The PMT should regularly review the register and response plans.

9.6.3 Action Tracking

The project action tracking database or spreadsheet should be updated. The PMT should also capture actions generated by their contractor(s), and ensure that all actions are progressively resolved, closed and documented. As a rule, all safety actions should be closed prior to handover to Operations, but any not closed should be communicated to the Operator.

9.7 PERFORMANCE TEST RUNS

The engineering design contractor and/or technology licensor for a specific process unit normally provide a performance guarantee in relation to production rate, product quality and/or efficiency parameters. Individual items of equipment may also have performance guarantees. These guarantees are typically conditional on operation in accordance with the design conditions, and the approved operating procedures and maintenance practices.

Performance test runs (a.k.a. performance guarantee test run (GTR) or acceptance test run) are carried out only after steady state operation has been achieved, in order to check the guaranteed production and efficiency parameters. It is essential that all startup activities are complete, any problems experienced have been resolved, and all instruments and process analyzers calibrated before commencing a test run. Test runs are normally conducted when facility operation has stabilized at full-load for a predetermined period of time. This may occur much later after startup and is also proof of quality of individual equipment and systems. This is often the last task in a turnkey contract.

A detailed procedure for each test run should be developed and agreed by the project, client, engineering design contractor and/or technology licensor. While more equipment focused than process focused, ASME has published a number of Performance Test Codes (PTCs) covering power production, combustion and heat transfer, fluid handling, and emissions, plus guidance on analytical techniques and measurement of process parameters (ASME).

Each test run should be conducted by a team representing all parties, and is typically led by an experienced process engineer. The team records the necessary data, takes feed and product samples for analysis, evaluates the results, and, if necessary, makes recommendations for improvements to address data discrepancies, instrument errors, control system and operational adjustments.

Upon completion of all test runs, and assuming guarantees are met, the facility is declared ready for commercial operation.

Further information and guidance is available from the following publications: *Chemical and Process Plant Commissioning Handbook*, (IChemE, 2011); *Achieving Success in the Commissioning and Start-up of Capital Projects* (CII, 2015).

9.8   HANDOVER

Following completion of performance test runs and successfully meeting performance guarantees, the facility is ready for commercial operation. At this point the facility is normally handed over to the Operator providing that they agree that the project's process safety, EHS, technical, operational and quality specifications have been met.

Many companies have a formal process for handover that typically includes a detailed procedure, checklists, and a transfer of responsibility form. The latter is normally signed by the project manager and a senior Client manager to transfer authority for the facility(s). All personnel, whether members of the project, operations or support teams, should understand the changing boundaries of their respective responsibilities, especially for large capital projects where handover of individual process units may be phased over time.

Arrangements for engineering, technology licensor, and vendor support may be appropriate for a limited period of initial commercial operation, and are often a condition of handover.

The PMT should have compiled extensive documentation over the lifecycle of the project, and agreed on the core information that the Operator requires for ongoing operation of the facility. The format of the documentation should also have been agreed with the Operator. Examples of the documents that should be handed over to the Operator are:

- Information required for ongoing operation, maintenance and further development of the facility,
- Documentation of design intent/criteria, verification, and quality certificates to be retained for statutory purposes and in event of future changes,
- Notifications, requirements and obligations for regulatory compliance,
- Commercial agreements, and commitments to stakeholders,

- Operations Case for Safety (if applicable),
- Contractual and financial documentation to be retained in respect of legal liabilities, warranties/guarantees, financial audits, and tax requirements,
- Other project documentation including risk registers, incident reports, action tracking data, punch-lists, and any technical standards developed by the project.

The first two items above contain all the process safety information (PSI) including as-built drawings, equipment datasheets, operations/maintenance manuals, operating procedures, and much more. The voluntary Operations Case for Safety explains all the safety considerations, including why and how certain safeguards were specified. Documentation is discussed in greater detail in Chapter 12.

The facility may not be 100% complete when handed over to the Operator. Some issues that may need to be resolved are: outstanding punch-list items, and outstanding as-built drawings and documents. The Project and Operator should agree who has responsibility for completing any outstanding items.

9.9   PREPARATION FOR ONGOING OPERATION

The Client/Operator should have developed and refined plans for commercial operation during the project execution stages. The plans should have addressed, but not limited to, the following activities:

- License to operate and regulatory approvals,
- Site management system including EHS and process safety,
- HIRA (including any SIMOPS),
- Recruitment and training of the workforce,
- Operating and maintenance procedures,
- Maintenance management system (baseline data, ITPM tasks, etc.),
- EHS and process safety procedures (e.g. safe work practices, MOC, emergency response, incident investigation, etc.),
- EHS equipment (e.g. ambulance, fire truck, fire extinguishers, etc.),
- Spare parts, consumables, etc.,
- Feedstock, chemicals, lubricants, catalysts, etc.,
- Engineering, technical and vendor support,
- Other business support functions (e.g. production planning, HR, accounting, etc.)

Many of these activities fall within the elements of risk-based process safety (CCPS, 2007), and completion of some activities (e.g. EHS and process safety procedures) are required prior to commencement of commissioning.

The ongoing operation is discussed in greater detail in Chapter 10.

9.10 PROJECT CLOSE-OUT

While close-out of construction contracts should have already been completed, final project close-out involves the process of completing all remaining tasks and documentation to finalize the project. Overall project close-out, in some instances, may not be fully complete until a year or so after handover, when actual equipment performance can be compared against any contract warranties. Planning for close out should commence early in the project schedule to ensure the appropriate resources are available.

Key close-out tasks involve, but are not limited to:

- Resolve outstanding invoices and claims from contractor(s),
- Dispose or handover surplus project materials to the Operator,
- Prepare a Close Out Report,
- Capture lessons learned for future projects,
- Evaluate the performance of the contractor(s),
- Complete project documentation for handover to Operator or archive,
- Project audit (to satisfy stakeholders on financial and technical issues),
- Reassign or terminate project team members.

9.10.1 Close Out Report

A common practice among large companies is to prepare a close out report to assist future projects estimate costs of equipment, materials, labor, and design services. For example, the report should include data on specifications, cost, delivery, weight, spares, etc. for major equipment. Labor should be addressed by man-hours for each craft, area, supervision, cost per hour/ton/foot, etc. Data on design and management services should also be costed by discipline and system plus expenses for travel, accommodation, offices, etc.

9.10.2 Post-Project Evaluation

The final step in the project should be a post-project evaluation (a.k.a. post-project appraisal). Inevitably there are areas in which all projects can be improved upon, and this review captures lessons learned (good and bad) that could be positively applied to future projects. While the motivation and minds of project personnel may be on the next project, time spent reporting and disseminating lessons can be enormously beneficial to the project manager of the next project. Typically, the remaining PMT members, sometimes with a representative(s) of the Client/Operator, discuss and document the lessons.

A key issue to address is whether the project matched the original scope, and Client and other stakeholder expectations and how well this was achieved. The lessons learned may cover any aspect of the project, such as leadership, competencies, tools, techniques, procedures, planning, progress reporting, deliverables, problem analysis, teamwork, communication, interface management, contractor performance, etc. From a process safety perspective, lessons on competencies, tools, techniques, and contractor performance may have implications for improved implementation of fundamental process safety principles and practices.

9.11 SUMMARY

The Startup stage comprises two main steps: commissioning and startup. The project's primary objective is turning a collection of process equipment into an operational facility meeting the client's requirements. To do so requires various process safety activities culminating in an operational readiness review verifying that the facility meets the design intent, hazards and risks are understood and actions completed, personnel are trained and competent, adequate procedures (EHS, process safety, operating, asset integrity) are in place, and the future Operator is prepared for ongoing operations. In this latter regard, successful projects involve the Operator throughout, and especially during Startup stage activities.

10 OPERATION

Following commissioning, startup and handover to the Client organization, the project moves from Execution into the Operation stage. Entry to the Operation stage is normally contingent upon completion of pre-commissioning, commissioning, startup and achievement of steady state operating conditions. Assuming that essential documentation and data have been provided to the Operator (see Chapter 12) and that all necessary preparations are complete (see Chapter 9 Section 9.8), the facility(s) is now ready for commercial operations. Figure 10.1 illustrates the position of Operation in the project life cycle.

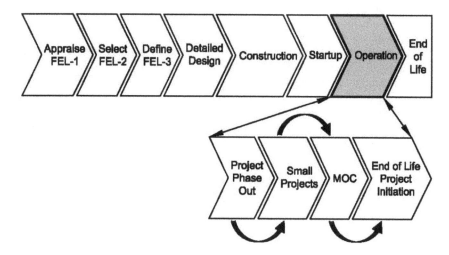

Figure 10.1. Operation

The Operation stage is the prime responsibility of the Operator, although some companies may delay full handover until the Operator formally accepts that the design production rate and product quality has been achieved. The Operator's focus will be on achieving safe and reliable operation at the design production rate. In essence the Operator's initial objectives are to evaluate the facility(s) to ensure that performance meets specifications and to maximize return on investment. Excellence in process safety and EHS is fundamental to achieving these objectives.

Some performance tests may be conducted during initial operation to verify product quality at the design production rate. Some process safety and technical studies are also performed periodically throughout the Operation stage to ensure performance specifications are met, maximize return to shareholders, and protect license to operate.

Project Management Team

Although the Operator is ultimately responsible, the PMT may have a continuing, albeit declining, involvement until at least the achievement of stable operation at the design production rate. PMT activities may include, but are not limited to, the following:

- Provision of engineering and technical support for a period of initial operation,
- Completion of outstanding punch list items,
- Full transparency and communication of any outstanding items turned over to the Operator,
- Formal acceptance of facility(s) by the Operator,
- Participation in stage gate and performance reviews,
- Comparison of project performance vs. project objectives,
- Completion and turnover of project documentation,
- Project close-out and demobilization,
- Reorganization of project team to deliver activities above.

Operator

The Client/Operator should have developed and implemented plans for commercial operation during the project execution stages. Ideally the Operator's team should have been involved throughout the development, and especially during the Startup stage activities. This should have helped to build operating competence and create 'ownership' of the new facilities.

Implementation of these plans should have delivered, but is not limited to, the following activities:

- License to operate and regulatory approvals,
- Site management system including EHS and process safety,
- HIRA (including any SIMOPS),
- Recruitment and training of the workforce,
- Operating and maintenance procedures,
- Maintenance management system (baseline data, ITPM tasks, etc.),
- EHS and process safety procedures (e.g. safe work practices, MOC, emergency response, incident investigation, etc.),
- EHS equipment (e.g. ambulance, fire truck, fire extinguishers, etc.),
- Initial inventory of spare parts, consumables, etc.,
- Feedstock, chemicals, lubricants, catalysts, etc.,
- Engineering, technical and vendor support,

- Other business support functions (e.g. production planning, HR, accounting, etc.)
- Transition of any remaining low risk items from Project to Operator, e.g. punch list, PSSR, stage gate review

Some of these activities fall within the elements of risk-based process safety (CCPS, 2007), and are discussed below in Section 10.1.

Environment, Health & Safety

From an EHS perspective, the Operator management need to demonstrate a strong and visible commitment to EHS by setting and enforcing high EHS standards, and provide adequate resources to deliver positive EHS performance. The project EHS risk register should be updated to address the transition to Operations, and the risks managed appropriately. A documented EHS management system with robust EHS procedures should be in place and being implemented.

Process Safety

The key process safety objectives in the Operation stage include:

- Recommendations from operational readiness and stage gate reviews have been satisfactorily resolved and implemented,
- Process safety risk register updated to address the transition to Operations, and the risks managed appropriately, and any remaining action items tracked to resolution,
- Process Safety Plan implemented to address transition to Operations,
- Process safety management system with robust procedures in place and properly implemented,
- Thorough investigation and timely response to any process safety incidents, process upsets, and asset integrity problems,
- Lessons learned from operation of the new facility(s) are captured and widely shared.

These and other process safety activities during the Operation stage of the project are discussed below:

- Process Safety Management System (Section 10.1)
- Other Project Activities (Section 10.2)
- Technical Support (Section 10.3)
- Operation Stage Gate Review (Section 10.5)

This chapter also briefly discusses some other project activities that typically occur during early operation:

- Performance Test Runs (Section 10.4)
- Post Operational Review (Section 10.6)
- Project Close-Out (Section 10.7)

10.1 PROCESS SAFETY MANAGEMENT SYSTEM

The Operator should have developed plans for a Business Management System (BMS) during the Execution stages of the project. The BMS should comprise local site procedures, the content of which are likely to be influenced by any corporate policies and standards. A key component of the BMS is a process safety management system (PSMS) that is integrated with other requirements, such as manufacturing operations, EHS, HR, engineering, procurement, etc. For example, both the PSMS and HR system are likely to address competence and training practices, and should therefore ensure consistency.

The PSMS should be fully documented and implemented prior to the Operation stage. It is should be designed on a thorough evaluation of the operational hazards and their associated risks. All hazards and risks are not created equal, so more resources and requirements should be focused on the higher hazards and risks. However, all hazards and risks should be managed by *doing whatever is necessary* to ensure safe and reliable operations. Compliance with any local/national regulations **alone** is no guarantee that hazards and risks will be satisfactorily managed to avoid process safety incidents.

One means of developing and implementing an effective PSMS is to follow the guidance of CCPS for a risk-based process safety (RBPS) management framework. The RBPS framework is based upon four accident prevention pillars:

- Commit to process safety
- Understand hazards and risk
- Manage risk
- Learn from experience

These four pillars contain twenty RBPS elements as shown in Table 10.1. If a site focuses its process safety efforts on these elements, then its process safety performance should improve, and the likelihood and severity of incidents should decline. Process safety performance and effectiveness can be optimized by varying the rigor with which each element is implemented commensurate with the level of hazard and risk. Each element is briefly discussed below.

Further guidance on process safety management systems is available from the following publications: *Guidelines for Risk Based Process Safety* (CCPS, 2007); *Guidelines for Implementing Process Safety Management,* 2nd edition (CCPS, 2016). *Guidance on Meeting Expectations of EI Process Safety Management Framework,* Energy Institute, London, UK, 2013.

Table 10.1. Risk-Based Process Safety Elements

RBPS Pillar	**RBPS Element**
Commit to Process Safety	Process Safety Culture
	Compliance with Standards
	Process Safety Competency
	Workforce Involvement
	Stakeholder Outreach
Understand Hazards & Risk	Process Knowledge Management
	Hazard Identification and Risk Analysis
Manage Risk	Operating Procedures
	Safe Work Practices
	Asset Integrity & Reliability
	Contractor Management
	Training and Performance Assurance
	Management of Change
	Operational Readiness
	Conduct of Operations
	Emergency Management
Learn from Experience	Incident Investigation
	Measurement & Metrics
	Auditing
	Management Review and Continuous Improvement

10.1.1 Process Safety Culture

Commitment to process safety is the cornerstone to a positive safety culture. This starts at the highest levels of the organization and must be shared by all. The quality of site leadership and commitment can drive or limit the culture. Leadership must care and lead by example in order for the entire organization to share the commitment. Leaders also need to understand how process safety activities are influenced by culture.

The essential features of a sound culture are:

- Enforcement of high standards, i.e. intervention to correct normalization of deviance,
- Maintain sense of vulnerability,
- Open and effective communication,
- Timely response to issues and workforce concerns.

Further guidance on culture is available from the following CCPS publications: *Essential Practices for Creating, Strengthening, and Sustaining Process Safety Culture* (CCPS, 2018e); *Building Process Safety Culture: Tools to Enhance Process Safety Performance* (CCPS, 2005).

10.1.2 Compliance with Standards

Site leadership should ensure that the organization is aware of and understands all applicable regulations, standards, codes, and other requirements issued by national, state/provincial, and local governments, consensus standards organizations, and the corporation. While compliance will not necessarily prevent process safety incidents, implementation of all applicable requirements should reduce risk and legal liability, and contribute to process safety practices.

10.1.3 Process Safety Competency

Process safety should be the responsibility of, and delivered by, most members of the workforce whether they are employed in operations, maintenance, engineering, EHS, other departments or by a contractor organization. The Operator should have developed and implemented plans for recruitment and training during the project execution stages. This should have included the process safety knowledge and skills in the right places necessary to (i) understand hazards and risks, and (ii) manage the risks, i.e. at least two of the pillars of RBPS.

Further guidance on process safety competency is available from the following CCPS publication: *Guidelines for Defining Process Safety Competency Requirements* (CCPS, 2015).

10.1.4 Workforce Involvement

As indicated above, process safety should be the responsibility of most employees and contractors. While workers involved in operating and maintaining the plant are most exposed to the hazards, they are potentially the most knowledgeable in day-to-day operations. Therefore broad involvement of operating and maintenance personnel in process safety activities is essential. It also ensures that lessons learned by the people closest to the process are considered and addressed.

10.1.5 Stakeholder Outreach

The PMT should have started outreach to stakeholders, such as the local community and government authorities. The site leadership should continue outreach activities to build trust and support the license to operate. This will also help external responders and the public to understand the plant's hazards and potential emergency scenarios, and how to address these scenarios in the event of an incident.

10.1.6 Process Knowledge Management

The PMT should have handed over to the Operator all information needed to perform process safety activities. This should include 'as-built' technical documents and specifications, engineering drawings and calculations, assumptions, studies, and other relevant documentation concerning technology, process equipment, and process chemicals and materials. Any outstanding redline drawings should be updated to 'as built' status as soon as possible and handed over to the Operator.

Site leadership needs a document management system to archive this information for future reference, maintain it up-to-date, and provide ready access to whoever needs this information to safely perform their work. Process safety studies and reviews, such as relief and flare, facility siting, HIRA, operational readiness (PSSR), and associated supporting information and closure of action items, should also be archived.

Documentation is discussed in detail in Chapter 12. Further guidance on documentation is available from the following CCPS publication: *Guidelines for Process Safety Documentation* (CCPS, 1995).

10.1.7 Hazard Identification and Risk Analysis

Hazard identification and risk analysis (HIRA) uses the information discussed in process knowledge management (Section 10.1.7 above) and is the foundation of other process safety activities necessary for managing process risk. The Project should have handed over HIRA studies conducted during design and construction, but the Operator should periodically revalidate these studies. In the event of new facilities or modifications to existing facilities, the Operator should conduct new HIRA studies. All HIRA studies should be periodically revalidated or redone.

In the case of large capital projects, it is likely that some process units may be handed over to the Operator while construction and/or commissioning activities continue in close proximity on other parts of the project. In these circumstances, the Operator should participate in JSAs for work permits and studies to identify and manage the hazards and risks of simultaneous operations (SIMOPS). Close liaison and communication between the Project and Operator is essential as, in the event of a process upset, it may be necessary to rapidly shutdown adjacent construction and commissioning activities.

Further guidance on HIRA is available from the following CCPS publications: *Guidelines for Hazard Evaluation Procedures*, 3rd edition (CCPS, 2008); *Guidelines for Chemical Process Quantitative Risk Analysis*, 2nd edition (CCPS, 2000); *Revalidating Process Hazard Analyses* (CCPS, 2001); *Layer of Protection Analysis* (CCPS, 2001); *A Practical Approach to Hazard Identification for Operations and Maintenance Workers* (CCPS, 2010).

10.1.8 Operating Procedures

Operating procedures should have been developed during the execution stages of the project. These provide detailed instructions for the safe startup, shutdown and normal operation of each process unit in terms of the sequence of steps, hazards, and protective equipment for each task. The consequences of deviation from procedures, safe process limits, key safeguards, and any special situations and emergencies should also be covered.

As a result of experience gained during commissioning and early operation, it may be necessary to capture any lessons learned and modify the operating procedures, especially if modifications to the process equipment or operating conditions are necessary. Thereafter the Operator should ensure that the procedures are periodically reviewed, and maintained accurate and up-to-date.

Further guidance on operating procedures is available from the following publications: *Guidelines for Writing Effective Operating and Maintenance Procedures* (CCPS, 1996); *Guidance on Meeting Expectations of EI Process Safety Management Framework, Element 8: Operating Manuals and Procedures*, Energy Institute, London, UK, 2013; HSE, *COMAH Guidance, Technical Measures, Operating Procedures*, Health & Safety Executive, Bootle, UK, http://www.hse.gov.uk/comah/sragtech/techmeasoperatio.htm accessed October 2017.

10.1.9 Safe Work Practices

The Operator should establish safe work procedures, which may be supplemented by work permits, to safely manage non-routine work. Local regulations and corporate standards may define certain requirements.

Typical safe work practices include:

- Site access control
- Hot work
- Energy isolation
- Line breaking
- Working at height
- Excavation and trenching
- Confined space entry
- Heavy lifts
- Electrical systems
- Hazard communication, e.g. safety data sheets
- Personal protective equipment (PPE)

JSAs should support each work permit. Daily toolbox meetings should be held to cover the day's job tasks, hazards, required safeguards, and adjacent activities. In the event of continuing construction and/or commissioning on adjacent facilities, the Operator should participate in SIMOPS studies to identify and manage potential interactions, and maintain close liaison and communication with personnel responsible for adjacent activities.

Further guidance on safe work practices is available from the websites of US OSHA and UK HSE.

10.1.10 Asset Integrity and Reliability

A strategy to ensure that process equipment remains fit for purpose throughout its life should have been developed during the execution stages of the project. Failure to maintain asset integrity could cause, contribute to, or fail to prevent or mitigate, a major incident. This strategy should be based on the criticality of equipment (including safety criticality), knowledge of potential damage and failure mechanisms, and one or more of the following approaches to managing failure: reliability, predictive, preventive or run-to-failure basis. Guidance on these alternative strategies is available in a recently published CCPS book: *Guidelines for Asset Integrity Management* (CCPS, 2017).

Based on the strategy, industry codes, and information from equipment vendors, the Operator should have established a program of systematic activities comprising periodic inspection, testing and preventive maintenance (ITPM) for all process equipment and piping. For example, each SIS should be maintained and functional tested per its Safety Requirements Specification (SRS). This program should be entered into a maintenance management system (MMS) to facilitate timely implementation of ITPM tasks. Procedures for ITPM and repair tasks, and quality management (QM) should be available to maintenance and contractor personnel.

Case Study: Corrosion of New Process Unit

A new crude distillation unit suffered serious corrosion and had to be shutdown for extensive repairs just weeks after the major expansion project started up.

Not long after startup, the unit was shutdown to fix a minor vapor leak. While the unit was shut down, sodium hydroxide continued to flow into the unit piping. Although the caustic pumps were shut down, system pressure resulted in caustic flowing through the pumps, which were not valve isolated, and into the process piping. When the unit restarted, the caustic was heated and vaporized, quickly corroding piping causing leaks that led to small fires and a ruptured heater system.

An inspection found caustic cracking of stainless steel piping. Repairs took 7½ months and the caustic system was redesigned to prevent unregulated flow into the unit.

Baseline inspections of key equipment and piping should have been conducted during construction activities. Thereafter initial inspection frequencies should err on the side of caution until operating experience gained, and, if appropriate, some frequencies may then be extended in line with industry good practice. Testing of key process equipment and controls, such as emergency shutdown systems, should be based on the required reliability. Once the facility has been operating for a few years and sufficient data has been gathered, an IEC stage 4 functional safety assessment (FSA) should be conducted of each SIF to review initial assumptions and validate them or update them to reflect operational experience.

If early operation and/or inspection and testing indicates any asset integrity concerns, a thorough investigation should be conducted to determine the root causes of failures or other deficiencies. Recommendations to resolve integrity problems should be implemented in a timely manner to prevent potential process safety incidents.

Further guidance on integrity management is available from the following CCPS publication: *Guidelines for Asset Integrity Management* (CCPS, 2017); *Guidelines for Safe Automation of Chemical Processes* (CCPS, 2017); *Guidelines for Writing Effective Operating and Maintenance Procedures* (CCPS, 1996).

Case Study: Fire During Early Operation of Fine Chemical Plant

A fire occurred during the first production cycle of a pharmaceutical process unit. A toluene leak from the cover joint of a solvent reception tank occurred while the unit was being cleaned with toluene. The tank drain pump failed to start due to an electrical defect (poorly secured lug) resulting in overfilling of the tank. A dozen employees sustained light burns from the ensuing flash fire when the leak ignited.

The new unit had several design and maintenance faults:

- undersized solvent tank meant drain pump had to start several times,
- non-redundant tank level alarm (sole overfill safety device),
- plastic tanks and pipes unsuitable for dielectric properties of toluene,
- tank made of fragile plastic,
- poorly secured cover (4-bolt design modified to 8 bolts but drawing not updated). Only 7 bolts installed and eighth bolt replaced with a clamp,
- lack of gas detection in facility.

Lessons were related to design hazard management, management of change, and asset integrity.

Reference: No. 14500 ARIA database; Ministry of Ecology, Sustainable Development and Energy, France.

10.1.11 Contractor Management

A number of contractors and vendor representatives may be required during early operation to provide support. Thereafter, depending on the site's contract strategy, various contractors may be employed for maintenance, engineering or other activities. The Operator should have developed a contractor management system to select, monitor, and review contractors to ensure that contract workers can perform their jobs safely, and that contracted services do not add to or increase risks.

Work crews should receive orientation training on the Operator's EHS and process safety expectations, rules and procedures, when they first access the site. They should then receive daily briefings on the hazards of their work and any hazards adjacent to the job site through pre-job toolbox meetings, participation in developing JSAs, or other means. A safety specialist and/or supervisor should monitor the work site(s) and intervene in the event of unsafe acts or conditions.

The Operator should regularly review each contractor organization's performance in meeting the EHS and process safety rules and procedures, and contract conditions, and intervene if performance improvement is required.

10.1.12 Training and Performance Assurance

The Operator should have recruited staff and employees based on their qualifications, knowledge, skills, and experience. Any training, where necessary, should have been conducted prior to startup. For example, process operators should have received instruction in operation of the process unit(s) and familiarization during commissioning and startup activities. All employees should also have received training on the EHS and process safety procedures.

Thereafter, the Operator should establish a system to ensure that new employees receive appropriate practical instruction in job and task requirements. The Operator should also verify that all employees perform proficiently in respect of the knowledge and skills they have been trained in.

10.1.13 Management of Change

The design and construction of the new facilities were subject to hundreds, if not thousands, of man-hours of technical and safety analysis by professional engineers to ensure the integrity of the facilities is fit for purpose. A hasty and ill-considered change could easily impair that integrity and cause, contribute to, or fail to prevent or mitigate, a major process safety incident.

During early operation it is not unusual to experience operating problems that may require changes to equipment or operating conditions and procedures. Later, the Operator may identify opportunities to debottleneck the facility(s) or make other production improvements. Therefore, the Operator should establish a formal management of change (MOC) practice to carefully review and authorize proposed changes to facility design, operations, organization, or activities prior to implementing them. The last stage of MOC should require documentation to be updated and communicated.

Further guidance on management of change is available from the following CCPS publications: *Guidelines for the Management of Change for Process Safety* (CCPS, 2008); *Guidelines for Managing Process Safety Risks During Organizational Change* (CCPS, 2013).

10.1.14 Operational Readiness

During early operation it is quite likely that the facility(s) may experience process upsets or equipment problems that result in a total or partial shutdown. Experience shows that the risk of an incident during transient operations such as startup is higher, especially if process conditions are not exactly as those intended. Therefore, the Operator should establish an operational readiness review practice to formally evaluate the plant before startup or restart to ensure the process can be safely started.

This practice should be applied to:

- Startup of a new facility or modified facility,
- Restart of a facility after being shut down or idled, e.g. power failure, maintenance, etc.

Further guidance on operational readiness is available from the following CCPS publication: *Guidelines for Performing Effective Pre-Startup Safety Reviews* (CCPS, 2007).

10.1.15 Conduct of Operations

An inadequate level of human performance can adversely impact operations and may cause, contribute to, or fail to prevent or mitigate, a major process safety incident. Site leadership should develop an effective program to ensure that workers are held accountable for performing their tasks flawlessly in a deliberate, faithful, and structured manner. Conduct of operations is central to culture, and managers should intervene to enforce high standards and prevent deviations from expected performance.

Further guidance on conduct of operations is available from the following CCPS publication: *Conduct of Operations and Operational Discipline: For Improving Process Safety in Industry* (CCPS, 2011).

10.1.16 Emergency Management

An emergency response plan for the site should have been in place for commissioning and startup activities. This plan should cover all possible emergencies and define the actions to be taken and the necessary resources to execute those actions.

As a result of practice drills and an emergency exercise during early operation, it may be necessary to update and improve the original plan. Thereafter, periodic exercises and drills should be held to continuously improve the plan and training of internal emergency personnel, and coordination with external resources. Employees, contractors, neighbors, local authorities, and other stakeholders should be informed of any changes that affect them in the event of an incident. Consideration should be given to occasionally conducting drills during shifts and weekends to test response in off-hour situations under realistic conditions, when limited resources are available.

Further guidance on emergency management is available from the following CCPS publication: *Guidelines for Technical Planning for On-Site Emergencies* (CCPS, 1995).

10.1.17 Incident Investigation

A system for reporting and investigating all incidents and near-misses that occur on the site (including those involving contractors) should have been in place for commissioning and startup activities. This system should identify root causes and corrective actions, track completion of actions and communicate lessons learned to the workforce. Incident trends should be periodically evaluated to determine if further management intervention is appropriate to reduce similar incidents.

Some companies also report and investigate asset integrity incidents (e.g. equipment failure) and process upsets, such as a demand on a protective system (e.g. PSV lifting), temperature or pressure excursion, or alarm flood. Some process upsets can be expected during early operation as the operations team gains experience operating the new facilities. Reporting and promptly investigating these upsets should result in smoother long-term operation.

Further guidance on incident investigation is available from the following CCPS publication: *Guidelines for Investigating Chemical Process Incidents, 2nd edition* (CCPS, 2003).

10.1.18 Measurement and Metrics

Process safety incidents tend to be high consequence/low frequency, whereas occupational safety incidents tend to be the reverse, i.e. low consequence/high frequency. Therefore, the Operator needs an *early warning system* to forewarn of declining and/or poor process safety performance, such that management has an opportunity to intervene before a major process safety incident occurs. This early warning system can comprise a number of information sources, such as:

- Listening to workforce concerns, i.e. bad news,
- Trend analysis of incident root causes,
- Audit findings,
- Unannounced field inspections,
- Learning from others' misfortune, e.g. other facilities or companies.

However, a set of carefully selected key performance indicators (KPIs) should be a major component of the early warning system. Leading and lagging indicators of process safety performance, including metrics that show how well key process safety elements are being performed, can be used to strengthen weak barriers and drive improvement in process safety. For example, reliability data on issues, such as seal failures and out of calibration sensors, can be used to improve asset integrity management. Site leadership should regularly review the metrics and intervene when weaknesses are highlighted (See Section 10.1.20 below).

KPIs for EHS and process safety may be stipulated by the local regulator, and are likely to include injuries (e.g. first aid, recordable, lost-time), environmental spills / emissions, and process safety incidents (fire, explosion, release of hazardous material).

Further guidance on measurement and metrics is available from the following CCPS publications: *Guidelines for Process Safety Metrics* (CCPS, 2009); *Process Safety Leading and Lagging Metrics... You Don't Know What You Don't Measure*

(CCPS, 2011); *Process Safety Leading Indicators Industry Survey* (CCPS, 2013); *Guidelines for Integrating Managemnt Systems and Metrics to Improve Process Safety Performance* (CCPS, 2016).

Additional guidance is available from the following publications: HSE, *Step-by-Step Guide to Developing Process Safety Indicators*, HSG 254, Health and Safety Executive, UK, 2006; API, *Process Safety Performance Indicators for the Refining & Petrochemical Industries, Part 2: Tier 1 and 2 Process Safety Events*, Recommended Practice 754, 2nd edition, American Petroleum Institute, Washington D.C., 2016; API, *Guide to Reporting Process Safety Events*, Version 3.0, American Petroleum Institute, Washington D.C., 2016.

10.1.19 Auditing

Site leadership should establish a program of periodic audits of EHS and process safety to provide a review of management system performance. These audits should be conducted by auditors not assigned to the site in order to provide an objective review. The audits should probe deeply to provide a critical review, as superficial 'check the box' auditing can lead to complacency and a loss of a sense of vulnerability.

Process safety audits of any of the elements discussed in Section 10.1 above can alert site management to gaps in performance that could give rise to a process safety incident, and identify improvement opportunities. Focus areas for auditing could be determined by any incident trends, metrics, observations, and employee concerns.

All audit findings, recommendations, and improvement opportunities should be recorded, and corrective actions tracked to closure. Follow-up audits should verify that corrective actions have resolved the original findings.

Further guidance on auditing is available from the following CCPS publication: *Guidelines for Auditing Process Safety Management Systems* (CCPS, 2011).

10.1.20 Management Review and Continuous Improvement

Site leadership should establish a regular forum to review management system performance. This forum should:

- Set process safety expectations and goals for staff,
- Review process safety performance by examining metrics, findings from incidents and audits, and other 'early warning system' information,
- Review progress towards process safety goals,
- Identify improvement opportunities and track to close.

A successful approach for the management review forum is a regular monthly meeting that examines (say) two elements of process safety, such that in the course of a year all elements are reviewed. If a particular element is perceived as being weak, it may be appropriate to review that element two or more times during the year. The forum should be chaired by a line manager, but a process safety engineer may prepare the performance data for review and facilitate the meeting.

10.1.21 EHS and Process Safety Procedures

In addition to operating and maintenance procedures, site policies and procedures are also required for EHS and process safety. These policies and procedures, as a minimum, are likely to address the elements within Section 10.1 above. Corporate standards and/or local regulations may set minimum requirements for these procedures.

10.2   OTHER PROJECT ACTIVITIES

In addition to the various process safety and technical activities needed for operation, there are a number of other activities that support project closure.

10.2.1   EHS and Process Safety Plans

The EHS Plan and the Process Safety Plan should be updated, if necessary, to reflect any specific activities for facility operation (Appendix B).

10.2.2   Risk Register

The Project Risk Register should be updated for any new or changed hazards/risks identified for facility operation (Appendix C). Individuals should be identified as responsible for developing a response plan to manage each item. The PMT and Operator should jointly review the register and response plans.

10.2.3   Action Tracking

The project action tracking database or spreadsheet should be updated. Some actions may be outstanding when the facility is handed over to the Operator. Some actions that may need to be resolved are: outstanding punch-list items, and outstanding as-built drawings and documents. Some obligations from regulatory requirements, EHS and process safety compliance, and commericial agreements may also require completion. The Project and Operator should agree who has responsibility for completing any outstanding items.

10.3   TECHNICAL SUPPORT

Technical support for the Operator is often appropriate for a limited period of initial commercial operation. This support may be provided by the project process engineer, engineering design contractor, technology licensor and/or vendor of specialized equipment, such as a compressor. Contracts and/or a condition of handover may specify a certain level and duration of technical support.

In the case of large capital projects, additional support for operations, process safety and EHS may be available for a limited period during early operation from other company facilities and offices.

10.4   PERFORMANCE TEST RUNS

While some performance test runs (a.k.a. performance guarantee test, acceptance test) may have commenced during commissioning and startup activities (see Chapter 9, Section 9.7), it is likely that one or more test runs to prove facility operation at full production rate will continue during early operation. These test runs are conducted to check the guaranteed production, product quality and efficiency parameters in respect of performance guarantees provided by the engineering design contractor and/or technology licensor.

A detailed procedure for each test run should be developed and agreed by all parties, who will also witness the test and evaluate the results. The procedure should also include data to be recorded, and process samples to be analyzed. Before commencing a test run, any equipment problems should be resolved, and all instruments and process analyzers calibrated.

10.5   OPERATION STAGE GATE REVIEW

A stage gate review should be conducted for larger projects to verify that process safety (and EHS) performance during early operation meets the design intent and that lessons learned from early operation are shared. This stage gate review should be conducted approximately 12 months after steady state operation is achieved.

The stage gate review team may use a protocol and/or checklist, such as the detailed protocol in Appendix G. A typical process safety scope for a pre-startup stage gate review is illustrated in Table 10.2.

Table 10.2. Operation Stage Gate Review Scope

Scope Item
Confirm that an adequate Process Safety and EHS management system has been properly implemented
Confirm that Process Safety and EHS performance of the operating facility(s) meets design intent
Verify the adequacy of response to any process safety and EHS incidents, and process upsets that have occurred during early operation
Verify the adequacy of programs to address any asset integrity problems that have occurred during early operation
Confirm the rationale for any changes/modifications made during early operation vs. the original design intent
Confirm that lessons learned from early operation of the facility(s) are documented and shared

The stage gate review team should be independent of the project, familiar with similar facility/process/technology, and typically comprise an experienced leader, operations representative, process engineer, process safety engineer, and EHS specialist. The project's process engineer, who is most knowledgeable in the design intent, should also attend. At the conclusion of the review, the review team will make recommendations for any improvements needed in the subject project and lessons for future projects.

10.6 POST-OPERATIONAL REVIEW

Some companies perform an external benchmarking review of large capital projects (typically >$50M) approximately one year after achieving steady state operations. This operability review compares a facility's production performance in the first year of operation with that of comparable industry facilities to identify practices that affected production performance (both technical and market-constrained). This provides targets for realistic project improvement and an understanding of changes required to achieve improvement. The information in this feedback to project management may be used for planning future projects.

Other companies may perform an internal review with differing scope, but the overall objective is similar, i.e. improving future projects.

10.7 PROJECT CLOSE-OUT

Project close-out activities should have started during commissioning and startup (see Chapter 9, Section 9.9). However, overall project close-out, in some instances, may not be fully complete until a year or so after handover, when actual equipment performance can be compared against any contract warranties. In particular, resolution of outstanding invoices and contractor claims, and a final audit to satisfy stakeholders on financial and technical issues, often delays closure. If applicable, the Operation stage gate review (see Section 10.5) and post operational review (see Section 10.6) also delay close-out by 12 months.

10.8 SUMMARY

Once the facility is handed over and accepted by the Operator, it is ready for commercial operation. The Operator's primary objective is achieving safe and reliable operation at the design production rate, for which a business management system comprising all the elements of *Risk Based Process Safety* is essential. Robust process safety and EHS procedures should be in place and properly implemented. Some engineering and technical support from the Project may be appropriate for early operation. Thereafter, the responsibility is on the Operator who should periodically perform various process safety and technical studies to maintain safe, efficient operation and asset integrity.

11 END OF LIFE

Eventually all process facilities come to the end of their useful life. The circumstances that determine this situation are often related to economic, socio-political, regulatory, process safety and EHS, and aging asset pressures. For example, the process unit may have inferior yield, productivity, product quality, emissions and waste, energy demand and/or excessive maintenance requirements compared to newer larger-scale plants, sometimes designed to a different technology. Production could be relocated to a new geographically remote site due to changes in product supply and demand. Major damage from a catastrophic incident may be uneconomic to repair, rebuild or modernize. Sometimes supply and demand issues can also extend the life of the plant beyond its original design intent, but finally a decision will be taken to shutdown permanently. Figure 11.1 illustrates the position of end of life in the project life cycle.

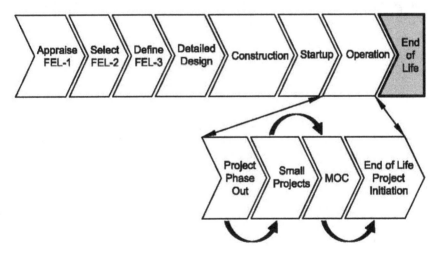

Figure 11.1. End of Life

Once the decision has been taken to decommission, there are various options on what to do with the facility that reflect the nature of the decommissioning. These options, all of which have process safety implications, include:

- Mothballing, i.e. process unit or equipment *may* be potentially re-commissioned at a later date,
- Deconstruction, i.e. process unit is dismantled and individual items of equipment may be re-used,
- Demolition, i.e. process unit is essentially destroyed for scrap and potential material recycling.

Irrespective of which option is adopted, decommissioning should be treated as a project in its own right. Thus, many of the issues discussed in earlier chapters are equally applicable. Decommissioning a facility in the chemical, oil and gas, pharmaceutical, and similar process industries can be challenging, due to the nature of the hazardous materials handled in these facilities. A rigorous approach embracing careful planning, hazard evaluation, and risk management throughout all phases of decommissioning are particularly important to avoid injury and environmental damage.

Project Management Team

The PMT is likely to be different from the team that managed the design, construction and commissioning of the facility when it was new. Many companies have leadership and engineers skilled in designing and building new facilities, but few have expertise in decommissioning. Most companies therefore require consultants and contractors experienced in decommissioning to assist and/or manage all stages of the project from confidential pre-project strategic studies to deconstruction/demolition and site remediation.

The PMT also requires some support from the site operations team, including production, engineering, maintenance, EHS and process safety departments, as their expertise and knowledge of the facilities is invaluable. The site administration personnel may also be required to advise and handle regulatory permits, financial and taxation, HR and community issues.

The PMT's primary focus should be on safely decommissioning, and removing if appropriate, a collection of vessels, tanks, pumps, compressors, valves, piping, controls and structures, while doing so within cost and schedule. Typical project objectives for decommissioning include:

- No injuries or process safety incidents,
- No environmental non-conformances,
- Facilities handed over to decommissioning contractor in decontaminated and safe condition,
- Completing project on budget and schedule,
- Maximizing value of any reusable items of equipment and recyclable materials,
- Site restoration and successful implementation of remediation measures,
- Compliance with regulatory and company requirements.

Environment, Health & Safety

From an EHS perspective, the EHS risks of decommissioning should be identified, understood, and managed to reduce risk. A project EHS plan should be developed and implemented including robust EHS procedures and an emergency response plan suitable for decommissioning activities.

Process Safety

The key process safety objectives during decommissioning include:

- Process Safety Plan to address preparedness for decommissioning,
- Identification and evaluation of decommissioning hazards, and understanding associated risks,
- Procedures and practices to manage decommissioning risks,
- Competent decommissioning workforce,
- Asset integrity management during late-life operation,
- Credible emergency response plan for decommissioning,
- Operational readiness of mothballed or re-used equipment.

These and other process safety activities during decommissioning are discussed below for:

- Design for Decommissioning (Section 11.1)
- Planning for Decommissioning (Section 11.2)
- Decommissioning Procedures (Section 11.3)
- Deconstruction and Demolition (Section 11.4)
- Process Safety for Decommissioning (Section 11.5)
- Other Project Activities (Section 11.6)

11.1 DESIGN FOR DECOMMISSIONING

Historically, little if any thought was given to decommissioning at the end of a facility's life cycle when the original design was developed. With many of the fixed platforms installed in the 1970's and 1980's in the North Sea, Gulf of Mexico and other offshore areas aging, making provision for their eventual decommissioning has assumed increasing importance for their Operators. Decommissioning offshore installations is governed by national and local regulations, but the industry faces a substantial technological challenge in their removal. Ease of decommissioning wasn't prioritized during their design over 30 years ago, decommissioning costs are reportedly rising, and significant safety precautions are required when working on older offshore installations and pipelines. The problem isn't unique to the offshore industry, as design of decommissioning issues for new chemical plants also has the potential to reduce life cycle costs, lower risk, and reduce safety and environmental impacts.

Today there is a growing recognition that design engineers need to incorporate the end of an offshore platform's life into the early stages of platform design. However, design for decommissioning is not widely embraced by onshore process industries, and it is left to individual client companies to emphasize and specify the requirement to engineering and construction contractors. Primary concerns are de-inventorying the plant, structural integrity during dismantling/removal, and process safety and EHS risks, which can be eliminated or reduced by selection of appropriate facility design. Risk management and the other process safety considerations discussed below, if considered at the concept and engineering design stages of a project, would significantly reduce risks during decommissioning.

11.2   PLANNING FOR DECOMMISSIONING

Decommissioning involves many of the hazards associated with construction (see Chapter 7), but involves additional hazards due to unknown factors that make the work potentially more hazardous. These unknown factors may include:

- Changes from the facility's design introduced during construction that may or may not have been approved (i.e. Project change management system),
- Subsequent modifications that altered the original design that may or may not have been approved (i.e. Operator's management of change system),
- De-inventorying and disposing of all process fluids, catalysts, and other materials,
- Residual hazardous materials within process vessels, piping, insulation, and structural members, such as process chemicals, asbestos, lead, heavy metals, etc., requiring special handling,
- Unknown strengths or weaknesses of construction materials due to aging,
- Hazards created by the tasks necessary for the decommissioning methods used (e.g. hazards to adjacent process units, third party installations, and local community).

These decommissioning tasks are often challenging and can include, but are not limited to:

- Handling, storage and disposal of hazardous materials,
- Handling and storage of explosives,
- Removal of heavy equipment and structures,
- Integrity of partial and/or damaged structures,
- Working at height,
- Working near/over water,
- Presence and/or removal of overhead/underground/subsea pipelines and utilities.

These and other issues require careful planning in order to perform decommissioning safely and efficiently, and the workforce must be fully knowledgeable of the hazards and the appropriate safety measures to mitigate the hazards. Demolition, rather than deconstruction / dismantling, has the potential for even greater hazard and proper planning is essential to avoid incidents and injuries.

The PMT should also observe corporate policies, standards and practices, including process safety and EHS, when planning facility decommissioning. Appendix E is an example of a site-specific decommissioning checklist / questionnaire that can be used during planning and execution of decommissioning tasks.

Case Study: Oil Refinery Decommissioning

An oil refinery went into liquidation, shutdown, and laid off most of the workforce. Eventually a decommissioning team was appointed, but 12 months after production ceased and in the absence of any maintenance, the condition of the process units was unknown. Steam and power services were out of commission. An extensive flare system, known to have heavy pyrophoric deposits, was under a nitrogen blanket, but posed a fire risk if the nitrogen was not maintained.

A detailed work program was developed and implemented to safely decommission the refinery after thorough planning, including an engineering survey, hazard identification and risk analysis, provision of temporary utilities, and design of temporary piping (0.9 mile) for decontamination cleaning of process equipment. The crude oil atmospheric and vacuum fractionation towers were successfully cleaned and declared free of pyrophoric material, isolated from the flare system, and opened to atmosphere.

11.2.1 Engineering Survey

As part of the planning exercise, an engineering survey should be conducted by a competent person(s) to identify potential hazards and thoroughly evaluate the condition of structures and buildings, and the possibility of an unplanned collapse. The impact of deconstruction and/or demolition on surrounding facilities should also be evaluated. Recommendations for safeguards to prevent incidents should be included.

The survey will require original construction and structural drawings including isometrics, and design information and calculations, if available. Some facilities may not have accurate drawings due to poor or non-existent management of change practices. Documents may also have been lost or mislaid. Under these circumstances it is essential to walk the facility and revise drawings to reflect as-built conditions. Operations personnel who have worked at the site for many years may be able to assist. However, site manning levels are likely to drop rapidly as the project progresses, so this support could be short lived. It is also likely that staff motivation may be poor if the facility is closing. While detailed facility knowledge is indispensable for decommissioning projects, it may not be readily available without incentives for experienced personnel to remain to the end of the project.Existing damage to the subject facility structures should be identified as this may impact the original design integrity. Existing damage to nearby structures should be documented, supplemented by photography, as this may influence the choice of decommissioning method, and mitigate potential liability if a nearby structure is inadvertently damaged further.

A typical content for an engineering survey report is illustrated in Table 11.1.

Table 11.1. Typical Content of Engineering Survey Report

Typical Content*
Structure / building characteristics • Construction type • Structure size • Height / number of stories • Structural hazards • Confined spaces / basements • Bracing / wall tie locations • Shoring requirements for adjacent structures • Type of shoring & location
Protection requirements for adjacent structures
Decommissioning method(s) to be used • Demolition, deconstruction /dismantling, and/or mothballing • Explosives handling / storage • Cutting (air-arc, mechanical, etc.) • Heavy lifting, toppling • Temporary support requirements, etc.

Typical Content*
Security / workforce and public protection
• Barricades/fencing and personnel access control
• Warning signs
• Relocation / protection of pedestrian walkways or roadways
• Lighting and housekeeping
• Special controls or procedures if portion of structure is occupied

* Additional content may be required dependent upon project and site specific circumstances

National/local regulations and/or industry standards may set requirements for certain decommissioning methods, and thereby influence the selection of decommissioning method. For example, the following apply in North America:

- *Explosives and Blasting Agents*, 29 CFR 1910.109, Occupational Safety & Health Administration, Department of Labor, USA, (OSHA 1972, and subsequent amendments).
- *Safety and Health Program Requirements for Demolition Operations*, ANSI/ASSE A10.6 - 2006, American National Standards Institute, USA.
- *Safety Requirements for Transportation, Storage, Handling and Use of Commercial Explosives and Blasting Agents*, ANSI/ASSE A10.7-2011, American National Standards Institute, USA.
- *Safety Nets Used During Construction, Repair, and Demolition Operations*, ANSI A10.11-1989 (R1998), American National Standards Institute, USA.
- *National Guidelines for Decommissioning Industrial Sites*, CCME-TS/WM-TRE013E, Canadian Council of Ministers of the Environment, Canada.

Other international regulations and standards include:

- ISO 7518:1983 *Technical drawings - Construction drawings - Simplified Representation of Demolition and Rebuilding*, International Organization for Standardization, Switzerland.
- *The Construction (Design and Management) Regulations* (CDM, 2015), Statutory Instruments, 2015 No. 51, Health And Safety, UK.
- *Code of Practice for Full and Partial Demolition*, BS 6187:2011, British Standards Institute, UK.

11.2.2   Hazard Evaluation

Once the engineering survey has defined the decommissioning method, HIRA studies, such as HAZID or What If/checklist, should be performed to identify hazards and appropriate safeguards to avoid adverse impacts to people, adjacent property, and the environment. The HAZOP methodology is not ideally suited to decommissioning work as it is focused on deviations from design intent of process systems depicted on a P&ID. There may be limited HAZOP application to temporary lines and connections necessary for de-inventorying and decontamination purposes.

Irrespective of the HIRA methodology employed, the PMT may need to develop additional guidewords that are more relevant to decommissioning tasks. A similar checklist of guidewords may be required to supplement the issue of work permits. Key issues to address are hazardous materials likely to be present, deconstruction and/or demolition methods and procedures, and other inherent site specific hazards, such as use of explosives, cutting equipment, cranes, movement of heavy machinery/vehicles, temporary systems, recontamination, etc.. Some hazardous materials may pose fire, explosion, toxicity and other health hazards. The HIRA studies should be documented including actions taken to resolve recommendations.

Hazards with potentially significant consequences should be compiled in a risk register with an individual assigned to develop risk reduction options for the PMT to periodically review.

Further guidance on decommissioning hazards is available from the following publication: *Demolition Man, Expert Observations of Demolition Dangers and How to Avoid Them* (IChemE 2018).

Further guidance on HIRA studies is available from the following CCPS publication: *Guidelines for Hazard Evaluation Procedures*, 3rd edition (CCPS, 2017).

11.2.3   Hazardous Materials

Most hazardous materials should be identifiable prior to commencement of decommissioning. However, there is always a risk that another hazardous material is found during work, and plans should include safety personnel alert to the possibility.

Aging facilities may include the presence of:

• Asbestos as insulation, fireproofing and in some building materials,
• Polychlorinated biphenyls (PCBs) in electrical equipment (e.g. capacitors and transformers), heat transfer systems, and some coatings.
• Lead in some pipe systems and lead-based paints.

Other hazardous materials that may be present include, but are not limited to, process fluids (e.g. hydrocarbons, reactive chemicals, acids, alkalis, etc.), heavy metals (e.g. mercury, chromium, arsenic, vanadium, etc.), naturally occurring radioactive material (NORM), waste streams, and other health and environmentally sensitive materials.

Plans should be developed for proper de-inventorying, removal, safe handling, and disposal of all hazardous materials present. National regulations may control handling and disposal of some materials. Process vessels should be cleaned to remove residual materials.

11.2.4 Process Safety Plan

A process safety plan should be developed that is tailored to the specific decommissioning activities to be undertaken (Appendix B). This plan may be combined with the EHS plan, and should cover, but not limited to, the following:

- Ongoing HIRA studies, especially SIMOPS, in event of task and/or sequence changes,
- Engineering method statements and associated task risk assessments / JSAs,
- Handover of areas/systems to demolition contractor,
- Step-by-step decommissioning procedures,
- Safe work practices, including work permits,
- Specialized dismantling equipment (e.g. crane, ROV, etc.),
- Safety equipment,
- Certification (workforce, equipment),
- Workforce orientation/training,
- Site perimeter barricades/fencing and warning signs,
- Safety oversight,
- Fire prevention and protection,
- Incident reporting and investigation,
- Emergency response plan.

11.2.5 Utilities

The decommissioning contractor will require some utilities, such as electricity, water, sewer and telecommunications, even if the facility is being totally deconstructed/demolished. Decommissioning plans should take this into account or provide alternative temporary supplies. Lines should be identified, protected, shut-off and/or relocated before work commences. In the USA, the National Association of Demolition Contractors recommends color-coding utility lines: green if a line is to be removed, and red if not.

Underground lines/cables are less obvious, and drawings may not be available or to as-built standard. Nevertheless, underground utilities should be identified using the local one-call system or equivalent and marked. Overhead utility lines should also be shut-off and isolated or protected.

11.2.6 Re-Engineering

Some existing systems may need to be re-engineered or replaced by new systems to allow decommissioning to progress safely. For example, a large offshore oil platform can take years to decommission, and may require fuel supply changes (diesel instead of production gas) and installation of associated storage tanks and piping for power generation and firewater pumps. Other changes could involve alternative firewater pumps (diesel instead of electric), and increased demand for service water, HVAC, and platform cranes. Consideration needs to be given to the reliability and capacity of existing generators, pumps and cranes, and whether replacement is necessary.

New systems, although temporary and likely to be skid-mounted, must be reliable and their design fit-for-purpose. These changes will need to be managed through the project's change management process (see Section 11.5.10 below), and may involve deviations from existing company standards that are based on a much longer asset life. Such standards may not be cost-effective for temporary operations.

11.3 DECOMMISSIONING PROCEDURES

Detailed step-by-step procedures need to be developed for decommissioning each item of equipment ranging from above and below-ground storage tanks to tall fractionation towers and structures. The procedures should reflect the decommissioning method(s) identified in the engineering survey, engineering method statements, and safeguards recommended in HIRA studies. A task sequence should also be developed with the aim of facilitating easy access to each item of equipment to be deconstructed or demolished.

Decommissioning a large facility may take several years to complete. For example, a large oil refinery or offshore oil platform may take up to five years from stopping production until the facility is deconstructed/demolished and physically removed. In the case of an offshore platform, it may take even longer to remove associated subsea equipment.

Each facility is different, but typically decommissioning procedures should as a minmum address the following stages of varying durations:

- Late-life operations,
- Cessation of production,
- Cleaning and decontamination,

- Deconstruction and/or demolition,
- Removal and remediation.

Some jurisdictions tightly regulate permitting, waste disposal, and site remediation for these stages, and therefore plans and procedures should address compliance with requirements for abating environmentally sensitive process systems.

The procedures need to address many tasks including, but not limited to: depressurization, deinventorying, isolation, purging, inerting, cleaning, decontamination, deconstruction/dismantling, demolition, and removal. These activities are discussed below.

11.3.1 Late-Life Operations

During late-life operation, there may be an opportunity to progressively deinventory some feedstock, intermediate process streams and products in preparation for final decommissioning. For example, an oil refinery may change to sweet crude oil to allow shutdown of sulfur recovery units and associated amine systems, or a chemical plant may decommission certain pipelines and/or storage tanks and transfer the line contents/tank bottoms to other tanks for processing or export. Following deinventorying, the equipment should be isolated, purged and cleaned. The equipment may also need to be inerted and gas-tested if hot work is necessary for removal.

Normal operations and maintenance should be maintained for other equipment, although there may be opportunities to relax some preventive maintenance (PM) tasks for non-critical equipment. PM tasks for safety-critical equipment should not be deferred or stopped. PM tasks should also be maintained for any equipment that may be mothballed and/or re-used.

Some equipment may be required to be functional during decommissioning and PM tasks should maintain its reliability. For example, effluent treatment systems will be required to handle disposal of chemicals within discharge consents, but eventually some residual fluids (e.g. cleaning streams) may require offsite disposal to regulated waste disposal sites to avoid breaching discharge consents. Planning should include (i) sampling residual fluids to ascertain compositions, and (ii) attempting to minimize cross-contamination and volumes of cleaning fluids to reduce waste disposal costs.

Further guidance on aging plant considerations is available from the following CCPS publication: *Dealing with Aging Process Facilities and Infrastructure* (CCPS, 2018d).

11.3.2 Cessation of Production

Shutdown procedures for process facilities that will never re-start are unlikely to be the same as normal shutdown or procedures for making equipment available for maintenance/turnaround. Every effort should be made to progressively process all (or as much as possible) feedstock and intermediate streams prior to ceasing production. Systems should be depressured to flare. Management of change and temporary operating procedures may be required for portable diesel-driven pumps and flexible hoses to recover fluids from the bottom of various storage tanks and pressure vessels, and from low points in pipelines. All remaining process fluids that are not processed need to be removed from process vessels and tanks, and disposed of properly.

11.3.3 Cleaning and Decontamination

Following deinventorying, equipment should be isolated, purged and cleaned. Positive isolation of all systems and interconnections is normally required for equipment containing hazardous materials. Some equipment may need purging with steam, nitrogen, and gas-tested after inerting if hot work is necessary for removal. Positive isolation, preferably with an air gap between vessels, is essential to eliminate fire and explosion hazards in connected equipment. Large storage tanks sometimes present a problem in gas-freeing and may require lengthy purging and ventilation. Any asbestos present should be removed before other decommissioning activities commence.

Cleaning and decontamination may be as simple as flushing with water (and detergent in some cases) and disposal to the effluent system. However, some equipment may require heating, chemical treatment and/or mechanical cleaning to avoid personnel exposure, and remove undesirable chemicals from equipment to be salvaged for potential re-use. For example, some systems may need neutralizing acids or alkalis, and/or careful removal and disposal of deposits/sludge containing heavy metals or NORM. Vessels and piping may retain some naturally occurring radioactivity after cleaning and should be identified.

Decontamination should proceed on a system by system basis to avoid cross-contamination, and progress in phases to achieve the desired 'clean' state. Typically the initial chemical cleaning phase is followed by water flushing, and then steam, nitrogen, and compressed air. Between each phase, samples of the residual contamination are analyzed. If contamination is still high, the phase is repeated. The next phase proceeds when contamination levels meet acceptable criteria.

11.3.4 Mothballed Facilities And Equipment

For various reasons a facility or equipment may be taken out of service with the possibility of future use. These so called 'mothballed' facilities and equipment require special decommissioning procedures including depressurization, deinventorying and cleaning, plus additional measures for preservation and any ongoing inspection, testing, and preventive maintenance (inspection) tasks that need to be performed to maintain integrity and a state of near-readiness. These measures may include maintaining a proper internal atmosphere (e.g. dry nitrogen) to prevent corrosion, and should be labelled to warn the workforce of unsafe atmospheres inside vessels and other equipment.

Some facilities have an area where items of equipment that have been removed (or are surplus to current requirements) are stored for possible use at a later date. These 'boneyards' present similar asset integrity management challenges, and design and inspection documentation should be retained for these items of equipment.

Some mothballed facilities and boneyard equipment never start up again, and eventually progress to permanent decommissioning. However, if a mothballed facility or equipment does start up at a later date, special recommissioning procedures should be prepared and ITPM tasks completed to ensure asset integrity. The facility or equipment should also be subject to an operational readiness review prior to startup.

11.4 DECONSTRUCTION AND DEMOLITION

Following cleaning and decontamination, the facility is ready for deconstruction and/or demolition. Deconstruction and demolition activities require a variety of specialized equipment including, but not limited to: cranes, man lifts, trackhoes, trackloaders, reinforced buckets and trailers, cutting equipment (hydraulic shears, torches, plasma arc, air arc, etc.), grapples, magnets, and concrete processors. These and other demolition equipment require experienced workers for safe operation. Deconstruction and demolition share some common hazards requiring rigorous implementation of some elements of RBPS, which are covered in subsequent sections below.

11.4.1 Deconstruction

Although deconstruction may appear to be the reverse of construction, the task sequence is likely to be different to merely reversing the original construction sequence and involves numerous additional hazards (IChemE 2018). It involves dismantling and segregating potential equipment for re-use and materials suitable for recycling. Therefore deconstruction is more labor intensive and likely to be more time consuming than demolition. It is the preferred means of decommissioning when one process unit close to adjacent units is at its end of life and has to be removed.

Equipment that is dismantled should be checked internally for the presence of contaminants. If samples do not meet acceptable criteria for contaminants when analyzed, the equipment should be cleaned again with chemicals, water, and/or steam until acceptable criteria are achieved.

Following this secondary decontamination, the equipment may be inspected and tested to determine its value for future re-use, and, if appropriate, disassembled into its component parts. If the equipment is declared fit for re-use it may be sold or used at another company facility.

Case Study: Decommissioning of Process Unit

A process unit with a very tall fractionation tower was decommissioned while surrounded by other in service process units. Although the contract specified dismantling, the decommissioning contractor requested permission to topple the fractionation tower. The plant manager and work permit authority refused to give permission, and emphasized the contract requirements.

The following weekend when the day staff were not present, the contractor collapsed the tower without any work permit to do so. The tower fell across a plant road that had not been barricaded to prevent the passage of vehicles and pedestrians. Fortunately no one was injured, and the other process units survived the vibration shock without disruption to operations or damage. However, an oily water sewer was damaged. In slightly different circumstances, the consequences could have been much more severe. Plant management were called in by shift staff and disciplined the decommissioning contractor.

Equipment that may be re-used elsewhere should be removed, labelled, segregated, inspected, and stored with appropriate preservation measures that may involve ongoing ITPM tasks until disposal. Design information and inspection records for this equipment should be retained. Packing and additional preservation measures may be required for transportation. Further guidance on re-using process equipment is available from the following CCPS publication: *Guidelines for Asset Integrity Management* (CCPS, 2017).

Following completion of all required dismantling operations, the remaining structures may be demolished.

11.4.2 Demolition

Demolition is less labor intensive than deconstruction, and involves a significant amount of heavy equipment, specialized machinery and recycling equipment. The percentage of materials suitable for recycling are reportedly generally similar for demolition and deconstruction methods at up to 90%, which can be used to offset the cost of decommissioning.

Most demolition activities involve use of low energy cutting with hydraulic shears and heavy equipment for excavating underground pipelines, and cutting and razing structures. A long-reach trackhoe/excavator fitted with hydraulic shears allows steel piping, tanks, and structures to be cut into small pieces safely with the operator protected inside the cab. Air arc (a.k.a. air carbon arc) cutting and other hot work cutting equipment may also be used. Monitoring for the presence of decontaminents, especially flammable vapors during storage tank demolition, is important even when using hydraulic shears. Some projects also require monitoring for explosive dust.

Use of explosives to topple or implode structures is generally a relatively minor fraction of the demolition process. However, it can be effective and timesaving, although it requires a competent person and care not to damage any adjacent facilities, sewers, and other underground utilities that may be required for decommissioning. Explosives are unlikely to be suitable for selective demolition of a portion of an operating facility, e.g. a single process unit surrounded by other units.

A competent person should conduct ongoing inspections during demolition to identify any hazards due to weakened process equipment, structures, and buildings. Safety personnel should also provide continuous oversight to identify other hazards, and ensure work permit requirements and other safety measures are in place. Housekeeping to remove demolished steel, concrete, and other debris should be monitored to keep walking surfaces and other work areas clear.

11.5 PROCESS SAFETY FOR DECOMMISSIONING

Implementation of a process safety management system is essential to ensure that decommissioning activities, whether deconstruction or demolition, are conducted safely without harm to people, property or the environment. While this system should be built around hazard identification, safe work practices, and detailed decommissioning procedures, a number of other process safety elements are also important, as discussed below.

11.5.1 Contractor Management

As previously discussed, most companies do not have expertise in decommissioning activities, and so deconstruction/demolition of a process facility should start with selection of a contractor(s) with the required competency and experience, and resources (workforce, specialized equipment) necessary for the project. The contractor(s) competencies should ideally include, but not limited to:

- Conduct of engineering surveys,
- Preparation of as-built drawings,
- Knowledge of how to decontaminate, dismantle, and demolish the relevant types of facilities, structures and buildings (e.g. chemical process plant, steel tanks, pipelines, transformers, etc.),

- Knowledge of how to safely decontaminate unusual hazards (e.g. asbestos, heavy metals, NORM, etc.),
- Preparation of deconstruction and demolition procedures,
- Performing deconstruction and demolition tasks safely, including operating heavy/specialized equipment,
- Dismantling, segregating and tracking equipment for re-use/sale,
- Recovering, segregating and selling recyclable materials,
- Generation, storage and proper disposal of liquid and solid waste,
- Environmental remediation of surface and sub-surface contamination.

A pre-mobilization meeting should be held with contractor leadership to discuss EHS and process safety expectations, rules and procedures. When mobilizing on site, the decommissioning work crews including any sub-contractors, should receive orientation training when they first access the site. Thereafter, work crews should be briefed daily on the hazards of their work and any hazards adjacent to the job site. This may be accomplished at pre-job toolbox meetings, participation in developing JSAs, or other means. Regular safety meetings should reinforce procedures, and share lessons learned from any incidents that occur.

An adequate number of safety specialists and deconstruction/demolition supervisors employed by the project and contractor(s) should maintain constant viligance around the site(s) to ensure that contract workers perform their jobs safely, and that contracted services do not add to or increase risks. A key aim should be that contractor vehicles and heavy equipment meet project's safety standards, are maintained in safe working order, and are operated by competent operators at all times. There should be a culture of zero tolerance for not following safety policies, rules and procedures.

The PMT should regularly review contractor(s) performance in meeting the EHS and process safety expectations, rules, and procedures, and rapidly intervene if performance improvement is required. This management review process should also ensure that the contractor(s) is complying with contract conditions.Further guidance on contractor management is available from the following CCPS publication: *Guidelines for Risk Based Process Safety* (CCPS, 2007).

11.5.2 Safety Culture

At the commencement of decommissioning activities, project management and contractor leadership should establish a positive environment where contractor employees at all levels are committed to safety. Conduct of operations is closely related to safety culture, and leadership should set expectations for deconstruction/demolition tasks to be carried out in a deliberate, careful, and structured manner that follows EHS and process safety procedures. Managers should set a personal example, ensure that workers perform their tasks properly, and enforce high standards if deviations from expected performance occur.

Further guidance on safety culture and conduct of operations is available from the following CCPS publications: *Essential Practices for Creating, Strengthening, and Sustaining Process Safety Culture* (CCPS, 2018e); *Guidelines for Risk Based Process Safety* (CCPS, 2007).

11.5.3 Workforce Involvement

The broad involvement of the workforce in reporting hazards and improving decommissioning activities can assist in driving a positive safety culture. Leadership should listen to workforce concerns, and make sure that lessons learned by the people closest to the deconstruction/demolition are considered and addressed.

This can be a time of great concern for some members of the workforce whose employment is directly linked to the process unit(s) being decommissioned. As such, their focus may not be on the job at all times, and it is important that managers, supervisors and safety personnel regularly walkabout during decommissioning to observe and motivate employees.

Further guidance on workforce involvement is available from the following CCPS publication: *Guidelines for Risk Based Process Safety* (CCPS, 2007).

11.5.4 Stakeholder Outreach

Before and during decommissioning, the PMT should hold regular meetings with their key stakeholders to keep them informed, understand concerns, seek alignment, and attain regulatory approval (e.g. permits) in order to smooth the progress of deconstruction and demolition. For example, the local community may have concerns relating to disturbance due to noise and heavy traffic to/from the site on a daily basis. In addition to the decommissioning workforce's vehicles, there is likely to be a large number of trucks disposing of equipment for re-use, recyclables, and liquid and solid waste. Route planning should attempt to avoid disturbance to the local community. Key project stakeholders are likely to include the local community, regulatory agencies, emergency services, employees, unions, partners, and contractors.

Further guidance on stakeholder outreach is available from the following CCPS publication: *Guidelines for Risk Based Process Safety* (CCPS, 2007).

11.5.5 Hazard Evaluation

Decommissioning activities, such as chemical cleaning, working at height, working near/over water, heavy lifts, hot work, confined space entry, use of explosives, excavation, and use of multiple vehicles and mobile machinery, involve many hazards. Simultaneous activities in close proximity to one another add further complexity, and a SIMOPS study should be conducted to identify and manage potential interactions.

Project's initial HIRA studies during the planning phase may have already identified some of the decommissioning hazards and recommended safeguards. However it is unlikely that HIRA studies recognized all deconstruction/demolition hazards. The project should ensure that safe work practices are rigorously implemented, and each work permit should be supported by a task hazard assessment, such as a job safety analysis (JSA). The JSA (a.k.a. job hazard analysis (JHA) and task hazard analysis (THA)) should involve the work crew and preferably a safety specialist (e.g. process safety, occupational safety, industrial hygienist, etc. as appropriate for the specific tasks), identify potential hazards at each step of the permitted job, and determine safeguards to manage the hazards.

All hazards and required safeguards must be communicated to the relevant job crew(s) including any hazards adjacent to the job site. Work crews should also report hazards and unsafe conditions to their supervisor for the attention of project management.

Further guidance on HIRA is available from the following CCPS publications: *Guidelines for Risk Based Process Safety* (CCPS, 2007); *Guidelines for Hazard Evaluation Procedures*, 3rd edition (CCPS, 2008).

11.5.6 Safe Work Practices

Rigorous enforcement of safe work practices is critical. All decommissioning work crews are responsible for following the approved safe work practices that may be regulated and/or required by the client/project. The client/project may require more stringent practices than local regulations.

The safe work practices may cover, but not be limited to:

- Site access control
- Work permitting
- Hot work (air arc and plasma cutting, grinding, naked flames/sparks, etc.)
- Energy isolation (LOTO), including underground utilities,
- Line breaking
- Working at height (scaffolding, man-lifts, fall protection, etc.)
- Excavation (underground cables/pipes, shoring, sloping, etc.)
- Confined space entry (including excavations, sumps, sewers, etc.)
- Heavy lifts (cranes, lift plans, signalers, forklifts, etc.)
- Electrical systems (high voltage, overhead/buried cables, etc.)
- Vehicles and mobile heavy machinery (excavators, trucks, banksman, etc.)
- Hazard communication (SDS, chemical cleaning, etc.)
- Working near/over water,

- Unusual hazards (e.g. asbestos, PCBs, heavy metals, naturally occurring radioactive materials (NORM), etc.),
- Handling and storage of explosives,
- Waste disposal,
- Personal protective equipment (PPE), including eye, face, head, hand, foot, respiratory, hearing, fall protection, etc.

Every member of the decommissioning workforce will require some form of orientation training in the detailed safe work practices, emergency response plan, and critical safety rules to be employed on the site. Thereafter leadership should establish daily monitoring and periodic auditing (by a number of project supervisors and safety specialists) to ensure that the safe work practices are being implemented, and, if not, intervene to enforce their implementation. Repeated failure to follow approved practices and procedures should be subject to disciplinary action including dismissal.

It may be appropriate to issue 'blanket work permits' in certain areas (e.g. fenced area under control of the decommissioning contractor), where the only hazards are associated with the contractor's deconstruction/demolition heavy machinery and vehicles. It may be appropriate to have a formal handover process to the contractor for areas or systems that the Operations and/or Project teams have isolated and deinventoried. Any handover process employed should include documentation on process chemicals, process equipment, and drawings. Blanket permits may be renewed on a regularly providing the hazards have not changed. JSAs should support each work permit. Daily toolbox meetings should be held to cover the day's job tasks, hazards, required safeguards, and adjacent activities. Permits, JSAs and meetings should be communicated in the workforce's native language(s).

Project safety specialists and those employed by contractor(s) should maintain high safety standards, including good housekeeping and enforcing exclusion zones behind barriers for work such as heavy lifting and excavation. A competent person should prepare a detailed lifting plan for each heavy lift to manage its hazards and risks.

Further guidance on safe work practices is available from the following CCPS publication: *Guidelines for Risk Based Process Safety* (CCPS, 2007). US OSHA and UK HSE also provide guidance through their websites.

11.5.7 EHS and Process Safety Procedures

Some EHS and process safety procedures may be mandated by local regulations, while the client and/or project may require a higher standard and additional procedures to meet their EHS and process safety expectations. These procedures may cover any of the elements discussed in this Section 11.5, but the most important that should be required by all decommissioning projects are:

- Hazard evaluation (Section 11.5.5)
- Safe work practices (Section 11.5.6)
- Asset integrity management (Section 11.5.9)
- Change management (Section 11.5.10)
- Emergency management (Section 11.5.12)
- Incident reporting and investigation (Section 11.5.13)

Each project should carefully determine whether any other EHS and process safety procedures are relevant to their decommissioning activities.

11.5.8 Training and Competence Assurance

While the contractor and sub-contractor organization(s) would typically be selected on the basis of their competency and capability, sometimes a contractor's resources become over-stretched and/or lose key personnel. The project should ensure that the contractor(s) have the skills and resources necessary to perform their scope of work. Review of craft skill certifications, audits, and less formal interviews can verify whether the mobilized resources have the necessary skills and experience. Any deficiencies discovered should be addressed with the contractor(s) concerned, and could have contract consequences.

In addition to contractors being responsible for providing trained and competent work crews, the project should ensure that each contract employee (including sub-contractors) receives some form of orientation training appropriate to their job tasks before accessing the construction site(s). This orientation training should cover, but not be limited to:

- Client/project process safety and EHS expectations,
- Site safety rules,
- Site safe work practices (see Section 11.5.6 above),
- Site emergency response plan (see Section 11.5.12 below),
- Site specific hazards associated with decommissioning tasks.

In rare circumstances, a project may decide to provide additional training, especially when it is necessary to employ a less experienced contractor.

Further guidance on training and competence assurance is available from the following CCPS publications: *Guidelines for Risk Based Process Safety* (CCPS, 2007); *Guidelines for Defining Process Safety Competency Requirements* (CCPS, 2015).

11.5.9 Asset Integrity Management

Potential re-use of some equipment was briefly introduced in Section 11.4.1 above. In order to verify if items of equipment are suitable for sale or re-use, they should be inspected, which may involve disassembly, after cessation of production, cleaning/decontamination, and removal. Thereafter, ITPM tasks are likely to be necessary to preserve, and maintain quality and integrity, while the equipment is in storage until sale or re-use. Decommissioning a large facility may take an extended duration, and manufacturer's preservation recommendations should be observed to avoid degradation due to age-related mechanisms, such as corrosion, fatigue and embrittlement.

Prior to re-use various quality management practices (e.g. management of change, recommissioning procedures) are likely to be required to verify that the used equipment is suitable for the new service. As a minimum, consideration should be given to:

- Length of time the equipment was out-of-service/in storage,
- Program of ongoing ITPM tasks while the equipment was in storage.

Further guidance on aging plant considerations is available from the following publications: *Dealing with Aging Process Facilities and Infrastructure* (CCPS, 2018d); *Plant Ageing, Management of Equipment Containing Hazardous Fluids or Pressure* (HSE, 2006c); *Managing Ageing Plant: A Summary Guide* (HSE, 2010b).

Further guidance on asset integrity management and re-use of equipment is available from the following CCPS publication: *Guidelines for Asset Integrity Management* (CCPS, 2017).

11.5.10 Change Management

Some changes during decommissioning can be expected. Some changes could be due to discovering previously unidentified hazards or unavilability of certain demolition machinery (e.g. breakdown). Whatever the reason, all changes to decommissioning plans and procedures should be subject to a change management process, similar to that for the construction stage (see Chapter 7, Section 7.4.15). The primary focus should be on understanding and managing risks, and approval of the change at the appropriate line management level.

Further guidance on management of change is available from the following CCPS publications: *Guidelines for Risk Based Process Safety* (CCPS, 2007); Guidelines for the *Management of Change for Process Safety* (CCPS, 2008).

11.5.11 Operational Readiness Review

Some process companies have instituted a 'safety stop' before certain critical and/or hazardous decommissioning tasks to substantiate that risks have been properly assessed and mitigated. The practice is similar to the 'go/no go' decision-making prior to startup of a chemical plant, and relies on development of an appropriate What If checklist. An example of a checklist for fluid transfers in illustrated in Table 11.2.

Table 11.2. Example of Safety Stop Checklist[8]

What If / Checklist Questions for Fluid Transfers
Are the conditions (inerting, earthing, etc) correct for the transfer?
What is the consequence of any mixing which might take place?
Is the resulting mixture safe (thermochemistry, flammability, combustability, etc.) and acceptable for disposal?
Can effluent streams be safely sent to drain and will they be compliant with site discharge consents?
What other transfers may be in progress at the same time?

11.5.12 Emergency Management

The emergency response plan for the decommissioning site, and the necessary resources, should have been finalized during the pre-mobilization phase, and a table-top or emergency drill conducted during mobilization or early execution to test its effectiveness. The plan should address similar issues as for the construction stage (see Chapter 7, Section 7.4.16), plus some specific decommissioning factors, such as, but not limited to:

- Site evacuation,
- First aid and medivac,
- Fire and explosion,
- Toxic chemical release,
- Hydrocarbon/chemical spills including prevention of groundwater contamination,
- Rescue from height/confined space/water,
- Vehicle/mobile heavy machinery accident,
- Electrocution,
- Injury due to slips/trips/falls/struck by/crush,
- Unstable structures/buildings,
- Security incident (e.g. trespass).

---

[8] Dixon-Jackson, K., Lessons Learnt from Decommissioning a Top Tier COMAH Site, Symposium Series No. 154, IChemE, Rugby, UK, 2008.

A major concern at some sites could be limited resources for emergency response. For example, utilities supplies, especially firewater systems, could be isolated or inadequate at some phase of decommissioning work. Similarly, the workforce may be significantly reduced and unable to support internal fire and EMT teams. Prior to commencement of decommissioning, external agencies including medical personnel, fire department, utility companies and local authorities should be notified of the deconstruction/demolition and that their services may be required.

Another concern during decommissioning could be maintaining access around the site. Some area and road closures within the site are inevitable due to deconstruction/demolition activities, especially if explosives are being used. A system should be established to permit the emergency services to respond to any incident without delay.

Further guidance on emergency management is available from the following CCPS publications: *Guidelines for Risk Based Process Safety* (CCPS, 2007); *Guidelines for Technical Planning for On-Site Emergencies* (CCPS, 1995).

11.5.13 Incident Investigation

The project should set up a system for reporting all incidents including, but not limited to, injury, illness, fire, chemical spill, and property/vehicle damage occurring within the decommissioning site. All contractors and sub-contractors should be required to use this system to immediately report incidents.

The project should also ensure that all incidents and near-misses are investigated to identify root causes, and make recommendations to prevent recurrence. Corrective actions should be tracked to completion, and lessons learned documented and communicated to the workforce.

Further guidance on incident investigation is available from the following CCPS publications: *Guidelines for Risk Based Process Safety* (CCPS, 2007); *Guidelines for Investigating Chemical Process Incidents*, 2nd edition (CCPS, 2003).

11.5.14 Auditing

The PMT should consider periodic EHS and process safety audits to probe deconstruction/demolition activities in more depth than day-to-day safety oversight. This is particularly important for lengthy decommissioning work, i.e. a multi-year undertaking. These audits should preferably be conducted by independent auditors not assigned to the site in order to provide an objective review.

All audit findings, recommendations, and improvement opportunities should be recorded, and corrective actions tracked to closure. A follow-up audit should verify that corrective actions have resolved the original findings.

Further guidance on auditing is available from the following CCPS publication: *Guidelines for Auditing Process Safety Management Systems* (CCPS, 2011).

11.5.15 Disposal

During deinventorying, decontamination and decommissioning there is a variety of vapor, liquid and solid materials to be disposed of. Most vapors should be flared if within emission consents, but some may need to be incinerated.

Most liquids should be routed to the site effluent treatment system, but only after analysis and providing that discharge consents are not exceeded. If the characteristics of effluent and waste liquids are outside the standards specified in the consent order, they should be transported to a licensed industrial hazardous waste disposal site. These sites are able to treat (e.g. incineration, oxidation, etc.), store, and dispose of hazardous wastes (e.g. injection well, surface impoundment, etc.). The PMT and decommissioning contractor should try to minimize materials sent to the waste disposal facility, and keep a record of all shipments.

The disposal process for solid materials may consist of various options, including:

- Complete transfer of the dismantled facility to another location for re-use,
- Complete demolition of the facility and disposal for recycling and/or landfill,
- Partial sale/re-use and partial demolition/disposal for recycling and/or landfill.

Carbon steel, alloys, copper cables, concrete and other solid materials should be segregated and recycled if possible. Some solid waste may have to be transported to a licensed landfill site and/or incinerated. The PMT and decommissioning contractor should try to minimize materials sent to the landfill site, and keep a record of all shipments.

11.5.16 Remediation

The final stage of the project may involve site remediation in order to restore the site for a future use that may be different than the original decommissioned facility. Environmental remediation of surface and sub-surface contamination may require a combination of in-situ and ex-situ bio-remediation techniques to remove hazardous substances.

The decontamination of chemicals that may have leaked into the ground over many years is likely to be an expensive cost comprising groundwater protection, excavation of contaminated material, remediation, backfill and site grading. The extent of contamination may not be known until the facility has been removed and the clean-up commences.

Case Study: Decommissioning of Oil Refinery

The decommissioning and environmental clean-up of an oil refinery required extensive field activities, including asbestos abatement, hazardous waste removal and disposal (including tetraethyl lead and catalysts), facility demolition, and materials recycling. Remediation of soil and groundwater contaminated by hydrocarbons required a vapor extraction system comprising over 40 wells, air sparge blowers and compressors, and flare. Over 1300 tons of hydrocarbons were removed from beneath the refinery and destroyed.

11.6 OTHER PROJECT ACTIVITIES

In addition to the various process safety and technical activities needed for decommissioning, there are a number of other activities that support project execution. Some of these activities continue throughout decommissioning, which if lengthy, should be periodically updated. This requires good interface management between the PMT and the contractor.

11.6.1 EHS and Process Safety Plans

The EHS Plan and the Process Safety Plan, that were developed before decommissioning commenced (see Section 11.2.4), should be periodically updated to reflect any changes in hazards and/or decommissioning activities (Appendix B).

11.6.2 Risk Register

The Project Risk Register that was compiled at the planning phase (see Section 11.2.2) should be updated for any new or changed hazards/risks identified during deinventorying, cleaning/decontamination, and decommissioning (Appendix C). Individuals should be identified as responsible for developing a response plan to manage each item. The PMT should regularly review the register and response plans.

11.6.3 Action Tracking

The PMT should compile an action tracking database or spreadsheet to include all activity relating to, but not limited to, any legally binding, regulatory or contractual requirements/commitments, and recommendations from specialist studies, incident investigations, and peer reviews and other assurance processes. The PMT should also capture actions generated by their contractor(s), and ensure that all actions are progressively resolved, closed and documented.

11.6.4 General Decommissioning Management

In addition to measurement of progress and expenditure, a number of other general management activities should continue throughout decommissioning, including, but not limited to, the following:

- Administration of contractor personnel
- Control of contracts
- Regular progress meetings (typically weekly) with contractor(s)

The PMT should also keep a daily diary/logbook as a record of decommissioning progress detailing significant areas of activity. This logbook should include dates and details of: documents issued to contractor, areas and systems released to contractor, workforce numbers and equipment on site, accomplishments, pictures, labor disputes, and any weather or other delays. This information is particularly important in settling or challenging contractor claims at the completion of the contract.

11.6.5 Stage Gate Reviews

One or more stage gate reviews should be held to assess whether the project team has adequately addressed technical, process safety and EHS aspects of decommissioning. Common industry practice is to conduct independent reviews:

- after planning to assess preparations for decommissioning, and
- during the early phase of deconstruction/demolition to assess implementation of decommissioning procedures.

These reviews are broadly equivalent to the stage gate reviews at the Detailed Design (see Chapter 6 Section 6.8) and Construction (see Chapter 7 Section 7.9) stages (Appendix G). Table 11.3 is a typical scope for an End of Life Stage Gate Review. However, the review scope should be tailored for the site specific nature of the decommissioning project.

Table 11.3. End of Life Stage Gate Review Scope

Scope Item
Confirm that project plans for decommissioning are adequate
Confirm that the Operations Team is involved as necessary in preparation for decommissioning activities.
Confirm that the HIRA study(s) is complete and recommendations are being satisfactorily addressed
Confirm that appropriate specialist reviews have been carried out and their outcomes are being satisfactorily addressed, including engineering controls and checks are in place
Confirm that a Process Safety and EHS management system including a Process Safety and EHS Plan(s) is being implemented effectively
Confirm that an emergency response plan(s) has been developed and that it addresses relevant process safety risks associated with decommissioning
Confirm that Process Safety and EHS aspects have been adequately considered and are appropriate for decommissioning
Confirm that decommissioning workforce training, competency, and performance assurance arrangements are adequate and being implemented
Confirm that the decommissioning project team has a robust process to manage the interface with contractor(s)
Confirm that asset integrity management processes including quality management are sufficient to maintain structural and equipment integrity

Actions to address recommendations from these reviews should be tracked to resolution.

11.7 SUMMARY

Decommissioning a process facility, structure or equipment can be a complex process, and demands a structured approach. In fact, it should be treated as a project in its own right. There are many hazards and opportunities for process safety incidents and regulatory non-compliance. To complete a decommissioning project without incident requires the right culture and commitment of the site, project and contractor's workforce, detailed planning, rigorous hazard identification and risk analysis, competent contractors, and disciplined implementation of detailed deconstruction and demolition procedures and practices. Each facility to be demolished may have unique characteristics that require specific procedures that are likely to be much more complex than simply 'knocking it down'.

12 DOCUMENTATION

All projects, whether large or small, require, use and generate copious quantities of information. This information is both hard copy and increasingly electronic, and takes many forms including, but not limited to:

- Memoranda, letters
- Procedures and Practices
  - o Project procedures (administration, HR, etc.)
  - o Process safety and EHS procedures
- Technical
  - o Reports
  - o Specifications
  - o Drawings
  - o Codes and standards
  - o Quality certificates
- Legal and Contractual
  - o Permits
  - o Contracts (and contract amendments)
  - o Purchase orders
- Databases
  - o Action tracking
  - o Risk register
  - o Incidents

The development and assembly of this information starts early and continues throughout the project life cycle. Eventually, a significant proportion of this information has to be handed over to the Operator for the ongoing operation, asset integrity management, and future development of the facility.

12.1 DOCUMENT MANAGEMENT

Document management encompasses both the equipment and procedures required to effectively handle the vast amount of information and documentation developed by projects. Most projects have a system comprising a combination of hardcopy and electronic information, although increasingly the trend is towards electronic. Large capital projects use CAD and 3D modelling systems for design and drawings, and future advances are likely to integrate greater technical information into these software systems. Ultimately, system selection depends on the scale of the project, types of records, number of users, user preference, and existing system(s).

Nevertheless, whether paper-based, electronic or a combination, all document management systems require a good file indexing system. The format of documents needs careful consideration, especially in the case of brownfield developments where the Operator has an existing information and document management system. In this case, the format and indexing (reference numbering/coding) should be compatible with that system. Consequently, the project information and document management system requires *early* design, including hardware, procedures, and catalogue/archive format. Once the project team, or contractor on their behalf, starts collating and coding documents, it will be progressively more expensive and disruptive to change.

The documentation must also be readily available to those who need it to safely perform their jobs. In a project context, process safety information (PSI) and other essential documentation must be defined *early* in the project, i.e. the 'right information' is available at the 'right time', so that the 'right decisions' can be made in order to optimize the facility design from process safety and technical perspectives. Some information may be need to be accessed frequently by multiple engineers, while other information may be of more historical value.

Case Study: Drawings Unavailable for Plant Expansion

A major plant expansion involved new trays/packing for several distillation towers. Five years later a training simulator project required the expansion project documents and drawings for the expansion project. However, the relevant drawings were not in the plant engineering equipment files.

The original project team was contacted and confirmed that the drawings were archived. Two personnel spent several weeks searching for the drawings, and found about half of the drawings. Most of these drawings were 'approved for construction' rather than 'as-built', and were variously archived in files related to (i) contractor who installed the trays, (ii) inspection folders, and (iii) equipment folders. None of the drawings were in folders related to the tray manufacturer. When contacted, the manufacturer was able to provide their most up-to-date drawings (not 'as-built' status) for the missing drawings.

The simulator project was not delayed, but, in addition to the small cost related to the searches, in the event of an emergency (e.g. process upset or loss of performance) troubleshooting could have been compromised resulting in a product quality and/or production impact.

Many project design activities, especially in front end loading (FEL), are iterative, meaning that PSI and other documentation is subject to frequent change. Changes can also occur when new or updated regulations, and industry or corporate standards are issued. It is therefore vitally important to verify the PSI is accurate and up-to-date before its use. Project change management procedures are required to control document/drawing changes, and confirm that the latest information is always available to users who have authorized access.

Another aspect of document control relates to the need to duplicate and distribute multiple paper copies. Late design changes during construction and pre-commissioning can present difficulties when multiple superseded hardcopy documents/drawings are being used in the field. Special procedures may be required, such as controlled distributions with numbered copies, to confirm that the latest information is provided to a potentially large number of users.

The Project, and later the Operator, should establish a retention policy for every type of information. National/local regulations may set a minimum retention period for some information. In respect of the Project, much of the documentation will be handed over to the future Operator. Some of the information may also need to be retained for other purposes, such as:

- Statutory and/or possible audit, e.g. confirmation of design criteria,
- Satisfying legal liabilities and financial/tax audits,
- Reference and estimating by future projects, e.g. project close-out report/data.

The documentation that the Operator receives in the handover package will be required for the ongoing operation, asset integrity management, and potential future development of the facility. As such, most, if not all, of this information should be retained for the life of the facility. The Operator will also generate updated and additional documentation related to the Operator's process safety and EHS management system, e.g. HIRA revalidation, management of change, incident investigation, etc. This updated and additional information should also be subject to the Operator's retention policy.

Many of the PSI and other project documents required substantial effort to create, and, in the event of loss, might be difficult to re-create. Therefore, Project and Operator documentation should be protected from inadvertent loss, which could occur in various ways:

- Inadvertent/unauthorized change,
- Physical removal or misfiling,
- Environmental damage, e.g. water and smoke damage,
- Total loss, e.g. fire,

- Electronic loss from lack of back-up or incompatible computer database changes.

Hardcopy documents should be duplicated and stored in separate locations and/or protected in fire-proof safes or buildings (e.g. in accordance with NFPA 232 *Protection of Records*, 2017 edition). Electronic documents should be regularly backed up on a redundant server at a remote location. Controls should be established to protect against unauthorized change, physical removal and misfiling.

Further information and guidance is available from the following CCPS publication: *Guidelines for Process Safety Documentation*, 1995.

12.2   PROCESS KNOWLEDGE MANAGEMENT

In line with the objective of this book, the remainder of this chapter will concentrate on the process safety documentation that should be developed, assembled, and managed throughout the project life cycle.

Process knowledge is essential for understanding the hazards and risks inherent in a project. As such, PSI is the foundation for risk-based process safety, and is needed to perform, and is generated by, process safety activities at each stage of the project. Identifying hazards early in the process and developing the scope to manage the risks of these hazards is a critical step in achieving a balance between safety, capital cost, operability, and life cycle cost.

The scope and management of two projects are rarely the same, and this influences the timing, requirements for, and generation of, PSI. Appendix F contains a comprehensive list of PSI, some of which may be relevant depending upon the scope of a project.

The following discussion illustrates typical PSI at each stage of the project life cycle, but some projects may require or produce the same information one or more stages earlier or later in the project life cycle.

12.2.1   Front End Loading 1 Stage

The front end loading (FEL) 1 stage is essentially an appraisal or feasibility stage of a potential project (see Chapter 3). Until a broad range of development options in line with corporate strategy have been evaluated, the commercial viability of the project is unknown. Therefore, alternative technologies, processes, and locations are normally assessed in terms of value, risk (threats and opportunities), and uncertainty.

Process safety considerations must begin early in this conceptual stage in order to optimize objectives, such as safety and risk reduction. Each alternative development option is typically reviewed at a high level (as detailed information is unlikely to be available) to identify potential hazards and inherently safer technology. This requires a range of PSI, including chemical hazards (e.g. flammability, toxicity, reactivity), hazardous inventories, applicable codes and standards, and other data needed for preliminary HAZID and conceptual risk analysis (CRA) studies, such as operating parameters (e.g. temperature, pressure) for each alternative process, and location specific information (e.g. topography, meteorology). Table 3.1 lists various issues and information that typically should be reviewed.

Some of the development options considered are likely to be rejected, but the most promising development options identified (if any) in terms of safety, technical and commercial viability proceed to the FEL 2 stage for further development. Nevertheless, all options considered and the reasons for their selection or rejection should be documented, along with all the PSI used, assumptions made, and results obtained. Finally, a stage gate review that appraises all of the options should be reported.

12.2.2 Front End Loading 2 Stage

During the FEL-2 stage, the most promising development options are further refined and evaluated to maximize opportunities and reduce threats/uncertainties to the point where a single option is selected (see Chapter 4). Development of this option continues with creation of a preliminary development plan, including the site, facilities, and infrastructure requirements, to take forward into FEL-3.

Normally the main compilation of PSI and other documentation commences in FEL-2. While the main focus is on the single option selected, documentation on any rejected 'promising option(s)' should be retained. As the development continues, some early PSI will need to be revised and updated.

Typical documentation in FEL-2 includes, but is not limited to:

- Corporate policies, standards and practices (including process safety, EHS, document management, insurance requirements),
- National/local regulations,
- Engineering codes and standards,
- Chemical hazards (flammability, explosivity, toxicity, reactivity, corrosivity, etc.),
- Process technology (chemistry, hazardous inventories, block flow diagram, PFD, mass/energy balance, etc.),
- Process parameters (safe upper/lower limits, consequences of deviation, etc.),
- Process equipment (key items, preliminary materials of construction, plot plan, etc.),
- Protective systems (for major hazards),

- Design hazard management strategy/process (see Chapter 4, Section 4.2.1),
- Design philosophies (e.g. blowdown, pressure relief & flare system, fire & gas, etc.),
- Location characteristics (topography, meteorology, population, infrastructure, etc.),
- Preliminary studies (HAZID, CRA, ISD, facility siting, blowdown/relief, fire & gas, etc.),
- Preliminary plans (development plan, process safety plan, EHS plan, etc.),
- Hazard/risk register,
- Action tracking database,
- Audit/review reports (technical peer review, stage gate review, etc.),
- Deliverables (statement of requirements, technology plan, outline basis of design, procurement plan for long lead items, cost estimate, project schedule),
- Preliminary strategies (project organization, HIRA, commissioning, operations and maintenance),
- Community outreach strategy/plan, and zoning buffer (if relevant),
- Other information for the Design Case for Safety (if applicable),
- FEL-2 stage gate review report and follow-up records.

Much of this documentation will be high level, preliminary and subject to change. However, as development progresses it should be updated.

12.2.3 Front End Loading 3 Stage

Further definition of the selected development option occurs in the FEL-3 stage with the objective of confirming the business case and achieving financial sanction for project execution (see Chapter 5). This involves completion of a design package for final engineering of the project that contains all the essential information, such as details of major equipment, materials of construction, piping/tie-ins, structural steelwork, wiring, buildings, etc. To do so requires information from FEL-2 to be updated and finalized, and preliminary drawings (e.g. layout, P&IDs, cause & effect) and process equipment datasheets prepared.

The compilation of PSI and other documentation, including calculations and design assumptions, continues throughout FEL-3 and into project execution. As the design evolves, the early information will often need to be revised and updated. As the level of project definition increases, the evaluation of major hazards involves more detailed, quantitative HIRA studies than was feasible earlier. This permits optimization of residual risk by applying ISD principles and a diverse range of passive and active design safety measures.

Typical documentation in FEL-3 includes, but is not limited to:

- Revised/updated information from FEL-2 (see Section 12.2.2 above),
- Preliminary P&IDs,
- Design calculations/assumptions,
- Procurement specifications of major equipment and protective systems (SCE including SIS),
- Datasheets for process equipment and protective systems,
- Performance standards (safety critical equipment (SCE), safety instrumented systems (SIS), other design safety measures),
- Deviations/exceptions and associated waivers from engineering codes and standards,
- FEED package,
- Project change management process,
- Commercial agreements,
- Contracts for main equipment,
- Strategies/plans
  - o Contracting/procurement,
  - o Resourcing/training,
  - o Integrity management/engineering assurance,
  - o Quality management,
  - o Partner management,
  - o Regulatory approval management,
  - o Preliminary emergency response,
- FEL-3 stage gate review report and follow-up records.

Although this documentation is increasingly more detailed, it is subject to change and should be updated as the development progresses.

12.2.4 Detailed Design Stage

Following FEL and financial sanction, the project moves into the first stage of execution, i.e. Detailed Design (see Chapter 6). Much of the FEL work requires refining and updating to achieve design completion prior to procurement and construction.

Thus, the compilation of PSI and other documentation, including calculations and design decisions, continues throughout detailed design. Contractor, vendor and supplier activities must be monitored to ensure production of relevant documentation. Some of this documentation will be required by the Operations team as it is essential for the ongoing operation and maintenance of the facility.

Typical documentation in detailed design includes, but is not limited to:

- Revised/updated information from FEL-3 (see Section 12.2.3 above),
- Project execution plan,
- Project performance records (expenditure, schedule, progress, quality, process safety, EHS, actions, etc.) and reports to client,

Design

- Design studies and follow-up (in addition to FEL, e.g. SCE vulnerability, RAM, SVA, SIMOPS, human factors, corrosion, structural, electrical system protection, pipeline integrity monitoring, decommissioning, emergency response, etc.),
- Design limitations for safe operation,
- Design assumptions on how the facilities will be operated,
- Design review reports and action follow-up (e.g. P&ID, 3D model, operability, inter-discipline, value engineering, etc.),
- Process equipment inspection and test plan,
- Design package for construction (e.g. procurement details for equipment, systems, buildings, structures, etc., and construction drawings, such as isometrics, P&IDs, electrical one line, cause & effect, etc.),
- Detailed design stage gate review report and follow-up records,

Procurement

- Contracts/purchase orders for initial construction (e.g. demolition/site clearance, grading, access roads, foundations, temporary buildings/camp, services, etc.),

Preparation for Construction

- Permits from local authorities and regulators,
- FAT for long lead equipment,
- Constructability report and follow-up,
- Construction plan (e.g. task sequence, manpower, construction equipment, SIMOPS, heavy lifts, transportation, area to system transition, engineering support, etc.),

- Construction contractor records (e.g. EHS/PS performance metrics, contractor management system, safety plan, safe work practices, sub-contractors, supplied equipment, etc.),
- Pre-mobilization plan (e.g. meetings with contractors, EHS expectations, hazards / risks, procedures, bridging documents, etc.),
- Construction site organization plan (e.g. offices, housing, security, utilities, telecoms, waste disposal, catering, lighting, fuel, design information storage, laydown area, warehousing, etc.),
- Construction contractor administration, orientation/training, and safety oversight plans,
- Construction emergency response plan (e.g. first aid, fire, rescue, access/egress, procedures, etc.),
- Pre-commissioning plan (e.g. hydro-testing, flushing/cleaning, mechanical completion certification, punch-lists, 'as-built' documentation, PSSR, etc.),

Preparation for Commissioning and Startup
- Commissioning/startup plan including test runs (if any),
- Plan for commissioning and operating procedures,
- Plan for asset integrity management (e.g. ITPM procedures, software, spares, etc.),
- Operator's plan for EHS/PS management system and document management,
- Operator training plan,
- Functional test plan,
- Handover plan,
- Preliminary Operations Case for Safety (if applicable).

While the design documentation should be finalized and a design freeze initiated before the construction stage commences, some of the plans for construction and commissioning/startup will be preliminary and subject to change. Any requests for design changes should be challenged and documented.

12.2.5 Construction Stage

The goal of the construction stage is to safely build a facility that will start up and operate safely, i.e. it is constructed as the detailed design intended to safely manage the inherent hazards and risks of the facility (see Chapter 7). The project should establish a formal system for managing receipt, storage, retrieval, and updating of design and technical specification information at the construction site.

Late design changes can present complications, especially when multiple hardcopy drawings and documents are being used in the site and must be replaced. Failure to replace all relevant site documentation with updated information could potentially result in a process safety incident or costly rework.

The project document management system should also track any outstanding deliverables, especially documents required prior to commissioning. A key activity during construction is the compilation of all documentation required for commissioning and subsequent handover to Operations.

Typical documentation required or generated during construction includes, but is not limited to:

* Late changes/revisions to information from detailed design (see Section 12.2.4 above),
* Engineering design drawings and technical specifications (including design intent, codes & standards, performance standards, etc.),
* Project execution plan,

Procurement

* Contracts/purchase orders for equipment, materials and services,
* Quality management (QM) plan to vendors defining tests and QC for major and critical equipment,

Construction and Pre-commissioning

* Construction plan,
* Construction EHS plan,
* Change management records and DCNs (e.g. late design changes, field changes),
* Engineering queries/RFIs,
* Construction contractor records (e.g. EHS/PS performance metrics, management system bridging documents, sub-contractors, supplied equipment, etc.),
* Equipment/material preservation procedures,
* Project quality plan (including an inspection and test plan),
* Positive material identification (PMI) records,
* Fabrication quality records (including FAT, NDT, weld radiographs, non-conformances, certificates, QA reports, technician/inspector qualifications, etc.),
* Pre-commissioning quality records (including SAT, NDT, field weld radiographs, hydro-tests, flushing/cleaning/drying, checklists, non-

conformances, baseline data, certificates, QA reports, technician/inspector qualifications, etc.),

- Mechanical completion certificates/dossier,
- Punch-lists,
- EHS and process safety procedures:
  - o HIRA and SIMOPS studies,
  - o Safe work practices, work permits, JSAs,
  - o Incident reports and follow-up records,
  - o Audit reports and follow-up records,
  - o Contractor competency, and orientation/training records,
  - o Emergency response plan, procedures, drills and follow-up records,
  - o Operational readiness review/PSSR.
- Construction metrics (e.g. progress, financial, EHS/PS, change, rework, etc.),
- Stakeholder outreach records, including commitments to third parties (e.g. regulator, NGO, local community, etc.),
- Construction stage gate review report and follow-up records,

Preparation for Commissioning and Startup

- Permits from local authorities and regulators,
- Recruitment and training records (e.g. operators, technicians, engineers, EHS, admin., etc.),
- Operator's EHS and process safety management system (e.g. procedures, plans, etc.),
- Operator's document management system,
- Operating and maintenance manuals from suppliers and vendors,
- Commissioning and operating procedures,
- Asset integrity management (e.g. master equipment list, SCE list, ITPM tasks and frequencies, maintenance management system build, etc.)
- Contracts/purchase orders (EHS equipment, vendor support, spare parts, consumables, chemicals, lubricants, catalysts, etc.),
- 'as-built' documentation (e.g. drawings, technical information).

The generation of 'as-built' drawings and technical information should commence as soon as possible, and preferably be completed for inclusion in the handover package to Operations. If this is not possible, red-line drawings should be provided until such time as the final CAD drawings are supplied.

12.2.6 Commissioning and Startup Stage

The final phase of project execution, commissioning and startup, commences after mechanical completion of the facilities (see Chapter 9). Some deviations from intended normal operation should be expected during commissioning and initial startup, and operating logs, shift handover notes, and records should be reported in greater detail to assist with future troubleshooting.

Typical documentation required or generated during commissioning and startup includes, but is not limited to:

- Late changes/revisions to information from construction (see Section 12.2.5 above),
- Permits from national/local government agencies,
- Pre-startup stage gate review report and follow-up records.
- Stakeholder outreach, including community liaison, reports, meeting minutes, etc.,
- Commissioning and operating procedures,
- Commissioning (with safe chemicals) report,
- Commissioning team operating logs, shift handover notes, and records for each step in the commissioning procedures,
- Comprehensive file for each system and item of equipment showing status of each commissioning step performed,
- Startup (with process chemicals) report,
- Startup team operating logs, shift handover notes, and records for each step in the startup procedures,
- Comprehensive file for each system and item of equipment showing status of each startup step performed,
- Performance test run procedures, including operating parameters, sample analysis, etc.,
- Performance test run results (by each equipment item and each system), data, sample analyses, and follow-up records,
- ESD, trip, and SIS and IPL activation records,
- Temporary operations, e.g. blind and strainer lists,
- Outstanding punch-list items to be inherited by the Operator,
- Outstanding action items to be inherited by the Operator,
- 'as-built' documentation (e.g. drawings, technical information) in preparation for handover.

Following commissioning and startup, the project is ready for handover to the Operator.

12.2.7 Handover

The facility is ready for commercial operation and handover to the Operator, after meeting any performance guarantees (verified by test runs) and other technical specifications, including process safety (see Chapter 9). The PMT should have compiled a vast amount of documentation over the project lifecycle, and agreed with the Operator:

- Information required for the ongoing operation of the facility,
- Format and content of information, i.e. hardcopy/electronic, coding system, etc.,
- Number of copies of hardcopy information, e.g. operating and maintenance manuals.

In particular, contracts with technology licensors should clearly indicate all relevant documentation that the Operator requires.

Good industry practice involves a formal handover process for the core information comprising a detailed procedure, checklists, and a transfer of responsibility form. These procedures should address responsibility for (i) any outstanding action items, and (ii) future change and updating of documentation, including HIRA and technical controls.

Typical documents that should be included in the handover package to the Operator are:

- Information required for ongoing operation, maintenance and further development of the facility, including:
  o Process chemicals/materials, safety data sheets (SDS), reactivity matrix,
  o Process technology,
  o Process equipment, e.g. equipment datasheets, calculations, codes & standards,
  o Operating and Maintenance manuals,
  o Operating procedures,
  o 'as-built' drawings and technical information,
  o Equipment quality certificates (retain for statutory purposes, future changes),
  o Initial/baseline inspection reports,
  o SCE and other important safeguards, performance standards,
  o Equipment/system commissioning/test run reports,
  o ITPM procedures, tasks and frequencies,
  o ESD and other trip activation during commissioning records,

- Documentation of design intent/criteria, verification, assumptions,
- Final HIRA reports and follow-up records,
- Final safety/technical study reports and follow-up records,
- Operational readiness review and follow-up records,
- Startup stage gate review report and follow-up records,
- Commissioning operation with safe chemicals report and follow-up records,
- Startup with process chemicals report and follow-up records,
- Notifications, requirements and obligations for regulatory compliance,
- Commercial agreements, e.g. licenses, feedstock, product, consumables, etc.,
- Contracts, e.g. engineering, technology licensor, and vendor support,
- Commitments to stakeholders,
- Operations Case for Safety (if applicable),
- Contractual and financial documentation to be retained in respect of legal liabilities, warranties/guarantees, financial audits, and tax requirements,
- HR, training and performance assurance records for any Project staff seconded to the Operator,
- Other project documentation, including:
  o Blind and temporary strainer lists,
  o Risk registers,
  o Incident reports,
  o Action tracking data, including outstanding actions inherited by the Operator,
  o Punch-lists, including any items inherited by the Operator,
  o Technical standards and approved waivers, if any, developed by Project,
- List of any unclosed action items / elevated risks that Operator will inherit.

Appendix F comprises a more comprehensive list of information that, depending on the scope of the project, may be appropriate for inclusion in the handover package.

12.2.8 Operation Stage

Assuming that startup achieved steady state operation in line with BOD expected production, the facility is now ready for commercial operation (see Chapter 10). At this point, the Operator takes responsibility from Project, and should have received all essential documentation in the handover package. Any outstanding technical information should be handed over to the Operator as soon as possible. Outstanding redline drawings should also be updated to 'as-built' status and handed over, unless the Operator has agreed to assume responsibility for updating.

The handover documentation should be archived in the Operator's document management system for future reference, protected from inadvertent loss (e.g. fire), and ready access to whoever needs the information in order to safely perform their job. During the operation stage of the project lifecycle, the Operator must maintain the accuracy of this information by keeping it up-to-date.

Therefore, typical documentation during operations will be the same as that in the handover package (see Section 12.2.7 above), with, but not limited to, the following additions:

- Process knowledge management program, policy,
    - o Document management system, control documents, retention policy, loss/fire protection, etc.,
    - o Revisions to information from handover (see Section 12.2.7 above), including, but not limited to:
        - Resolution of outstanding actions inherited by the Operator,
        - Resolution of punch-list items inherited by the Operator,
        - Changes in chemicals, process technology, process equipment,
        - Changes to operating procedures as result of commissioning, startup, and operating experience,
        - Changes in ITPM procedures, tasks, and frequencies due to operating experience,
        - Changes due to debottlenecking projects and modifications,
        - Project PS/EHS risk register to address transition to Operation,
- Compliance with standards program, corporate policies and standards, national/local regulations, citations/improvement notices,
- Process safety and EHS management systems (e.g. procedures, safe work practices, plans, objectives, etc.),
- Process safety culture program, culture assessments and follow-up records,
- Workforce involvement program, roles/responsibilities, records,
- Stakeholder outreach program, objectives, meeting minutes, records,
- Risk management program, including HIRA, methodology procedures, facilitator/team member qualifications, risk management philosophy, risk tolerance criteria, risk register, revalidation reports and follow-up records, communication records,
- Operating procedures program, format/content, temporary procedures, checklists, periodic review records, etc.
- Safe work practices program, procedures, work permits, JSAs, permit authorizer qualifications, permit reviews/audits,

- Asset integrity program, procedures, maintenance management system,
  - o Master equipment list,
  - o Criticality analysis, SCE list,
  - o Reliability analysis reports,
  - o Quality management program,
  - o ITPM task plan, task records, technician/contractor qualifications, data analysis and plan update, inspector recommendation follow-up records,
  - o Process equipment deficiencies, failure analysis reports and follow-up records, repair/replace/re-rating procedures/records, technician / contractor qualifications,
  - o Spares, preservation procedures,
  - o Control software (DCS, PLC, interlocks, etc.),
- Contractor management program, screening/selection procedures, contractor EHS/PS performance records, contractor qualifications, pre-qualified contractor list, orientation/training materials, oversight, end of contract evaluations, etc.
- Training and performance assurance program, employee qualifications, training matrix, training materials, trainer qualifications, verification test/observation data, assessment of program effectiveness,
- Management of change program, procedures, files including scope, design information, HIRA studies, technical reviews, authorization, link to operational readiness review, etc.,
- Operational readiness review program, procedures, reports, checklists, etc.,
- Conduct of operations program, procedures, operator logs, shift handover notes, checklists, equipment labeling/warning signs, housekeeping, etc.,
- Emergency management program, response plans, drills/exercises and follow-up records, liaison/communication with stakeholders, ITPM for emergency facilities, equipment and PPE, etc.,
- Incident reporting and investigation program, procedures, reports, forms, checklists, investigation/facilitator qualifications, root cause analysis methodology, trend analysis, recommendation follow-up records, etc.,
- Measurement and metric program, procedures, KPI records, periodic review/analysis and follow-up records, communication, etc.,
- Auditing program, procedures, plans, audit protocols, periodic self-assessment/audit reports and follow-up records, etc.,
- Management review and continuous improvement program, policy, procedures, plans, review meeting information, meeting minutes and follow-up records, etc.,
- Operations stage gate review report and follow-up records.

Appendix F comprises a more comprehensive list of information that, depending on the scope of the project, may be appropriate for documentation during the operation stage.

12.2.9 End of Life Stage

Documentation at the end of a facility's lifecycle (see Chapter 11) depends to some extent on the nature of the decommissioning that can involve:

- Mothballing, i.e. process unit or equipment may be potentially re-commissioned at a later date,
- Deconstruction, i.e. process unit is dismantled and individual items of equipment may be re-used,
- Demolition, i.e. process unit is essentially destroyed for scrap and potential material recycling.

Typical documentation at end of life may involve, but is not limited to, the following:

- National/local regulations and/or industry standards (e.g. 29 CFR 1926, Subpart U, *Blasting and the Use of Explosives;* ANSI A10.6. *Safety Requirements for Demolition;* BS 6187:2011 *Code of Practice for Full and Partial Demolition*),
- Normal operation and maintenance procedures/practices in late-life for large facilities, while deinventorying raw materials, intermediates and products in preparation for decommissioning,
- Shutdown procedures for facilities that will not re-start,
- Structural engineering survey report to identify potential hazards (e.g. premature collapse, cave-in, etc.), original structural drawings, calculations, etc.,
- Engineering/safety study of impact of deconstruction and/or demolition on surrounding facilities,
- HIRA studies of deconstruction and demolition hazards, and follow-up records, risk register,
- Safety plan/report for deconstruction and/or demolition, including oversight,
- Safe work practices, including work permits, positive isolation of all energy sources – especially underground utilities, unusual hazards (e.g. asbestos, PCBs, heavy metals, naturally occurring radioactive materials (NORM), etc.), handling explosives, heavy equipment operations, waste disposal, etc.,
- Security plan, site perimeter barricade, warning signs, etc.,
- Emergency response plan for deconstruction and/or demolition,

- Decommissioning procedures, task sequence, depressurization, deinventory remaining materials, cleaning, decontamination, purging, inerting, etc.,

- Asset integrity management, preservation procedures, ongoing ITPM tasks (if necessary) to maintain assets in a state of readiness or near-readiness, etc.,

- Contractor management, contractor qualifications/experience of deconstruction and demolition, contractor selection, orientation/training, oversight, sub-contractors, etc.,

- Deconstruction and/or demolition procedures, task sequence, groundwater protection, segregation of equipment and materials for re-use, recycle, and/or scrap, etc.,

- Incident reporting and investigation procedures, reports, records,

- Remediation procedures,

- Recommissioning procedures for mothballed process unit/equipment,

- Operational readiness review for mothballed process unit/equipment.

If any equipment is to be re-used, the original or modified design and inspection documentation should be retained. Decommissioning of offshore oil/gas production facilities requires additional documentation related to the hazards of working and heavy lifting over/under/near water.

12.3 SUMMARY

Information is the life blood of projects, which use and generate large quantities of documentation. A significant proportion of these documents are process safety information (PSI) that is critical to the design and residual risk of the completed project.

In liaison with the future Operator, Project should define *as early as possible* the documentation to be retained, and that to be handed over to the Operator. When seeking tenders, contracts should define the information, including formatting and coding, to be produced by contractors. Above all else, the timing of availability of certain documents is critical for risk management decision-making and meeting the project schedule.

APPENDIX A. TYPICAL PROCESS SAFETY STUDIES OVER PROJECT LIFE CYCLE

Table A-1 is intended as a guide to some of the process safety studies and reviews that *may* be appropriate for a project to undertake in order to develop and deliver a facility that is safe and reliable to operate. Generally, no two projects are the same. They may vary in strategy, scope, complexity, location, design basis, local laws and regulations, and various other factors. When reviewing the table below, the user should consider all project-specific factors before determining which process safety studies and reviews apply.

As an example, a small, relatively simple modification project *may* only require the following:

- Employ a competent and experienced project team,
- Identify hazards inherent to the modification (e.g. conduct HAZOP and/or What If study),
- Understand the process safety risks associated with the hazards (e.g. use risk matrix/risk ranking),
- Manage the risks to meet regulations and corporate policy (e.g. select appropriate engineering standards; contractor management; construction safe work practices and quality management; update all operating, AIM, process safety and EHS procedures; conduct MOC; train workforce in project/changes; conduct ORR; etc.)

Conversely, a major project for a greenfield chemical plant *may* require many of the process safety studies and reviews in the table depending upon its scope.

The timing of certain studies and reviews can also vary between projects based upon the project strategy (i.e. traditional, fast track, insource, outsource, etc.) and/or corporate/contractor preference. The timing (i.e. stage of project life cycle) of each study shown in the table is based upon a traditional strategy (i.e. develop a plan and work the plan) for a major capital project.

Again, the user of the table should consider what timing is appropriate for the intended studies and reviews required for their project. Studies may be conducted *early* compared to the table below, if the required input data and information are available, and have value in aiding stage appropriate decision-making. Any study, irrespective of timing, should be updated if new data and information become available.

Table A-1. Typical Process Safety Studies over Project Life Cycle

Warning: It is essential that the guidance above is understood and followed before using the information in this table for a specific project.

Process Safety Studies Category	Front End Loading (FEL)			Project Execution				
	Appraise FEL-1	Select FEL-2	Define FEL-3	Detailed Design	Construction	Commissioning Startup	Operation	Decommissioning
Process Safety Plan / Risk Register / Action Tracking List			Updated continuously throughout capital project stages					
Hazard Identification	Prelim HAZID	HAZID	Prelim HAZOP/What If/Checklist	Final HAZOP/What If/Checklist*			Revalidate/ReDo HAZOP/What If/Checklist	HAZID/What If/Checklist
				DCN after Final HAZOP/What If #	Change Mgmt. for DCN	Change Mgmt. for temp piping, etc.	MOC	Change Mgmt./MOC
					JSA	JSA	JSA	JSA
					ORR	ORR*	ORR*	
Consequence Assessments		Prelim FSS/FEA	FSS/FEA	FSS/FEA*			Revalidate FSS	
		Prelim FHA	FHA	FHA*				
			SGIA/SIP	SGIA/SIP*				
Safety Assessments	Prelim ISD	Update Prelim ISD	ISD	ISD*				
		Prelim DHM	DHM	DHM*				
			HFA	HFA*				
		Prelim BD/PR/Flare	BD/PR/Flare	BD/PR/Flare*				
			Prelim HAC	HAC*			HAC*	
			RAM	RAM*				
			SCE	SCE*				
			SCEVA	SCEVA*				
			EERA	EERA*				
			DOA	DOA*	DOA*			
			Prelim Design Case for Safety	Design Case for Safety	Operations Case for Safety		Update Operations Case for Safety	
			Prelim SVA	SVA	SVA*			
			Prelim SIMOPS	SIMOPS	SIMOPS* Audits	SIMOPS* Audits	SIMOPS Audits Mgmt. Reviews	SIMOPS
Transportation Studies		Prelim Pipeline HCA	Pipeline HCA					
		Prelim Road Anal.	Road Analysis					
		Prelim Rail Analysis	Rail Analysis					
		Prelim Marine/WSA	Marine/WSA					
Risk Assessments	CRA	Update CRA						
			Prelim QRA	QRA			Periodic IRE Revalidate QRA	
			Prelim HRA	HRA	HRA			
				LOS	LOS*			
				TRIA				
				LOPA	LOPA*		Revalidate LOPA	
				SRS/SIL	SRS/SIL*			
Risk Mitigation			Prelim F&G	F&G	F&G*			
			Prelim ESD	ESD	ESD*			
			Prelim FP	FP	FP*			FP
			Prelim PW	PW	PW*			
			Prelim ER	ER	ER*		ER	ER
Regulatory Studies†				RMP (USA) Design Safety Case (UK) Preconstruction Safety Report (UK)	Pre-operational Safety Case (UK) Pre-operation Safety Report (UK)		Periodic RMP (USA)	Dismantlement Safety Case (UK)
Stage Gate Reviews	Concept Review	Selection Review	Technical Definition Review	Detailed Design Review	Construction Review	Pre-Startup Review	Post Operation Review	

* Update process safety study, as necessary

† Examples of country-specific studies. User to determine any country-specific studies required for their project.

(Please access the CCPS website for better resolution of this table.)

APPENDIX B. PROJECT PROCESS SAFETY PLAN

A Project Process Safety Plan evolves throughout the stages of the project lifecycle, and should describe all the process safety activities and their timing necessary to deliver a safe and reliable operating facility. For a large project these activities comprise a plethora of studies, assessments, competencies and training, documentation, reviews and inspections covering all the elements of CCPS Risk Based Process Safety, plus Inherently Safer Design and Design Hazard Management.

Some companies combine the process safety plan with the Project EHS Plan, while others have separate plans. Nevertheless, it is important that each discrete process safety and EHS activity is included in the overall project plan for FEL and/or Execution, as appropriate. The generic content of a typical project process safety plan described below includes various overlaps with health and occupational safety activities, although environmental issues have been ommitted as they are not the focus of this book.

Appendix A provides the timing of *key* activities by project stage for a typical major project, although some of the activities may not be relevant depending upon the scope of the project. Other process safety activities for specific project stages are discussed in Chapters 3 (FEL 1), 4 (FEL 2), 5 (FEL 3), 6 (Detailed Design), 7 (Construction), 9 (Startup), 10 (Operations), and 11 (Decommissioning). It should be noted that the timing of these typical activities may vary between projects of differing scope, complexity, strategy, and corporate preference.

The list below is not meant to be all inclusive nor to imply that every item should be included for every project.

GENERIC CONTENT OF A TYPICAL PROJECT PROCESS SAFETY PLAN

- Project description
- Key project milestones and timing of key process safety activities
- Roles and Responsibilities for managing process safety activities
- Communication, meetings, etc. for integrating process safety into overall project plans, and for promoting safety
- Process Safety Culture (activities to promote positive culture)

- Compliance with Standards (regulations, engineering codes/standards, variances, etc.)
- Workforce Involvement (activities to involve the workforce (i.e. Project, contractors, Operator, etc.) in process safety and EHS)
- Stakeholder Outreach (regulator, emergency services, local community, NGO's, etc.)
- Process Knowledge Management (chemicals/materials, process technology, process equipment, etc.)
- Hazard Identification and Risk Analysis (HIRA)
  - o HAZID, HAZOP, What If, etc.
  - o Consequence analysis (fire, explosion, toxicity, reactivity, equipment vulnerability, etc.)
  - o Risk matrix, LOPA, QRA,
  - o Facility Siting (location, occupied buildings, off-site impacts, etc.)
  - o Human Factors (ergonomics, human performance, human error, etc.)
- Inherently Safer Design (ISD) and Design Hazard Management (DHM)
  - o Design strategies and philosophies,
  - o Technical studies (process safety input on siting and layout, corrosion, geotechnical, resistivity, reliability, electrical, constructability, etc.),
  - o Risk reduction measures (prevent/detect/control/mitigate devices),
  - o Performance standards for risk reduction measures.
- Operating Procedures (normal operation, startup, shutdown, operating limits, etc.)
- Conduct of Operations (standing instructions, routines, shift handover, inhibit/override/bypass, etc.)
- Safe Work Practices (for office, fabrication, installation/construction, commissioning, startup, operation, decommissioning, SIMOPS, etc.)
- Asset Integrity (ITPM tasks, SCE, reliability/availability, quality management, condition monitoring, corrosion protection, etc.)
- Process Safety Competency and Training & Performance Assurance (process safety & EHS, task requirements, assessment, training, SME's, etc.)
- Management of Change (hazard/technical reviews, approval, DCN, late design changes, organisation changes, etc.)
- Contractor Management (pre-qualification, evaluation, oversight, interface management, etc.)
- Operational Readiness (pre-startup reviews, go/no go decision-making, etc.)
- Auditing (stage gate reviews, contractor performance, work permits, etc.)

- Metrics and Measurement (injury/illness, spills, releases, etc.)
- Emergency Plan (procedures, alarms, crisis management, evacuation, shelter-in-place, drills, etc.)
- Accident/Incident Reporting and Investigation (injury, loss of containment, property damage, near-miss, high potential, etc.)
- Management Review & Continuous Improvement (regular performance reviews by Project and/or Operator management)
- Action tracking and resolution (all studies, reviews, investigations, etc.)
- Risk Register (see Appendix C)
- Documentation (for all process safety elements, archive and field management, obsolete/superceded documents, handover package, etc.)

This list may be used by a gatekeeper as a guide to verify due diligence with regard to process safety during each end of stage review.

APPENDIX C. TYPICAL HAZARD & RISK REGISTER

In addition to process safety, a Project Hazard & Risk Register typically covers technical, EHS, contracting, commercial, administrative, etc. risks. Ideally the register should be reviewed at most project team meetings, and, for small projects, as a minimum at the end of each stage of the project lifecycle. The register should be updated frequently as new hazards/risks are identified, and existing risks are eliminated or reduced. While all project team members should have access to view the register, only a few project roles should have editing rights.

The risks are often documented using all or some of the following fields:

HAZARD IDENTIFICATION

HAZOP and other hazard identification studies identify and categorize hazards. Some hazards may be identified through brainstorming in project meetings or other means. It is common to designate a person responsible for managing the hazard (through prevention plans and contingency plans – see below). Typical register fields for hazard identification include:

- Description of the Hazard (with unique identifying number, and source of hazard e.g. HAZOP study)
- Categorization (i.e. type of hazard, such as project schedule/budget, safety, etc.)
- Responsible Person/Owner of the Hazard

RISK ANALYSIS

Some companies use a risk matrix to estimate an order of magnitude risk for specific consequences of hazards. In addition to risk matrices, quantitative techniques such as consequence analysis, LOPA and QRA can be used to estimate with greater accuracy the consequences and/or probability of the hazard/risk occurring. Most companies rank the hazards/risks by magnitude, i.e. the combination of consequence and probability.

Typical register fields for risk analysis include:

- Consequence/Impact
- Probability/Likelihood

- Risk Ranking

Consequences and probabilities may be expressed:

- qualitative (e.g. critical/high/medium/low; traffic lights; etc.), especially for non-process safety risks, such as impacts to project schedule and budget,

or

- quantitative (e.g. risk of thermal radiation/blast overpressure/toxic concentration levels resulting in potential injury/fatality/property damage/environmental damage).

RISK MANAGEMENT

Most projects develop plans to prevent non-process safety risks from being realized. While these plans focus on prevention, some projects also develop contingency plans just in case the risk does occur. For example, a delay in receiving equipment/materials from a specific supplier may trigger a contingency plan for alternative procurement.

It is likely that some process safety and EHS risks may exceed client policies or tolerance criteria. In these circumstances, a plan of how the project intends to respond should be developed, and the responsible person/owner for the risk should manage activities to reduce the risk, and record the residual risk after risk reduction measures have been implemented.

Typical register fields for risk management include:

- Residual Risk
- Response Plan
- Prevention Plan
- Contingency Plan (and possible causes/triggers for implementation)

REGISTER SPREADSHEET

The Hazard & Risk Register is often documented in a database or spreadsheet format. Large projects invariably use databases, while a spreadsheet, such as the example in Table C-1, may suffice for small projects:

Table C-1. Risk Register Example

PROJECT ABC HAZARD & RISK REGISTER									
No.	Description	Category	Consequence	Probability	Ranking	Residual Risk	Response Plan	Prevention/ Contingency Plans	Responsible Person
1	...	Safety	High	Medium	Medium High	Low	....	...	J. Smith
2	...	Schedule	Low	Low	Low	Low	N/A	...	S. Adams
3	...	Budget	Low	High	Medium	Low	...	...	A.N. Other

Alternatively the number and category could be combined. For example, safety risks could be designated as S1, S2, S3, etc., and budget risks as B1, B2, B3, etc.

Color coding can also be used to highlight ranking, and swiftly communicate the overall level of risk in the project.

Additional columns could be added for:

- Status of actions and/or implementation of plans,
- Key milestones (e.g. date risk added, due date for risk reduction implementation, etc.),
- Risk reduction/mitigation costs
- Overall item status (e.g. open, pending, in progress, closed, etc.).

A simplified executive summary version of the register may also be appropriate for key stakeholders (e.g. client corporate management) to provide a quick picture of overall project risk.

While a combined project hazard and risk register may be convenient, many EHS and process safety risks are likely to continue beyond the end of the project, whereas other project risks, such as schedule and budget issues, by definition end at the end of the project. If a combined register is used, any remaining open items should be communicated and passed to the Operator.

APPENDIX D. SAFETY CHECKLIST FOR PROCESS PLANTS

This Safety Checklist has been compiled from multiple references, and is not intended to be comprehensive. The checklist is a starting point and the user should think about any hazards related to the topics in the checklist, and consider additions appropriate to their specific project.

LOCATION

1. Accessibility. Avoidance of site cul-de-sacs, preference for ring roads, avoidance of road and rail bottlenecks and traffic congestion, provision of roadways round all process units. Alternative main entrances. Facilities near main entrance, for regular road traffic.

   – Alternative emergency access/egress to and from all areas.
   – Plant fences, barriers, etc.

2. Traffic - Vehicular and pedestrian. Need for barrier control.

3. Parking areas - Entrances, exits, drainage, lighting, enclosures.

4. Clearances - Buildings for railroad traffic and vehicles (overhead, width, turn-arounds), including fire engines.

5. Drainage.

6. Road locations, markings.

7. Entrances, exits - Pedestrian, vehicular, railroad.

8. Transformer location, to be in comparatively safe areas, least likely to be affected by fires, accidents, road traffic, mechanical equipment.

9. Location furnaces, units for heat transfer agents, flare stacks (exclusion zones).

10. Separation of hazards from people (general public outside boundary, employees within site). Dispersion of toxic and flammable releases both within and outside the site.

    – Separation of hazards from other hazards (domino escalation)
    – Separation of flammable/explosive hazards from ignition sources
    – Good siting and spacing in relation to other buildings or installations, effect of adjacent fire, etc. unusually hazardous areas. Separation of hazardous and occupied areas.

11. Important safety codes and standards that may be applied.

12. Hazards of inclined sites, tank rupture and overflow (liquids / heavy vapors), vehicle runaway.

13. Unnecessary low-level areas (flooding, collection of flammable liquids, collection of toxic/flammable heavy vapors)

14. Protection against flooding, e.g. for key utilities.

15. Climatic and meteorological conditions.

16. Adjacent internal and external site activities, external roads. Ignition sources, heavy activity, movement of machinery, personnel concentrations in hazardous areas.

17. Cooling tower location.

18. Possible equipment and vessel damage by vehicles.

19. Firewater intake location, possibility of oil spills, clogging by silt, weed, fish, shellfish, etc.

20. Possible operational restrictions by fire walls.

21. Local atmospheric conditions particularly affecting design, e.g. salt spray, freezing fog, ice loading, electric storms, hurricane/typhoon, etc. Also earthquake zone.

22. Rail facilities, possible damage to equipment.

23. Separate access for ship crews.

24. Strategic location of fire-fighting equipment. Fire pumps located away from major hazardous areas.

25. Prevailing wind direction.

26. Availability of scale models.

27. Effect of process, mechanical or control integration.

28. Need to segregate laboratories from general offices.

29. Location of major ignition sources, e.g. fired heaters, boilers, maintenance workshops.

30. Consider future expansion.

31. External factors, such as neighboring process facilities, proximity of flight paths, infrastructure, etc.

32. Location of exhausts, process vents, and HVAC inlets.

33. Helicopter landing area for emergencies.

BUILDINGS

1. Wind Pressure, snow loads, floor loads, earthquake design.

2. Roof material, anchorage.

3. Roof vents and drains, smoke dispersal.

4. Stairwells, ramps, lighting.

5. Elevators and dumbwaiters.

6. Fire walls, openings, fire doors.

7. Explosion relief, e.g. panels.

8. Exits - Fire escapes, identification, safety tread.

9. Record storage.

10. Ventilation - Fans, blowers, air conditioning, scrubbing of toxic vapors, location of exhausts inlets, smoke and heat ventilation dampers, fire curtains.

11. Lighting protection, structural and equipment grounding for electrical discharges.

12. Building heaters (hazardous or nonhazardous area), vents.

13. Locker rooms including need for separate lockers for work and street clothes, required number of each and air changes.

14. Building drainage - inside and out.

15. Structural steel and equipment fireproofing, including piperacks.

16. Access ladders to roofs from outside level, escape ladders, fire escapes.

17. Bearing capacity of subsoil.

18. Important safety codes and standards that may be applied.

19. Siting of control rooms, considering possible incidents.

20. Design of control rooms, including consideration of explosion (blast proof or blast resistant).

21. Siting of emergency services (fire, medical, response time) and emergency control room/communications.

SPRINKLERS, HYDRANTS AND MAINS

1. Water supply including secondary supplies, pumps, reservoirs and tanks.

2. Mains - adequate looping, cathodic protection, coated and wrapped when needed, sectional valves.

3. Hydrants - location.

4. Automatic sprinklers - occupancy classification, wet systems, dry systems, deluge systems.

5. Standpipes and tanks.

6. Type, size, location and number of fire extinguishers needs.

7. Fixed automatic extinguishing systems, CO_2, N_2, foam, dry powder, halon.

8. Special fire protection systems - rise in temperature alarms, sprinkler system flow alarms, photoelectric smoke and flame alarms, UV/IR, Hydrogen detection in battery rooms, etc.

9. Important safety codes and standards that may be applied.

10. Independence of firewater system, possible process connections.

11. Fixed fire protection equipment appropriate to risk, sprinkler systems, steam fire curtains, water curtains, fixed monitors.

12. Firewater pump drives.

13. Need for more than one fire pump station.

14. Protection of mains against freezing, mechanical or fire damage.

15. Need for permanent monitors.

16. Water curtains.

17. Periodic spray system checking and cleaning.

18. Mobile foam equipment, foam storage and stocks.

19. Steam curtains, steam lances.

20. Snuffing steam systems.

21. Portable extinguishers (type, locations, numbers)

22. External emergency services (facilities, equipment, response time, etc.).

ELECTRICAL

1. Hazardous area classification. (Use of dispersion calculations for unusual circumstances not covered by Code).

2. Accessibility of critical circuit breakers.

3. Polarized outlets and grounded systems.

4. Switches and breakers for critical equipment and machinery.

5. Lighting - hazardous or nonhazardous areas, light intensity, approved equipment, emergency lights.

6. Telephones - hazardous or nonhazardous areas.

7. Type of electrical distribution system - voltage, grounded or ungrounded, overhead, underground.

8. Conduit, raceways, enclosures, corrosion considerations.

9. Motor and circuit protection.

10. Transformer location and types. Need for fencing.

11. Fail safe control devices protection against automatic restarting.

12. Preferred busses for critical loads.

13. Key interlocks for safety and proper sequencing, duplicate feeders.

14. Accessibility of critical breakers and switch gear.

15. Exposure of process lines and instrument trays to fire damage.

16. Important safety codes and standards that may be applied.

17. Complete failure of electricity supply, need for standby Diesel generators.

18. Cable protection against fire damage, location, and flame retardants.

19. Need for alternative cable runs.

20. Marking of underground cables.

21. Fire pump power supplies.

SEWERS

1. Chemical sewers - trapped, accessible clear-outs, vents, locations, disposal, explosion hazards, trap tanks, forced ventilation automatic flammable vapor detectors and alarms.

2. Sanitary sewers - treatment, disposal, traps, plugs, cleanouts, vents.

3. Storm sewers.

4. Waste treatment, possible hazards from stream contamination including fire hazard from spills into streams and lakes.

5. Drain trenches - open, buried, accessible cleanouts, presence of required baffles, and exposure to process equipment.

6. Important safety codes and standards that may be applied.

7. Disposal of wastes, air and water-pollution safeguards.

8. Furnaces and flammable service drains, curbs, fire-traps, drainage system capacity.

9. Pits and gas pockets.

STORAGE

1. General

   a. Accessibility - entrances and exits, sizes.

   b. Sprinklers.

   c. Aisle space.

   d. Floor loading.

   e. Racks.

   f. Height of piles.

   g. Roof venting.

2. Flammable Liquids - Gases, Dusts and Powders, Fumes and Mists.

   a. Closed systems.

   b. Safe atmospheres throughout system.

   c. Areas to be equipped with sprinklers or provided with water spray.

   d. Emergency vents, flame arresters, relief valves - safe venting location including flares.

   e. Floor drains to chemical sewers properly trapped.

   f. Ventilation - pressurized controls, etc. and/or equipment.

   g. Tanks, bins, silos - underground, above ground, distances, fireproof supports, dikes and drainage, inert atmospheres.

   h. Special extinguishing systems, explosion suppression - foam, dry chemicals, carbon dioxide.

   i. Dependable refrigeration systems for critical chemicals.

   j. Separation of reactive materials from one another.

3. Raw Materials

    a. Hazard classification of material including shock sensitivity.

    b. Facilities for receiving and storing, segregation of incompatible materials, clear labeling.

    c. Identification and purity tests.

    d. Provisions to prevent materials being placed in wrong tanks.

4. Finished Products

    a. Identification and labeling to protect the customer.

    b. Conformance with ICC and other shipping regulations.

    c. Segregation of hazardous materials.

    d. Protection from contamination, especially in the filling of tank cars and tank trucks.

    e. Placarding of shipping vehicles.

    f. Routing of hazardous shipments.

    g. Data sheets for safety information for customers.

    h. Safe storage facilities, piling height.

    i. Safe shipping containers.

5. Tank Failure.

6. Important safety codes and standards that may be applied.

7. Adequate distances from operating plants.

8. Adequate dikes for storage tanks.

9. Remote isolating valves.

10. Location of intermediate storage vessels at ground level and away from process units.

11. Tank boil-over or material 'roll-over'.

12. Tank foundations and soil conditions.

13. Flooding round empty tanks.

14. Tank contamination, e.g. by water.

15. High-pressure considerations.

16. Differential movement of tanks and piping.

INERT GAS BLANKETING OF ALL HAZARDOUS PRODUCTS

1. Consider raw material, intermediates, and products.
2. Consider storage, material handling and processes.
3. Important safety codes and standards that may be applied.
4. Capacity of nitrogen and inert gas supply facilities.

MATERIALS HANDLING

1. Truck loading and unloading facilities.
2. Railroad loading and unloading facilities.
3. Industrial trucks and tractors - gasoline, diesel, liquefied petroleum gas.
4. Loading and unloading docks for rail, tank trucks and truck trailer - grounding system for flammable liquids.
5. Cranes - mobile, capacity marking, overload protection, limit switches.
6. Warehouse area - floor loading and arrangement, sprinklers, height of piles, ventilation.
7. Conveyors and their location in production areas.
8. Flammable liquid storage - paints, oils, solvents.
9. Reactive or explosive storage - quantities.
10. Disposal of wastes - incinerators, air and water pollution safeguards.
11. Important safety codes and standards that may be applied.
12. Pipe track cross-walls.
13. Ship loading and unloading facilities. Also bilge / ballast water reception facilities.
14. Transfer by pipeline of duct using pumps, fans or compressors.
15. Mechanical handling equipment, e.g. elevators, fork-lifts, trucks, etc.
16. Manual transfer arrangements.
17. Vapor recovery systems.
18. Weigh-bridges

MACHINERY

1. Accessibility for maintenance and operations.

    a. Provision to prevent over-heating, including friction heat.

    b. Possible damage to fire protection equipment from machine failures.

    c. Protection of pipelines from vehicles, including lift trucks.

2. Emergency stop switches.

3. Important safety codes and standards that may be applied.

4. Dispersion of gas or liquid leakage.

5. Compressor house ventilation.

6. Monitoring for compressor vibration and axial vibration.

PROCESS

1. Chemicals - fire and health hazards (skin and respiratory), instrumentation, operating rules, maintenance, compatibility of chemicals, stability, etc.

2. Critical pressures and temperatures.

3. Relief devices and flame arresters. Identification of HP/LP interfaces and design for worst case LP relief.

4. Coded vessels and suitable piping material.

5. Methods for handling runaway reactions.

6. Fixed fire protection systems - CO2, foam, deluge, halon.

7. Vessels properly vented, safe location.

8. Permanent vacuum cleaning systems.

9. Explosion barricades and isolation.

10. Inert gas blanketing systems - listing of equipment to be blanketed.

11. Emergency shutdown valves, switches and alarms, location from critical area, action time for relays. Need for high integrity systems.

12. Fireproofing of metal supports.

13. Safety devices for heat exchange equipment - vents, valves, and drains.

14. Expansion joints or expansion loops for process steam lines.

15. Steam tracing - provision for relief of thermal expansion in heated lines.

16. Insulation for personnel protection - hot process, steam lines and tracing.

17. Static grounding for vessels and piping.

18. Cleaning and maintenance of vessels and tanks - adequate manholes, platforms, ladders, cleanout openings and safe entry permit procedures.

19. Provisions for corrosion control.

20. Pipeline identification, during construction and in operation.

21. Radiation hazards including personal protection for fire fighters - processes and measuring instruments containing radio-isotopes, x-rays, etc.

22. Important safety codes and standards that may be applied.

23. Suitable materials of construction.

24. Fire insulation of vessels and vessel supports.

25. Adequate spacing of equipment items and plant sections, alternative 'separation' means, e.g. steam curtains of fire-walls.

26. Contamination through common-line usage.

27. Washing of piping and valves.

28. Diversion of hazardous materials to the wrong vessels.

29. Dumping or blowdown facilities in the event of an incident.

30. Total cooling water failure, cooling system reserve capacity.

31. Remote isolating valves.

32. Monitoring of materials of construction on low-temperature, high-temperature, high-pressure or corrosive service. Minimizing effect of stress corrosion, embrittlement and creep. Suitability for emergency conditions.

33. Possible stack explosions and need for purge gas.

34. Furnace safety controls.

35. Vessel depressurizing when overheated.

36. Joint leakage protection, e.g. steam quench rings.

37. Pump location and protection. Need for double seals.

38. Any need for high-integrity protective systems for unusual hazard.

39. Segregation of hazardous pipelines.

40. Unexpected presence of water in a high-temperature process system.

41. Pyrophoric-forming materials. Exposure and handling precautions. Safe storage / burial area.

INSTRUMENTATION

1. DCS - redundant data highways, cyber security, etc.

2. Safety instrumented systems (manuals, ITPM plan, etc.)

3. Instrumented IPLs (alarms, control loops)

4. Valves located outside fire hazard areas or fire-safe

5. Failsafe position of valves

6. Hazardous area protection for instruments

7. Lightning protection for instruments

8. Cables fire-proofed and/or failsafe circuits

9. Process safety response times identified

10. Operator response to alarms identified

11. Bypass/inhibit management system in place

12. UPS systems

SAFETY EQUIPMENT FACILITIES

1. Dispensary and equipment.

2. Ambulance.

3. Fire truck.

4. Fire alarm system.

5. Fire whistle and siren - departments, inside and outside.

6. Fire pumps - approved.

7. Sanitary and process waste treatment.

8. Snow removal and ice control equipment.

9. Safety showers and eye wash fountains.

10. Safety ladders and cages.

11. Emergency equipment locations - gas masks, protective clothing, fire blankets, inside hose streams, stretchers, etc.

12. Laboratory safety shields.

13. Watchmen stations.

14. Hose houses - type, location, hose and allied equipment.

15. Instruments - continuous analyzers for flammable vapors and gases, toxic vapors, etc.

16. Communications - emergency telephones, radio, public address systems, paging systems, safe location and continuous manning of communication center.

17. Important safety codes and standards that may be applied.

18. Adequate compressed air and instrument air supplies.

19. Protection of utilities supply, emergency supplies, e.g. power and water.

20. Full-time Chief Fire Officer, full-time firemen.

21. Adequate fire-fighting machines.

22. Foam-forming stocks.

23. Outside back-up facilities, times involved, e.g. in medical assistance.

24. Fire and smoke detectors.

25. Vapor leak detectors.

26. Diesel engine precautions in hazardous areas. Hot exhausts / surfaces, flame arrestors, over-speed trip, etc.

27. Air compressor intake location.

28. Protection of critically important areas or single equipment elements, e.g. by location etc. Consideration of utility failures, vapor travel, remote explosions, flooding, vehicles, etc.

CLOSED RELIEF SYSTEMS

1. Radiation hazards from flares, distance from other facilities, low-level flares. Also toxic emissions (personnel and environmental effects).

2. Need for separate high-pressure and low-pressure flare systems.

3. Hydrate formation. Need for separate wet and dry gas flare systems.

4. Relief system isolation valves, need for remote actuation.

5. Need for heat tracing and/or low temperature specification.

6. Need for methanol injection.

OPERATIONAL SAFETY PROCEDURES

1. Safety committees, regular meetings.

2. Regular plant inspections / audits. Formal response and follow-up to findings.

3. Housekeeping.

4. Maintenance.

5. Adequate operating instructions, availability to all operators, regular updating, emergency instructions.

6. Regular proof testing of automatic protective instrumentation.

7. Work permit system.

8. Reporting of incidents and near-misses, and procedures to investigate and act.

9. Updating of key plant documents, independent technical audits.

10. Adequate quality of labor.

11. Personnel training, refresher courses, operator fire and safety training, use of extinguishers, adequate first-aid firefighting, training in type of fire to be expected, familiarity with protective devices. Evacuation and escape training.

12. Adequate fire safety organization structure.

13. Fire safety strategy; written and displayed fire safety policy; manager training.

14. Serious contingency plans. Alarm procedure, rapid plant shutdown, systems, protective clothing, emergency equipment, nominated people and deputies, communications, nominated safe assembly points, nominated disaster control center, arrangements with all outside organizations, action drills, salvage and other contingency arrangements.

15. Strategic display of notices.

16. Test procedures and facilities to prevent equipment failures, e.g. hydraulic testing, ultrasonic thickness testing, radiographic flaw detection, internal viewing facilities, leak testing, etc.

17. Security; fencing, alert systems, entry checks, identification, adequate lighting.

18. De-matching arrangements.

TECHNICAL SAFETY PROCEDURES

1. Loss prevention. Avoidance of most hazardous processes (extreme parameters, toxic / explosive materials, etc.). Minimization of hazardous inventories. Minimization of personnel (unmanned, low manned, remote operation, etc.). Separation of hazards, personnel and ignition sources.

2. Hazard identification. HAZOP study (process & utilities), FMEA, checklists, etc. Formal follow-up of actions. Re-HAZOP of late design changes.

3. Hazard analysis. Consequence analysis, fault tree, event tree, etc. of major hazards (fire, explosion, toxic, dropped load, etc.)

4. Risk analysis, QRA, LOPA, risk matrix.

5. Human factors. Human error, human intervention, organizational structure, reporting relationships, communications, multi-tasking, multi-skilling, training, qualifications, shift rota and leave arrangements, contractor/staff ratios and recruitment, motivation and morale incentives, working environment, personnel continuity, ergonomics, etc.

6. Technical safety audit. Phase gate reviews, cold eyes reviews, project safety reviews, theme audits.

7. Quality control, quality assurance programs (design, procurement, fabrication, construction, materials handling, maintenance, inspection and testing.

APPENDIX D REFERENCES

Dow's Safety and Loss Prevention Guide. Hazard Classification and Protection. Chemical Engineering Progress - Technical Manual.

D. F. Drewitt. The Insurance of Chemical Plants. I. Chem. E. Course on 'Process Safety - Theory and Practice', Department of Chemical Engineering, Teesside Polytechnic, 12-15th July, 1976.

Commercial Union Risk Management Technical Report No. 2. Fire and Explosion Risk Control in the Petrochemicals Industry.

G. Armistead. 'Safety in Petroleum, Refining and Related Industries'. 2nd Edition 1959. John G. Simmonds & Co. Ltd.

Fawcett and Wood. Safety and Accident Prevention in Chemical Operations. John Wiley and Sons.

C. R. Spitzgo. Chem. Engr. P. 103, 27th September, 1975.

Safety and Reliability Directorate, Report No. R254

APPENDIX E.
EXAMPLE OF SITE-SPECIFIC DECOMMISSIONING CHECKLIST / QUESTIONNAIRE

Based upon checklist derived from CCPS *Guidelines for Safe Process Operations and Maintenance,* 1995.

The checklist below is intended as a guide to some of the issues that *may* be appropriate for a decommissioning project, and is not exhaustive. Generally, no two projects are the same, and site-specific factors are likely to differ. When using the checklist, the user should consider all project-specific factors before determining which checklist items apply, and may require additional items for their project.

REVIEW OF THE ADEQUACY OF THE PLANT PREPARATIONS

1. Who has surveyed the site, examined the condition of the buildings and structures? Has a report been written indicating problem areas? Do any equipment/ structures require further special advice about the method of removal?

2. Are floors, stairways, etc. safe to use?

3. What liaison has been established among all the parties concerned, i.e., Contractor EHS department, the decommissioning inspector, and plant personnel? Will there be inspections during various stages of decommissioning to ensure that safe practices are in fact being carried out?

4. Is the area for the decommissioning completely isolated? Are piping and services (electricity, etc.) disconnected and/or blinded? Who has made the physical check? Who has checked any rerouted lines, etc.? Are there Plant Modification Sheets? Are there any temporary live services in the area? Are these clearly identified?

5. What checks have been carried out of underground hazards in the area? Does drainage from one process area pass through another process area? What is underneath the soft ground where cranes, etc. are likely to stand?

6. Has a comprehensive list of chemicals, including intermediate compounds, processed or stored in the areas in the present or past been established? Is this list accompanied by all relevant flammable and toxicity data, etc.?

7. Is a safe work permit system in force for opening up each piece of equipment and piping for cleaning? Will there be a separate permit for actually removing each piece of equipment or piping?

8. Can all vessels and piping that have been cleaned, etc. be easily identified? Is there a clear identification system (color code) for equipment and piping to indicate its safety status? For example, may it be removed to the steaming out point?

9. Do any of the vessels have linings that could create process hazards?

10. Has the process supervisor responsible for issuing permits enough knowledge of the buildings and chemicals handled in them? Who will accept the permits? The contractor or company inspector?

11. Where will the demarcation come between work carried out by plant maintenance personnel and work carried out by the contractor?

12. Has all the asbestos lagging been removed?

13. Will safety showers and eye wash bottles be provided in the area?

14. Has a study been made of previous dangerous occurrences and minor accidents that have occurred during the decommissioning of similar process equipment? Will work procedures prevent similar incidents from happening again?

15. Will safety boots, helmets, gloves and goggles be standard issue to all contractor personnel? Will the wearing of this safety gear be mandatory?

16. Will safety harnesses, etc. be available at all times on the site?

17. Will the steaming out point be clearly segregated from the cutting and breaking area to prevent the possibility of confusion? How will equipment be moved from one area to another?

18. Where will the acetylene, propane, and oxygen bottles be stored? Who checks the flexible hoses on the equipment?

19. What steps have been taken to remove all the materials at present stored in the decommissioning area? Some of the materials may be hazardous?

20. Has the paint been tested for lead content? Will the decommissioning contractors' employees be given medical checks if necessary? Who decides?

21. Has the fire prevention officer been consulted about fire points, alarm points, hydrants, etc.? Is he/she satisfied that sufficient access is available for emergencies? Will fire station personnel make an inspection of the site at the end of each working day?

22. Has the electrical classification for the surrounding areas been considered for the selection of the site for hot work, etc.?

23. Is the lighting adequate for the hours of work?

24. Is the decommissioning area adequately signposted?

   a. Danger Decommissioning Area

   b. Danger Asbestos Stripping Ongoing

   c. Decommissioning Traffic Only

   d. Decommissioning Control Office

25. Where will equipment to be recovered and claimed by other company personnel be stored until required?

26. Are any special preparations needed for dismantling windows and cladding? Could the cladding be made of cement or asbestos?

27. Will equipment and structural drawings be available to the contractor and decommissioning inspector to identify loading within each building and the safe approach for dismantling?

28. Are any vessels or piping yet to be cleared of toxic or flammable chemicals? What special arrangements have been made for these items?

29. If certain items are being gas freed while the contractor is dismantling, what safeguards are being undertaken to prevent the exposure of workers to fumes?

30. How will the plant shift managers be kept informed of daily progress and problems?

REVIEW OF THE WORK METHODS AND SUPERVISION

1. Has the decommissioning inspector a thorough knowledge of demolition work and also of the principles of building construction?

2. Are details available showing the process safety management system (e.g., safe work permits and the coordination needed among those involved in the decommissioning)? Will a responsible plant maintenance supervisor be available at all times on the site?

3. Does a list exist of personnel involved in the decommissioning? Will there be any need for access to a decommissioning area? Will there be any need for access to a decommissioning area by persons other than those involved? Will such entry be controlled? What happens if one of the decommissioning specialists is off work?

4. Do all personnel involved in the decommissioning fully understand the process safety management system in force at the Plant?

   - Safe Work Permits
   - Hot Work Permits
   - Management of Change
   - Process Safety Information
   - Emergency Planning and Response
   - Are copies available on site? Will a special alarm system be provided for use in the decommissioning area?

5. Has a consultant structural engineer been appointed? Is there a procedure for obtaining this type of expert advice?

6. Are plant supervisors and the contractor conversant with the following regulatory standards?

   a. OSHA Demolition Standard 29 CFR 1926.850 - 860

   b. OSHA Crawler and Truck Crane Standard 29 CFR 1910.180

   c. OSHA Material Handling Standard 29 CFR 1910.176

   d. OSHA Construction Work Standard 29 CFR 1910.12

   e. OSHA Means of Egress Standard 29 CFR 1910.37

   f. OSHA Scaffolding Standard 29 CFR 1910.29 g. OSHA Medical Services and First Aid Standard 29 CFR 1910.151

   h. OSHA Cutting and Welding Standard 29 CFR 1910.252 i. OSHA Asbestos Standard 29 CFR 1910.1101

7. Who has checked to ensure that all these regulations are met?

8. Should all the relevant state / federal regulations and the plant standards be written into the contract?

9. Is the contractor adequately insured to cover all possible contingencies?

10. Has the contractor employed competent personnel, for example, a sling hand for crane work (thorough knowledge of signals, etc.), and a certified crane operator with previous decommissioning experience?

11. Is there a master plan showing the sequence of decommissioning? Will the results of each step be forecast accurately? Who is responsible for each step? Who is in charge overall?

12. Who will ensure that all construction equipment on site is in good condition and has been regularly inspected to meet all plant and regulatory requirements?

13. What is the standard of housekeeping expected to be on a job of this kind? Have safe access, a means of escape, tripping hazards, holes in flooring, etc. been taken into account? Will any doorways be allocated safe for entry/exit? Will they be protected?

14. Will all the windows and side cladding be removed first?

15. Will any flame cutting take place inside buildings? When within the scope of work will this be done? Are there potential sources of flammable materials inside buildings?

16. Will there be a set procedure for lowering large items of equipment? Will it be necessary to meet with other adjacent area and plant shift managers?

17. Will "long term" permits be issued for any purpose? If so, where and for what purpose?

18. Who will renew permits each morning to allow work to progress safely and without delay?

19. Where will the permits be kept? Does everyone understand the permit system?

20. Has safety clearance been given for using hoists within buildings?

21. When permits are issued, how will the contractors differentiate between equipment for:

    a. Scrap

    b. Removal for sale

    c. Retainment for use by the company at other locations

22. Is it possible to overload a truck with scrap? Will it pass a weigh station?

23. What decommissioning methods are going to be used on tanks and vessels?

24. What "incident" or "accident" documentation will be held by:

    a. Plant?

    b. Contractor?

25. Has the company's head office EHS staff been informed of this work? Will they be making regular inspections?

26. Will gas tests for flammability be conducted on vessels before removal and again before doing hot work?

27. Because it will aid the continuity of work, will preliminary atmosphere tests (flammability, toxicity) be made by plant operations shift supervisor before any unit or equipment is isolated? Such tests will help determine the care needed when isolating units or equipment and when preparing the plant (steaming out).

Obviously, a plant laboratory test must be done before any hot work or entry situation.

28. Will grinding wheels be used? If so, has a competent person been assigned to check and change the wheels?

STUDY OF THE EFFECTS OF THE WORK ON THE SAFETY OF OTHER PLANT ACTIVITIES

1. What effect will the decommissioning have on surrounding process units, rail tracks, tank farms, etc.?

2. What effect will the area electrical classification for the surrounding areas have on the decommissioning program?

3. How will personnel working around the decommissioning area be protected from falling debris?

4. Will scrap be removed off site as soon as a permit has been issued? What form of "Security Pass Out" will be required at the main gate with a plant signatory approving inspection and certification of the load?

5. How will access to firefighting equipment being blocked with scrap be prevented?

6. Roadways in the area may be narrow. What measures are anticipated to keep them clear? Will traffic not involved with decommissioning be rescheduled to outside normal working hours?

7. Will the decommissioning of a process unit affect truck loading?

8. Will all permits in the area be countersigned by operations shift supervisors responsible for the units, etc. in the surrounding areas?

9. How will any rail tracks and pipe racks close to decommissioning activities be protected against falling objects, etc.?

10. Have all plant departments been informed of the decommissioning as it may affect their operations in the area?

11. Will the decommissioning area be fenced off in one large area or will parts of the working area be fenced off separately (e.g., steaming point, scrap collection, and rubble collection areas)?

12. Will the movement of any materials (e.g., recovered chemicals from vessels placed in drums) be likely to expose persons inside the plant? If the drums are disposed of outside of the plant, will the general public be exposed to any hazards?

13. Will vehicle loads be inspected before leaving the plant?

14. Will all asbestos-containing material leaving the plant be suitably packaged and labelled? Will it be disposed of at an approved site?

15. Will any materials such as rubble be disposed of within the plant?

16. Will the routes for the contractors' vehicles be clearly defined within the plant?

APPENDIX F. TYPICAL PROJECT DOCUMENTATION

Project documentation comprises process safety information (PSI) and other information. All documents associated with the 20 elements of CCPS RBPS are considered PSI. Other project documentation that has a safety content may also be categorized as PSI, such as design studies and calculations, and stage gate review reports. Project documentation associated with commercial agreements and finances is unlikely to be classified as PSI.

Table F-1 represents typical documentation that *may* be appropriate for an engineering project, but it is unlikely that all documents will be necessary for a specific project. This table is not intended to be comprehensive, and additional documents may be required for a specific project.

Table F-1. Typical Project Documentation

Documentation Category	Typical Project Documentation

Project Documentation	• Project procedures (e.g. change management, document management, etc.)
	• Strategies/Plans:
	   o Development plan
	   o Technology plan
	   o Regulatory approval
	   o Design hazard management strategy
	   o HIRA plan
	   o Integrity management/engineering assurance
	   o Contracting/procurement
	   o Quality management, including process equipment inspection/functional test
	   o Functional safety management plan
	   o Construction plan, construction site organization plan, construction EHS plan
	   o Pre-commissioning plan
	   o Resourcing/training

Documentation Category	Typical Project Documentation

	o Commissioning/startup plan, including test runs
	o Process Safety Plan
	o EHS Plan
	o Security plan
	• Development Options:
	  o Process technology options
	  o Location specific information (e.g. infrastructure, population, meteorology, topography, etc.)
	• Deliverables:
	  o Statement of Requirements (SOR)
	  o Basis of Design (BOD)
	  o Cost estimate
	  o Project schedule
	  o FEED package
	  o Project Execution Plan (PEP)
	  o Design package for construction
	  o Handover package
	• Design philosophies
	  o Operations and maintenance
	  o Blowdown, pressure relief & flare system, fire & gas detection, fire protection, process control, alarm management, construction, etc.
	• Design studies, calculations, assumptions, design intent:
	  o ISD,
	  o Spacing & layout,
	  o Blowdown/relief/flare,
	  o Fire & gas detection and suppression,
	  o Firewater analysis,
	  o SCE vulnerability,
	  o RAM,
	  o SVA,
	  o SIMOPS,
	  o Safety instrumented system (SIS) assessment,
	  o Safety integrity level (SIL) determination/verification,
	  o Facility siting (blast, fire, toxic),

Documentation Category	Typical Project Documentation
	<ul><li>○ Temporary refuge/shelter-in-place impairment assessment,</li><li>○ Human factors analysis,</li><li>○ Corrosion,</li><li>○ Structural,</li><li>○ Electrical system protection, electrical loads, short circuit, earthing/grounding,</li><li>○ Pipeline integrity monitoring,</li><li>○ emergency response/evacuation, escape, and rescue analysis,</li><li>○ Dropped object,</li><li>○ Decommissioning, etc., follow-up records</li></ul><ul><li>Design reviews:<ul><li>○ P&ID, 3D model, operability, constructability, inter-discipline, value engineering, deviations from engineering standards, etc.), follow-up records</li></ul></li><li>Design Case for Safety</li><li>Contracts/purchase orders for equipment, materials, services</li><li>Contractual and financial documentation to be retained in respect of legal liabilities, warranties/guarantees, financial audits, and tax requirements</li><li>Commercial agreements</li><li>Construction:<ul><li>○ Engineering queries, RFIs</li><li>○ Punch-lists</li></ul></li><li>Commissioning:<ul><li>○ Commissioning (with safe chemicals) report</li><li>○ Commissioning team operating logs, shift handover notes, and records for each step in the commissioning procedures</li><li>○ Comprehensive file for each system and item of equipment showing status of each commissioning step performed</li><li>○ Startup (with process chemicals) report</li><li>○ Startup team operating logs, shift handover notes, and records for each step in the startup procedures,</li></ul></li></ul>

Documentation Category	Typical Project Documentation
	o Comprehensive file for each system and item of equipment showing status of each startup step performed • Project performance: o Expenditure, schedule, progress, quality, rework, process safety, EHS, reports, records • Permits from local authorities and regulators • Project Hazard/Risk Register • Action tracking database • Technical Peer Reviews, reports, follow-up records • Stage Gate Reviews, reports, follow-up records
Operator Documentation	• Corporate memory: knowledge / information gained from similar plant experience • Process safety and EHS management system, policies, procedures, objectives • Document management system, procedures, control documents, retention policy, loss/fire protection, etc., • Recruitment records, qualifications, etc. (e.g. operators, technicians, engineers, EHS, admin) • Outstanding punch-list items inherited by the Operator, follow-up records • Outstanding action items inherited by the Operator, follow-up records • Operations Case for Safety • HR, training and performance assurance records for any Project staff seconded to the Operator • Operations Stage Gate Review, report, follow-up records • Structural engineering survey report to identify potential hazards of deconstruction/demolition, original structural drawings, calculations, etc • Engineering/safety study of impact of deconstruction and/or demolition on surrounding facilities • Safety plan/report for deconstruction and/or demolition, including oversight

Documentation Category	Typical Project Documentation
	• Security plan, site perimeter barricade, warning signs, etc. • Deconstruction and/or demolition procedures, task sequence, groundwater protection, segregation of equipment and materials for re-use, recycle, and/or scrap, etc.
Process Safety Culture	• Process safety culture program • Culture assessments and follow-up records
Compliance with Standards	• Compliance with standards program • National/local regulations • Corporate policies, standards, practices • Industry codes and standards • Technical standards, if any, developed by Project • Variance/waiver approval, records • Citations/improvement notices
Process Safety Competency	• Process safety competency program, standards, records
Workforce Involvement	• Workforce involvement program • Roles/responsibilities • Plan, procedure, records
Stakeholder Outreach	• Stakeholder outreach program, objectives, plan • Commitments to third parties • Meetings with stakeholders, minutes, actions, follow-up records
Process Knowledge Management (includes Project design documents)	Process knowledge management program, policy Hazardous chemicals information: • Safety Data Sheets (SDS) (fire/explosion, human/environmental toxicological, corrosivity, thermal hazard, flammability, dust/powder hazard, etc.) • Reactivity matrix • Other sources

Documentation Category	Typical Project Documentation

	Process Technology:
	• Block flow diagrams
	• Process flow diagrams (PFD)
	• Process description
	• Process chemistry
	• Mass/energy balance
	• Inventory of chemicals
	• Records of evaluation of consequences of deviation from normal process conditions
	• Safe upper and lower limits for temperature, pressure, flowrate, level, composition and other key parameters
	• Limitations for safe operation
	Process equipment:
	• Plot plan
	• Piping and instrumentation diagrams (P&ID's)
	• Piping service index with piping service specifications
	• Piping isometric drawings
	• Instrument index and specifications including description of operating conditions, materials of construction, process fluids
	   o Loop diagrams
	   o Termination diagrams
	   o Location plans
	   o Control narratives
	   o Network diagrams
	   o Control programs
	• Cause and effect charts
	• Electrical/hazardous area classification drawings
	• Electrical one line diagrams
	• Equipment datasheets/specifications including materials of construction, reference to applicable codes, etc.
	• Relief system design and design basis, pressure safety valve size calculations
	• Control room and process buildings design, fire/explosion resistance

Documentation Category	Typical Project Documentation

	• Heating, ventilation, and air conditioning (HVAC) systems related to process safety (fume or dust controls)
	• Protective systems
	  ○ Safety critical equipment (SCE) list
	  ○ Safety instrumented systems (SIS)
	  ○ Safety requirements specification (SRS)
	  ○ Safety systems (e.g., interlocks, detection or suppression systems)
	  ○ Performance standards
	  ○ SIL verification calculations
	  ○ Safety logic programs
	  ○ Cause & effect charts
	• Design codes and standards used for design
	• Design basis documents, which refer to compliance with good engineering practice
	• Fire zones, passive/active fire protection
	• Equipment operating and maintenance manuals from suppliers and vendors,
	• 'As-built' drawings and technical information

Hazard Identification & Risk Analysis (HIRA)	• HIRA program, plan, reports, follow-up records
	• Hazard Evaluation:
	  ○ HAZOP
	  ○ HAZID
	  ○ What If / checklist
	  ○ FMEA
	  ○ SIMOPS
	  ○ Consequence analysis (e.g. fire/explosion, smoke/toxic gas, transportation, etc.)
	• Risk Analysis:
	  ○ Risk matrices
	  ○ LOPA
	  ○ Quantitative risk assessment (QRA) policies, practices
	  ○ Concept risk analysis (CRA) reports, follow-up records
	  ○ QRA reports, follow-up records
	  ○ Corporate risk tolerability criteria

Documentation Category	Typical Project Documentation
	o Risk management philosophy (e.g. ISD, engineering vs. administrative controls) • Facility siting study reports, follow-up records • Human factors analysis, reports, follow-up records (e.g. ergonomics, human performance) • Procedures for each HIRA methodology/technique used • HIRA facilitator qualifications • HIRA team members – records of qualifications, initial and refresher training • HIRA revalidation reports, follow-up records • HIRA studies of deconstruction / demolition hazards, and follow-up records • Information for HIRA studies (e.g. relevant accidents/incidents, process changes, etc.) • List of engineering controls, administrative controls • Communication records (e.g. results/changes to all affected employees, changes to training and procedures)
Operating Procedures	• Operating procedures program, format/content, etc. • Operating manuals from suppliers and vendors • Commissioning and startup procedures • Commissioning team operating logs, shift handover notes, and records for each step in the commissioning procedures • Startup team operating logs, shift handover notes, and records for each step in the startup procedures • Performance test run procedures, including operating parameters, sample analysis, etc., • Performance test run results (by each equipment item and each system), data, sample analyses, and follow-up records • Changes to operating procedures as result of commissioning, startup, and operating experience • Operating procedures: o Startup procedures

Documentation Category	Typical Project Documentation
	<ul><li>o Startup procedures after maintenance/turnaround</li><li>o Normal operation procedures</li><li>o Shutdown procedures</li><li>o Shutdown procedure for facility that will not re-start again</li><li>o Emergency procedures</li><li>o Temporary procedures</li><li>o Decommissioning procedures, task sequence, depressurization, deinventory remaining materials, cleaning, decontamination, purging, inerting, etc.</li><li>o Recommissioning procedures for mothballed process unit/equipment</li><li>o Checklists</li><li>• Supervisor/operator logs</li><li>• Shift handover procedures</li><li>• Preparation for maintenance procedures</li><li>• Periodic procedure review records</li></ul>
Safe Work Practices	<ul><li>• Safe work practices program</li><li>• Routine work procedures</li><li>• Safe work practice procedures, JSAs, records, including, but not limited to:</li><li>o Cold/Safe work permit</li><li>o Hot work permit</li><li>o Energy isolation (lockout/tagout)</li><li>o Confined space entry permit, rescue plan</li><li>o Line breaking</li><li>o Drainage and diking</li><li>o Excavation</li><li>o Heavy lift</li><li>o Mobile heavy machinery/vehicles</li><li>o High voltage electrical systems</li><li>o Working at height</li><li>o Radioactivity, NORM</li><li>o Working over/near water</li><li>o Diving</li><li>o Asbestos, PCBs, heavy metals</li><li>o Handling explosives</li></ul>

Documentation Category	Typical Project Documentation
	o Process unit/facility access by non-operations personnel • Work permit, JSA records • Work permit audit reports • Training/qualification records for work permit authorities/issuers

Asset Integrity & Reliability	Process equipment integrity:
	
	• Quality management program, procedures
	  o Fabrication quality records, including FAT, NDT, weld radiographs, non-conformances, certificates, QA reports, technician/inspector qualifications, etc.
	  o Manufacturer/supplier documents, certificates, mill tests, etc.
	  o Pre-commissioning quality records, including SAT, NDT, field weld radiographs, hydro-tests, flushing/cleaning/drying, checklists, non-conformances, baseline data, certificates, QA reports, technician/inspector qualifications, etc.
	  o Quality control (QC) records, certificates, positive material identification (PMI), non-conformances and follow-up
	  o Quality assurance (QA) reports, follow-up records
	  o Initial/baseline inspection reports, records
	• Installation records, mechanical completion certificates/dossier
	• Master equipment list, instrument index,
	• Criticality analysis, reports, SCE list,
	• Equipment datasheets/specifications including materials of construction, codes and standards, design calculations,
	• Control systems records for DCS, programmable logic controllers (PLC), SIS, interlocks, software, functional specifications, etc.
	• Performance requirements for independent protection layers (IPLs), including safety

Documentation Category	Typical Project Documentation

	instrumented functions (SIFs), SIF proof test procedures, SIL analysis reports, SIL verification
	• Functional safety assessments (FSA)
	• Alarm list, set points, alarm flood study
	• Relief and vent system records
	• Piping systems records
	• Comprehensive file for each system and item of equipment showing status of each commissioning step performed
	• Comprehensive file for each system and item of equipment showing status of each startup step performed
	• Reliability analysis, reports, follow-up records
	• ESD, trip, and protective system activation records
	• Asset Integrity Management procedures:
	   o Maintenance manuals from manufacturers, suppliers and vendors
	   o Equipment/material preservation procedures, records
	   o ITPM plan, procedures, tasks, frequencies, records, technician/contractor qualifications, data analysis and plan update, inspector recommendation follow-up records, changes due to operating experience
	   o Emergency maintenance procedures
	• Maintenance management system, procedures, software
	• Spare parts list, preservation procedures
	• Process equipment deficiencies,
	   o Equipment failure analysis, reports, follow-up records
	   o Deficiency correction records, repair/replace/re-rating procedures/records, technician/contractor qualifications
	• ITPM personnel:
	   o ITPM qualifications, training, materials, records
	   o Contractor ITPM qualifications records
	• Code and standard compliance records

Documentation Category	Typical Project Documentation
	• Mothballed equipment: ○ Preservation procedures, ongoing ITPM tasks (if necessary) to maintain assets in a state of readiness or near-readiness

Contractor Management	• Contractor management program
	• Pre-qualification procedures, screening/selection, records
	• Contractor qualifications/competency, EHS/PS performance, records
	• Pre-qualified contractor list
	• Construction/turnaround pre-mobilization plan
	• Contractor records:
	  ○ EHS/PS performance metrics, contractor management system, safety plan, safe work practices, sub-contractors, supplied equipment, etc.
	  ○ Bridging documents
	• Contractor administration, orientation/training, materials, records
	• Safety oversight plans, procedures, performance records
	• End of contract evaluation reports, records

Training & Performance Assurance	• Training and performance assurance program, procedures, training matrix/schedule, initial/refresher • Employee qualifications/competency records • Employee (and contractor) training records, verification (written test or other means) • Trainer qualifications • Training procedures, materials • Evaluation of effectiveness of training program

Management of Change	• Management of change program:
	  ○ Project change management procedure, DCNs, records
	  ○ Operator management of change procedure, files (including scope, design information,

Documentation Category	Typical Project Documentation
	HIRA studies, technical reviews, authorization, follow-up records, link to operational readiness review, etc.) • Communication records
Operational Readiness	• Operational readiness review program, procedures • Operational readiness review/PSSR reports, checklists, records

Conduct of Operations	• Conduct of operations program, procedures, checklists
	• Routine tasks:
	   o Operator rounds, checklists, logs
	   o Shift handover notes
	   o Housekeeping inspection records
	   o PPE audits
	• Periodic system evaluations
	   o Protective systems bypass/inhibit procedure, records
	   o Locked open/locked closed valve checks
	   o Interstitial pressure between rupture disc and relief valve checks
	   o Safety equipment checks (fire extinguishers, safety showers/eye wash, SCBA, PPE, etc.)
	   o Work schedules, shifts, hours, overtime to avoid fatigue records
	• Physical systems maintained
	   o Limits on operation, alternative safety measure records
	   o Use of flexible hose/connection/jumper records
	   o Equipment labeling/warning signs, lighting, etc. checks, records
	   o Building pressurization system checks, emergency shutdown procedures
	• Commissioning temporary operations, e.g. blind and strainer lists

Documentation Category	Typical Project Documentation

Emergency Management	• Construction site emergency response plan, procedures, drills and follow-up records
	• Emergency management program
	• Emergency response plan
	  o Reporting and alarms
	  o Evacuation/escape procedures, routes, muster/assembly points, headcount
	  o Emergency equipment, PPE, rescue, firefighting, etc.
	  o Site plan drawings for plant areas, muster points, evacuation routes, wind socks, control rooms, locations of rescue and firefighting equipment, etc.
	  o Response plan for small releases
	  o Key personnel organization chart and job descriptions
	• Shelter-in-place building procedures
	• ITPM for emergency facilities, alarms, lighting, equipment and PPE, etc. records
	• Emergency drills and table-top exercises, and follow-up records
	• Outside agency coordination plan
	• Community plot plan which indicates location of all support agencies with contacts information
	• Liaison/communication with stakeholders, e.g. local community
	• Emergency response plan for deconstruction and/or demolition

Incident investigation	• Incident reporting and investigation program • Incident/near-miss reporting procedures, reports, forms, records • Evidence preservation procedures • Incident investigation procedures, reports, follow-up records • Root cause analysis (RCA) methodology/technique • Investigation team members (records of training and qualifications)

Documentation Category	Typical Project Documentation
	• Incident and root cause trend analysis, follow-up records
Measurement & Metrics	• Measurement and metric program, procedures • KPI records • Periodic review/analysis and follow-up records • Communication records
Auditing	• Audit program, policy, plan, procedures • Audit protocols • Periodic self-assessment/audit reports • Audit action follow-up records
Management Review & Continuous Improvement	• Management review and continuous improvement program, policy, procedures, plans • Review meeting information, meeting minutes and follow-up records • Communication records

APPENDIX G. STAGE GATE REVIEW PROTOCOL FOR PROCESS SAFETY

INTRODUCTION

Many operating companies within the process industries conduct reviews at key milestones during the life cycle of capital projects. These reviews are variously known as stage gate reviews, 'cold eyes' reviews, peer reviews, project technical safety reviews, etc., and are normally conducted by an independent and experienced multi-discipline team familiar with the relevant facility/process and technology. The objectives and scope also vary between companies, but have a strong focus on process safety, although they may also include technical and EHS issues. A fuller description is included in Chapter 2, Section 2.10.

Typical process safety scopes for the reviews are addressed at each stage of the project life cycle in the appropriate chapter. The following tables represent typical issues that the stage gate review team may use as a protocol to cover the scope at each stage. No two projects are likely to be the same, and the protocol is not intended to be exhaustive. Users may omit issues that are irrelevant and add new issues based on their specific project.

STAGE GATE: FEL-1

Review the technology and process for potential Process Safety risks.

- Hazardous properties (toxicity, flammability, reactivity) of materials (feeds, products, intermediate streams and discharges).
- Complexity of processes and severity of operating conditions.
- Potential for major accident risk (toxic / flammable inventories, high pressure / temperature operations, logistics, location considerations such as potential impact on the public and the workforce).
- Incident history of similar technology.
- HIRA for similar technology.

Confirm all project options were assessed for inherently safer design (ISD).

- Substitution/minimization/moderation/simplification
- Hazard elimination/prevention/control/mitigation
- Passive/active/procedural risk reduction

Review all potential locations for possible Process Safety impacts on neighboring facilities, local community and environment.

- Confirm that the project has identified environmental risks (e.g. protected/ sensitive areas, etc.) and determined that the risks can be sufficiently mitigated to meet Company policies and comply with applicable regulations.
- Regulatory and permitting concerns.
- Geo-hazards
- Potential security risks (e.g. plant/pipeline security).
- Crisis management and emergency preparedness.
- Logistics/transportation risks for raw materials and products.

Examine project options for issues that can significantly influence Process Safety performance.

- Contractors, partners, joint ventures, other stakeholders, including security due diligence.
- Impact/conflicts with company policies/public commitments,
- Life cycle of the site, equipment, and products.
- Construction risks.

Identify Process Safety uncertainties/unknowns of each project option.

- The need for pilot plant testing.
- Reactivity and chemical instability risks.
- Corrosivity data.
- Toxicology data, chronic effects from exposure.

STAGE GATE: FEL-2

Confirm that Process Safety hazards inherent in the proposed development warranting special attention, or uncertainties that need further investigation, have been identified.

- Topics considered typically include:
  - i   Properties of the process materials (SDS required for all raw materials, products, and intermediate streams).
  - ii   Reactive chemicals mixing matrix and reactivity risks.
  - iii Processing conditions (normal, startup, shutdown, & excursion).
  - iv Process inventories, fire and explosion potential, stored energy, toxic release potential.
  - v   Impact on company policies, commitments, targets, and strategies.
  - vi   Project strategy for inherently safer design.
  - vii Transportation hazards / risks.

- Review project plans to collect information on previous incidents and lessons learned from asset integrity reports of similar processes and to address them in the design.

- Review the impact of the proposed development on existing facilities, and vice versa, both onsite and offsite. Review the interfaces between new and existing facilities.

- Review the proposed location to ensure that any characteristics of special Process Safety concern have been acknowledged; e.g., stakeholders, local community, local environment, reputation risks, security risks, geographical features, geo-hazards (tsunami, earthquake, etc.) hydrology, meteorological conditions, etc.

- Review the expected emissions profile and natural resource use. The following will typically need to be considered over the full range of possible conditions; i.e., steady state, startup, shutdown, emergency release, batch change, normal, maximum, end of run, etc. Review mitigation options and check that anticipated abatement technology is in line with company experiences and is sufficient to meet company policy and comply with all applicable regulations. Identify uncertainties with:

i   Vents, oily wastes, wastewater, hazardous & non-hazardous solids & sludge.
ii   Flares and relief discharges to the atmosphere.
iii   Noise, smell, visual impact.
iv   Point source & fugitive emissions.
v   Greenhouse gas, NOx, SOx, and VOC emissions.
vi   Capabilities for offsite treatment and disposal.
vii   Use of scarce/non-renewable resources (including sustainability of fresh water supply).

- Review draft construction philosophy including numbers in workforce, access to site, interactions with existing operations, location of temporary construction facilities, onsite and offsite traffic routes, significant hazards and environmental impact likely to be encountered during construction, including use of substances/materials hazardous to health and also the application of legislation.

Confirm that acceptable solutions for hazards and uncertainties are available or are capable of being developed within the timeframe and organization of the project.

Confirm that all Process Safety concerns relating to the characteristics of the full life cycle of the project, novel technology, and the nature of the location have been identified.

Confirm all applicable regulations, standards, and relevant company expectations have been identified.

- Review the Project's approach to asset integrity management (AIM), including the list of approved specifications, codes and standards. Confirm that these are appropriate, taking into account current industry conventions, company engineering and technical practices.

- Review the project's plan for quality assurance of the design, including clarity of business need, competencies, selection of the engineering contractor, key skills of the project team, specialist support, and verifications that will deliver a quality design.

- Review contractor selection/procurement strategy and alignment with company expectations.

- Review the strategy for assuring compliance with the company's Process Safety program and AIM standards.

- Review project's strategy for assuring safety of the project workforce including:

  i   Application of the company's safety expectations and rules,
  ii Work permitting procedures, and
  iii Other relevant company standards.

- Review local and national regulatory requirements, the project's outline proposals for acquisition of a "license to operate" and plans for permit consents.

Confirm an adequate Process Safety Plan has been established, communicated to the project team, and endorsed by management for subsequent stages.

- Review the project's Process Safety Plan for the following:

  i   Adequate external, internal, specialist and peer assist/reviews and Process Safety audits have been identified and scheduled.
  ii Identification of the need for additional studies, pilot plant tests, etc., in order to address Process Safety uncertainties.
  iii A plan has been developed for conducting stage gate reviews.
  iv The program and intended organization for the procurement, execution and review of Process Safety studies are in place.
  v   Studies are scheduled to be completed in good time to ensure that actions can be incorporated into the design and resources are available in order to carry out the work.

- Review the intentions for the tracking of actions arising from Process Safety studies including any commitments made.

- Review the intended arrangements for resourcing and training of the workforce for the development.

Confirm an adequate Process Safety risk management strategy, including future HIRA studies, has been established.

- Review the potential for significant Process Safety incidents and initiate a Concept Risk Analysis (CRA) if significant hazard potential is identified.

- Review the project's assessment for major accident risk and its plans for addressing identified risks. How does the project affect the site's risk profile vs. company risk criteria? How will the project demonstrate continuous reduction of risks?

- Review the design principles for safety and security risk control and mitigation features, including:

  i   Process containment

ii   Unit separation / layout
iii Land take and land use, without compromising safety
iv   Protective systems (i.e., SIS, relief systems, fire protection, etc.)
v   Isolation philosophy
vi   Protection of personnel
vii Emergency service provisions/emergency response plan
viii Physical security requirements – identification of critical / vulnerable points

- Review plans for development of the project's hazard/risk register.

- Review project strategy for assuring that inherently safer design (ISD) is addressed during design.

- Review Process Safety risk management strategy for outsourced facilities.

- Review project's strategy for ensuring a comprehensive and high quality HIRA (e.g. HAZOP) and SIL assessment review, including resourcing of the leader and team, competency/operational experience, quality of P&IDs & design information, process safety information, and documentation requirements.

- Review proposals for ensuring that the scope of the HIRA study includes all aspects of the development, including vendor packages, with a significant hazard potential, facility siting study, etc.

- Review project's strategy for resolution of findings, including assignment of responsibility and handling of recommendations outside of project responsibility.

- Review the proposed timing of the HIRA study, and in particular arrangements for incorporating findings in the specifications for early procurement items.

- Review project strategy for management of change and hazard review of post-HIRA changes.

STAGE GATE: FEL-3

Confirm that Process Safety studies, including specialist reviews, are being satisfactorily addressed and followed up.

- Review progress on addressing recommendations/commitments arising from previous Stage Gate Reviews and other Process Safety studies, including HIRA (e.g. HAZOP, CRA, SVA, etc.).

- Determine if a system is in place to register and track all Process Safety recommendations, including commitments and full documentation of actions taken to resolve recommendations.

- Review whether there have been any changes in project scope and confirm that the relevant Process Safety studies have been updated.

- Review any project studies on hazard identification. Confirm that all areas of the new facilities and interactions with adjacent facilities have been considered. This should include hazards that could arise during normal, startup, shutdown, maintenance, transient, and emergency operations and those associated with construction.

- Review the project's capability to resolve identified hazards within the timeframe and organization of the proposed project.

- Determine if appropriate, that Specialist Reviews have been initiated; e.g., instrumentation and control, logistics, security, human factors, ergonomics.

- Review project's plans for studies targeted to improve Process Safety performance; such as noise assessment, air emissions modeling, security by design, design for safe construction, ergonomics, and 3-D model reviews of layouts for personnel hazards, accessibilities, and locations of safety equipment.

Confirm that Process Safety related aspects of the engineering designs meet or exceed regulatory requirements, and that satisfactory project codes and standards have been identified, and design philosophies have been established.

- Review the Project's list of specifications, codes and standards to confirm that these are appropriate; take into consideration that engineering technical practices are consistent with site practices. Verify that the design is being developed in compliance with these specifications, codes, and standards.

- Determine if the basis of design (BOD) acknowledges both established good practice and "lessons learned" in the application of the technology concerned.

- Review the proposed BOD for all of the individual systems that are likely to influence Process Safety performance and satisfy regulatory requirements and company policies, and external commitments.

- Review the proposed means of maintaining operations within the design envelope, and identify and assess any special features.

- Determine if HP to LP interfaces and primary pressure protection accomplished completely or partially by instrumentation have been identified and appropriately considered in the design.

- Review project's application of inherently safer design (ISD) and design philosophies relevant to Process Safety performance. These will typically include over-pressure relief/over-temperature protection, isolation, control and shutdown systems, safety instrumented systems, fire and gas detection, fire protection, emissions control, pollution prevention, noise control/abatement, and methods for detecting and correcting emissions and discharges.

- Review the BOD for the flare system including layout, materials of construction, and instrumentation used to reduce flare loads.

- Review the project's methodology for determining Safety Integrity Levels (SIL) for Safety Instrumented Systems and if it received appropriate specialist input. Determine if the calculated SILs have been taken into account in proposed hardware configurations, and if a Safety Requirements Specification (SRS) has been prepared.

- Review safety critical equipment (SCE) identification.

- Review the project's assessment of major accident risks for potential impacts on the site's risk profile and the resolution of findings. Review design philosophy for all occupied buildings and confirm that the designs will provide adequate protection for personnel, and complies with all regulations and location building codes. Confirm that a thorough analysis of blast overpressure scenarios has been performed.

- Determine if security has been addressed in the design of the facility, including an SVA, cyber security for control systems, etc.

- Review the Project's Process Safety management system to verify compliance with the Company's program.

- Review progress on the proposed strategy for acquiring "license to operate".

- Review whether the methods of installation and construction to be adopted are being addressed in the development of the design, including an assessment of risks.

* Examine the project's sequencing of systems for detailed design and construction and determine if it takes into consideration the order for pre-commissioning, commissioning, and startup, in order to minimize risks.

Confirm that all Process Safety concerns relating to the characteristics of the full life cycle of the project, novel technology, and the nature of the location have been identified.

* Review the design for acknowledgement of all the Process Safety features of both the technology and the location, which demand special attention.

* Review in particular any novel features, which may affect Process Safety performance, and any special considerations, which need to be given to vendor packages.

Confirm that asset integrity management (AIM) / engineering assurance processes are in place.

* Review the engineering authority role for the project for execution in accordance with the Company AIM standards and determine if levels of technical authority for the design have been clearly established in project procedures, understood at all levels, and are being applied.

* Examine waivers to company standards / policies for an appropriate level of review, documentation and approval. Determine if a system exists whereby: when the design contractor or a vendor departs from the specified standards, that the departure is reviewed and approved by the engineering authority and is included in the project's design dossier.

* Review project's plan for Quality Management (QM) of design. Confirm that the competence of personnel key to design integrity has been verified. Confirm that a process has been established to ensure continuity of key personnel during design.

* Review the management of interfaces that exist between multiple designers, vendors, and multiple sites etc. Review the process for inter-discipline checks of the design.

* Review criteria for selection of contractors, subcontractors, major vendors, and outsourced facilities for Process Safety and design integrity considerations and check that due diligence assessments have been carried out on contractors.

Confirm that Change Management procedures are in place.

* Review the proposed change control procedures for application to all changes that might affect Process Safety performance and equipment integrity or the HIRA integrity of the intended development.

- Verify that changes to P&IDs, Cause and Effect diagrams, hazardous area classification (HAC) drawings, etc., post HIRA will be properly controlled and authorized.

- Review the process for inter-discipline checks of design changes.

Confirm that documentation requirements have been addressed.

- Review provisions for supplying of Process Safety Information (PSI) and other related documentation required by Client site operating and maintenance systems.

- Check proposals for data management and transfer into site systems, (e.g., maintenance management, engineering records.)

- Review project and engineering contractor plans to obtain from vendors recommended mechanical integrity procedures for all equipment.

- Examine documentation requirements for the relief and flare systems. Review proposals for the compilation of a Register of Safety Critical Equipment (SCE).

Confirm that a resourcing and training strategy is established.

- Review how the project will address competency requirements for the intended operation.

- Review the outline operating and training strategies for compatibility with safety-related design assumptions. This should include development of:

    i   operating, maintenance and safety procedures;
    ii   manning levels;
    iii training provisions (e.g., simulator);
    iv   preventive maintenance (PM) frequency;
    v   inspection frequencies; and
    vi   emergency response capability and emergency drills for Operations.

- Review resourcing plans for Process Safety support for construction, required training resources, and plans for getting the construction contractor into the right safety mindset.

Confirm that project plans ensure Process Safety preparedness for commencement of construction.

- Review project's plan to conduct a site pre-mobilization review to provide verification of project and contractors' preparedness prior to mobilization to the site for construction.

Confirm that a risk register has been established for the project and that the risks associated with Process Safety are followed up and formally reviewed by competent personnel.

- Review the project's hazard/risk register that documents hazards associated with the development and the safeguards that will be implemented as a part of project development to mitigate risks.

STAGE GATE: DETAILED DESIGN

Confirm that final HIRA (e.g. HAZOP) is complete and its recommendations are being satisfactorily addressed.

- Check that the HIRA study has been undertaken on all systems, including vendor packages, in a satisfactory manner and that recommendations are being resolved in a timely manner and that closure is properly documented.

- Review the justifications for HIRA recommendations that have been rejected.

Confirm that change control procedures are being applied and that appropriate hazard review of changes has been instigated to maintain Process Safety integrity.

- Verify that change control procedures have been implemented and are adequate by checking project documents, e.g., P&IDs and Cause and Effects post-HIRA.

- Evaluate the proposed change control procedures for capture of all changes that might affect HIRA integrity and the Process Safety performance of the intended development. Determine if there have been any changes in scope and confirm that the relevant Process Safety studies were updated.

- Review project plan and arrangements for managing field changes during construction.

Confirm that appropriate specialist reviews have been carried out and their outcomes are being satisfactorily addressed.

- Review status and resolutions of recommendations from previous stage gate reviews for projects and other Process Safety studies including physical security surveys and QRAs for completion and closure of actions.

- Determine if a site pre-mobilization review has been conducted and that all recommendations have been implemented.

- Determine if appropriate specialist reviews have conducted and the results incorporated. For example:

  i   instrumentation and control,
  ii   logistics,
  iii security,
  iv   human factors,
  v   ergonomics,
  vi   alarm management.

- Determine if 3-D model reviews for accessibility, layout, and ergonomic reviews have been conducted and results have been incorporated.

Confirm that engineering controls and checks are in place.

- Verify that the Project engineering authority role is being effectively executed and that levels of technical authority for the design are clearly established in Project procedures and understood / applied, to design changes.

- Check that the design is being developed using approved specifications, codes, and standards. Determine if deviations from the specified standards, including those by vendors, are identified, reviewed, and approved by the Project engineering authority and included in the Project's register of deviations from specifications.

- Determine if quality management (QM) procedures for design have been followed and that the competency of personnel key to technical integrity has been verified.

- Identify and review any interface control procedures that exist between multiple designers and vendors.

- Review the process for inter-discipline checks of the design and design changes.

- Review the procedure for control and issue of latest design documentation.

- Review accountability for identification, verification, and compliance with statutory requirements and obtaining the required consents to operate the facilities. Review progress of submissions.

Confirm that a Process Safety management system including a Process Safety Plan is being implemented effectively.

- Review the Project's Process Safety Plan and verify that it will provide an effective process safety management system (PSMS) for the project.

Confirm its implementation including resourcing, process safety support, contractor engagement, and alignment with the site.

- Review the Construction Safety Plan. Assess whether the plan has the necessary components to drive the required safety performance during construction.

- Check that a robust hazard identification and risk management process has been established for addressing and mitigating potential construction safety risks.

- Review the Construction safety plan for alignment with the site's safe work practices and procedures. Verify integration of Company safety rules in the plan.

- Review project plans for training and site induction of the construction workforce.

- Examine the criteria for the selection of contractors, sub-contractors, and major vendors and whether it is being applied and maintained, including an assessment of the competence of contractors whose activities are critical to technical integrity. Review any contractual commitments relating to use of local labor and materials supply for safety concerns.

- Review arrangements for safe transport of workers to and from the job site.

- Review the project's plan for security, including physical security, loss prevention, and integration with site and community security.

- Examine the measurement and monitoring plan and safety KPIs.

- Review arrangements for construction, such as location of laydown areas, transportation of large equipment, higher risk activities, etc.

- Examine project's process to assure safety preparedness of contractors and subcontractors prior to mobilization to the site.

Confirm that asset integrity management (AIM) programs are being satisfactorily addressed.

- Verify that the Project engineering authority role has assured integrity of design and has a plan/procedure to maintain design intent during the fabrication, construction, and installation.

- Review the Quality Management Plan for the project and confirm that all activities are subject to system and compliance audits and findings are followed up and closed out. Sub-contractor and vendor audit trails should be in place for all design, procurement, and fabrication activities.

Confirm that Process Safety aspects have been adequately considered in the products of detailed engineering and that they are appropriate for construction.

- Review P&IDs and isometrics for completeness and readiness for construction.

- Review the Register of Safety Critical Equipment (SCE) for update of relief/flare system calculations based on isometrics and vendor data.

- Review safety instrumented systems (SIS) for the required Safety Integrity Level (SIL), component testing frequencies and alignment with operating philosophy.

- Review HP to LP interfaces and determine if primary pressure protection (accomplished completely or partially by instrumentation) have been identified and appropriately considered in design.

- Review the project's Hazard / Risk Register and verify that the project has incorporated into the design safeguards which will mitigate the identified hazards.

Confirm that Project's planning for startup includes development of procedures, training, pre-commissioning and commissioning activities.

- Review arrangements for the collation and transfer of all information required to enable safe and efficient future operation of the facilities.

- Check that a strategy is in place for the integration of the new facilities' procedures in the Client site procedures and plans.

- Verify progress on the development of operating, maintenance, and Process Safety procedures.

- Review proposals for training and ensuring competencies of site personnel for safe startup and operation of the facility.

- Review proposals for inspection, testing, pre-commissioning, and operational readiness review to confirm that they provide for adequate process safety and asset integrity management.

- Review project strategy for managing pressure testing, pre-commissioning and commissioning, including the disposal of wastes from testing, proving, pickling, and cleaning activities.

Confirm that the scope of process safety information (PSI) is defined and that a plan is in place for formal delivery to Operations.

- Review plans for PSI and other essential documentation for handover to Operations.
- Review arrangements for providing as-built documentation prior to start-up.

Confirm that an emergency response plan(s) has been developed or updated and that it addresses relevant process safety risks associated with startup and the operation.

- Review the emergency response plan(s), procedures and equipment for the construction site(s), including training, evacuation/shelter-in-place, drills and exercises.
- Review plans for routine & emergency medical support during construction.

STAGE GATE: CONSTRUCTION

Confirm that construction workforce training, competency, and performance assurance arrangements are adequate and being implemented.

- Examine the implementation of the site induction process, including hazard identification training. Determine if it is being applied to all project-related personnel, contractors, and sub-contractors.
- Review adequacy, implementation, and application of competence assurance programs for all trades.

Confirm that a construction Process Safety management system is adequate and being implemented.

- Review emergency and contingency plans for the construction period and integration of it with existing site emergency procedures. Verify that the project emergency procedures have provisions for accounting of personnel.
- Check that a safe system of work is in place and that it fully addresses site safe work practices, including topics such as:

  a. Work permitting
  b. Working at height / scaffolding
  c. Heavy lifting
  d. Excavation / trenching
  e. Isolation of process and electrical equipment
  f. Confined space entry

g. Hot work / ignition controls
h. Heavy machinery / vehicles
i. Task-based risk analysis.

- Determine if the requirements for site specific procedures, such as simultaneous operations (SIMOPS) have been assessed and implemented as necessary.

- Check that the project has implemented a process to ensure safety preparedness of contractors and sub-contractors prior to mobilization to the site and that all issues have been satisfactorily resolved.

- Check that a Project procedure for assessing contractor safety management systems is being implemented.

- Review the safety management systems for effective implementation throughout the construction organization, including lower tier sub-contractors.

- Check that a system exists for the development and authorization of method statements for tasks with identified hazards and performance of JSAs. Confirm that the project has implemented safeguards for hazards identified in JSAs and risks are being effectively managed.

- Review the security plan for full implementation.

- Verify that adequate medical services are in place.

- Review procedures for ensuring that all appropriate construction regulations have been identified and are being applied.

- Review procedures for tracking and reporting safety performance. Confirm that appropriate assurance programs are in place, and are driving performance improvement. Review incident investigations and lessons learned.

- Review progress on regulatory submissions with respect to target startup date.

- Review procedures for equipment flushing, cleaning, steam-blowing, drying, etc., and whether safety considerations have been addressed, (e.g., disposal of fluids and availability of suitable means to treat emissions/discharges/wastes; noise control, etc.)

Confirm that Client, contractors and vendors have clarity in regard to their scope and responsibilities for the mechanical completion, and that the construction team have a robust process to manage all interfaces.

- Review understanding of roles and responsibilities for mechanical completion where sub-divided between between multiple parties. Confirm understanding is aligned with scope of work in contracts.

- Review project interface management of fabrication, construction, and installation contractor(s) and sub-contractor(s).

- Check oversight and procedures that exist between project and fabrication, construction, and installation contractor(s) and sub-contractor(s).

Confirm that asset integrity management (AIM) processes including quality management are sufficient to deliver the design intent and facility integrity.

- Verify that the Project engineering authority role has assured integrity of construction and that procedures have been implemented that maintain design intent during the fabrication, construction, and installation.

- Review the quality plans and audit programs for the Project and contractors and confirm that these are being implemented.

- Review the Project internal controls and procedures related to inspection, testing, and material control. Confirm that they provide assurance that construction, fabrication, and materials management activities are being carried out in accordance with the design.

- Review arrangements for punch listing, including resourcing, training, and procedures for handling of punch items.

Confirm that change management is being applied.

- Establish that a change control procedure is in place to manage deviations from codes, standards, specifications, commitments, and all changes affecting P&IDS in project and contractor / sub-contractor organizations.

- Review Project document control procedures.

- Check that engineering authority and the levels of technical authority for authorizing changes are established, are understood, and are being applied.

- Review the register of deviations from specifications.

- Review the availability of up-to-date design information at the "workface" and the systems in place to determine if up to date information is available at all times.

Confirm that project plans for pre-commissioning, commissioning, and pre-startup are adequate.

- Review project pre-commissioning and commissioning plans and procedures. Confirm that all aspects of pre-commissioning and commissioning have been identified and included in procedures.

- Verify alignment between construction schedule, planned hand-over sequence for mechanical completion, and the pre-commissioning/commissioning plans.

- Confirm that potential safety risks, including safety concerns during pre-commissioning, have been identified and are being managed.

- Review procedures for proper handling and disposal of effluents generated during pre-commissioning activities, such as hydro-testing, equipment pre-treatment, and cleaning..

- Determine if emergency procedures and operations drills are being prepared or updated to take into consideration lessons learned from the project major accident risk assessment and the project hazard/risk register.

- Review progress on completing recommendations from previous stage gate reviews, HIRA and other Process Safety studies. Assess capability to complete all actions prior to startup.

- Review project plans for the pre-startup stage gate review and operational readiness review (ORR). Confirm that the project's checklist effectively covers all aspects for assuring construction integrity and preparedness for startup.

- Review physical security technical specifications.

Confirm that progress on Operations training and development (or update) of operating procedures is adequate.

- Check whether a competency management system has been established or updated for the intended operation.

- Review status of training & development/update of operating procedures for the facility.

Confirm that the Operations Team is involved as necessary in preparation for pre-commissioning and commissioning activities.

- Review the arrangements to ensure that all personnel involved with pre-commissioning and commissioning activities are competent to do so and that the Operator has similar standards in place to ensure long-term safe operation of the plant.

- Review the level of resources, skills, and training that has been provided to commissioning personnel, including Operations.

Confirm that plans for a site Process Safety management system and procedures are adequate.

- Review plans for the site's management system to ensure that an update will be carried out to include the new facilities. Any changes introduced by the project should not affect the site's ability to meet the expectations of the Client's process safety program.

- Review plans for the facility's process safety management program for completeness, implementation, training of personnel, and where appropriate, integration into existing site programs. This includes all elements of the site's safety program, including management of change (MOC) procedures.

Confirm that a document management system has been implemented and is performing as expected.

- Verify that the documentation management system is in place and provides assurance that the plant has been built and pre-commissioned in accordance with the design.

- Review plans for handover of as-built documentation and Process Safety Information (PSI), and its subsequent control by the Operator.

STAGE GATE: PRE-STARTUP

Confirm that pre-commissioning has been satisfactorily completed and the facilities are ready for commissioning.

- Review for completion of pre-commissioning activities. Verify that all issues encountered during pre-commissioning have been satisfactorily resolved and that the facility is ready for commissioning.

- Examine the system for punch listing and whether it has been effectively implemented. Review all high priority (category A) items for completion before commissioning.

- Review project commissioning plans and confirm that safety concerns during commissioning are recognized and mitigated.

- Review the utility, protective safety devices and support systems that will be fully commissioned, including testing where appropriate, before the plant starts up.

- Review for proper disposal of effluents generated during commissioning activities.

- Determine if adequate spares are available for commissioning and the initial operation of the plant.

- Determine if physical security systems have been commissioned and that they meet original specifications.

Verify that project and/or site is implementing a comprehensive process to confirm preparedness (e.g. Operational Readiness Review) and obtain approvals for startup.

- Review project/site's detailed plan for operational readiness review (ORR) and assess its adequacy and effective execution.

- Verify that the project has confirmed that recommendations arising from all previous stage gate reviews, HIRA and other Process Safety studies have been addressed, actions implemented and closure documented, also including all commitments made during the project.

- Determine if the project has satisfactorily mitigated all hazards/risks identified in the project's hazard/risk register.

- Determine if the necessary approvals have been obtained for a license to operate.

Confirm that integrity of the design has been maintained, and deviations from design have been satisfactorily addressed and will not compromise Process Safety performance.

- Review the project change and engineering authority records and the register of deviations from specifications to confirm that all design changes have been properly assessed, approved, and recorded/updated in the operator's hand-over documentation.

- Assess whether aspects of the plant likely to influence safety performance have not changed significantly throughout the project.

- Determine how the project confirmed that the effects of a number of small changes did not introduce a significant risk when considered together.

- Determine if changes identified in as-built checks have been subjected to an appropriate hazard review process.

- Establish that a hand-over system is in place that provides assurance and documented evidence the plant has been built and pre-commissioned in accordance with the design.

- Review the Project's Register of Safety Critical Equipment (SCE) for completion and handover to the site.

- Review the delivery of as-built documentation and Process Safety Information (PSI), and its subsequent control by the site.

Confirm that the commissioning, startup and operations teams are adequately trained, equipped, and competent and that all necessary procedures are available.

- Review commissioning procedures. Confirm that adequate systems and procedures, both for commissioning and subsequent operations have been prepared for startup, shutdown, normal operation, emergency shutdown, maintenance, and testing of the facilities. This includes procedures for all phases of operation, including temporary operations.

- Review safe work practice procedures for completion and verify that the workforce has been fully trained on these procedures.

- Review the level of resources, skills, and training that has been provided to commissioning personnel, including safety induction and training specific to the site. Verify through interviews that the workforce has a clear understanding of operating and safety procedures and has the competencies and skills for safe startup and operation.

- Review the arrangements to ensure that all personnel involved with commissioning activities are competent to do so and that the Operator has similar standards in place to ensure long-term safe operation of the plant.

- Check that the necessary vendor support for all mechanical, instrument, electrical, and control equipment will be available.

Confirm that the Client / site have made adequate preparations for startup.

- Review the site's management system to ensure that an update has been or will be carried out to include the new facilities. Any changes introduced by the project must not affect the site's ability to meet the expectations of the Client's process safety program.

- Examine the facility's process safety management program for completeness, implementation, training of personnel, and where appropriate, integration into existing site programs. This includes all elements of the company's safety program, including management of change (MOC) procedures.

- Review handover of management responsibilities from Project to site, for example:

    a. MOC,
    b. Security,
    c. Contractor oversight,
    d. Permit systems, etc.

- Review commitments made during the project for being carried through to the site's management system.

- Review the Operator's maintenance management system, ensuring that all necessary maintenance data has been input and the system has been tested.

- Examine the availability of procedures to carry out the periodic testing of high integrity protection systems. Verify that this testing period is in accordance with any reliability analysis previously carried out.

- Determine if a system is in place to manage a Register of Safety Critical Equipment (SCE), including identification of a document owner, technical authority, and assignment of upkeep responsibilities.

- Check that arrangements have been put in place for an extension to contracted services, where appropriate.

- Review any procedures for special commissioning arrangements, such as phased commissioning, temporary features, etc.

- Review for compliance with regulatory requirements.

- Perform a thorough site inspection for hazards, housekeeping, safety equipment installation, and readiness for startup.

- Review how the site will address the Operation stage gate review.

Confirm that emergency response arrangements and procedures have been established.

- Verify that emergency plans and procedures are in place, and roles and responsibilities along with channels of communication are clearly identified.

- Check whether emergency response procedures and arrangements fully address the findings of HIRA studies.

- Review emergency procedures that assure facility has adequate staffing of operators to bring unit to a safe state during emergencies. Confirm that emergency drills for identified risks have been put in place prior to startup.

- Determine if a program is in place to test the effectiveness of emergency response procedures and to feedback the lessons learned.

- If appropriate, assess the availability of a high level interface emergency plan, which ties together the various contingency plans. Links to local and national emergency services should be included.

STAGE GATE: OPERATION

Confirm that an adequate Process Safety management system has been properly implemented.

- Determine if a management system for all elements of the site's Process Safety program has been established.

- Review the site's programs to address requirements of the Company's process safety standards.

- Verify that recommendations arising from all previous Project stage gate reviews and other Process Safety studies have been addressed, actions implemented and closure documented.

- Review considerations and strategy that may impact end of lifecycle safety performance and liabilities.

Confirm that Process Safety performance of the operating facility(s) meets design intent.

- Review site safety performance data including noise, ambient air quality,

dust, discharges, metrics, etc.
• Review the emissions/discharge/waste profiles and compare with design and regulatory consent values.
• Review security arrangements for adequacy and learnings.
Verify the adequacy of response to any process safety incidents, and process upsets that have occurred during early operation.
• Review any design, operations or maintenance related problems, excursions outside of safe limits, incidents, and the measures taken to overcome them.
• Review incidents and lessons learned.
Verify the adequacy of programs to address any asset integrity problems that have occurred during early operation.
• Identify asset integrity management (AIM) concerns regarding the installation and lessons learned.
• Confirm that the site has satisfactorily implemented programs to maintain ongoing integrity of the equipment.
Confirm the rationale for any changes/modifications made during early operation vs. the original design intent.
• Review the modification record post startup and potential cumulative effect on hazards and risks.
• Understand rationale for changes vs. original design intent.
Confirm that lessons learned from early operation of the facility(s) are documented and shared.
• Identify lessons learned from the development and ensure that they are widely shared through Company website(s) and documentation system(s).

STAGE GATE: END OF LIFE

Confirm that project plans for decommissioning are adequate.

- Review Engineering Survey of the facilities, structures and/or buildings to be decommissioned.

- Review procedures for ensuring that all appropriate decommissioning regulations have been identified and are being applied.

- Check that a system exists for the development and authorization of method statements for decommissioning tasks with identified hazards.

- Review arrangements for decommissioning, such as location of laydown areas for mothballed/re-sale equipment, transportation of any hazardous materials and large loads, higher risk activities (e.g. storage and use of explosives), etc.

- Review the project's plan for security of the decommissioning site.

Confirm that the Operations Team is involved as necessary in preparation for decommissioning activities.

- Review liaison between Operations, Project, and decommissioning contractor(s) and sub-contractor(s).

- Review Operations input to decommissioning plans, especially where simultaneous operations (SIMOPS) are likely.

Confirm that the HIRA study(s) is complete and recommendations are being satisfactorily addressed.

- Check that a robust hazard identification and risk management process has been established for addressing and mitigating potential decommissioning safety risks.

- Check that HIRA studies have been undertaken in a satisfactory manner on all systems, structures and buildings to be decommissioned.

- Check that recommendations are being resolved in a timely manner and that closure is properly documented.

- Review the justifications for HIRA recommendations that have been rejected.

Confirm that appropriate specialist reviews have been carried out and their outcomes are being satisfactorily addressed, including engineering controls and checks are in place.

- Determine if a site pre-mobilization review for the decommissioning team/contractor(s) has been conducted and that all recommendations have been implemented.

- Determine if appropriate specialist reviews have conducted and the results incorporated. For example:

  i   stability of facilities, structures and buildings,
  ii identification and disposal of hazardous materials,
  iii site security.

- Check that the decommissioning plan is being developed using approved specifications, codes, and standards. Determine if deviations from the specified standards are identified, reviewed, and approved by the Project engineering authority and included in the Project's register of deviations from specifications.

- Review accountability for identification, verification, and compliance with statutory requirements and obtaining the required consents to deconstruct/demolish the facilities. Review progress of submissions.

Confirm that a Process Safety management system including a Process Safety Plan is being implemented effectively.

- Review the decommissioning project's Process Safety Plan and verify that it will provide an effective process safety management system (PSMS) for the project. Confirm its implementation including resourcing, process safety support, contractor engagement, and alignment with the site.

- Review the Decommissioning Safety Plan. Assess whether the plan has the necessary components to drive the required safety performance during decommissioning.

- Review the decommissioning safety plan for alignment with the site's safe work practices and procedures. Verify integration of Company safety rules in the plan. Check that a safe system of work is in place and that it fully addresses safe work practices, including topics such as:

  i.   Work permitting
  ii.   Working at height / scaffolding
  iii. Heavy lifting
  iv.   Excavation / trenching
  v.   Isolation of process and electrical equipment
  vi.   Confined space entry
  vii. Hot work / ignition controls
  viii. Heavy machinery / vehicles
  ix.   Use of explosives.

- Determine if the requirements for site specific procedures, such as simultaneous operations (SIMOPS) have been assessed and implemented as necessary.

- Check that a Project procedure for assessing contractor safety management systems is being implemented.

- Check that a system exists for the performance of JSAs for tasks with identified hazards. Confirm that the project has implemented safeguards for hazards identified in JSAs and risks are being effectively managed.

- Review procedures for tracking and reporting safety performance. Confirm that appropriate assurance programs are in place, and are driving performance improvement. Review incident investigations and lessons learned.

- Review procedures for equipment decontamination (e.g. flushing, cleaning, inerting, etc.), and whether safety considerations have been addressed, (e.g. disposal of fluids and availability of suitable means to treat discharges/wastes; noise control, etc.)

Confirm that an emergency response plan(s) has been developed and that it addresses relevant process safety risks associated with decommissioning.

- Review emergency and contingency plans for the decommissioning period and integration of it with existing site emergency procedures. Verify that the project emergency procedures have provisions for accounting of personnel.

- Review arrangements for emergency medical support.

Confirm that Process Safety aspects have been adequately considered and are appropriate for decommissioning.

- Review engineering drawings and other PSI for accuracy, completeness and readiness for decommissioning.

- Review the project's Hazard / Risk Register and verify that the project has incorporated into the decommissioning safeguards which will mitigate the identified hazards.

Confirm that decommissioning workforce training, competency, and performance assurance arrangements are adequate and being implemented.

- Examine the criteria for the selection of contractors and sub-contractors, and whether it is being applied and maintained, including an assessment of the competence of contractors whose activities are critical to safety.

- Review project plans for training and site induction of the decommissioning workforce.

- Examine project's process to assure safety preparedness of contractors and sub-contractors prior to mobilization to the site.

Confirm that the decommissioning project team has a robust process to manage the interface with contractor(s).

- Review project interface management of decommissioning contractor(s) and sub-contractor(s).

- Check oversight and procedures that exist between project and decommissioning contractor(s) and sub-contractor(s).

Confirm that asset integrity management (AIM) processes including quality management are sufficient to maintain structural and equipment integrity.

- Verify that the Project engineering authority role has assured integrity of facilities, structures and buildings before and during decommissioning.

- Review the Project internal controls and procedures related to inspection, testing, and preservation of mothballed/re-sale equipment.

- Review arrangements for material handling (i.e. segregation) of recyclable materials.

REFERENCES

ACC et al, *Site Security Guidelines for the U.S. Chemical Industry*, American Chemistry Council, Chlorine Institute, and Synthetic Organic Chemical Manufacturers Association (SOCMA), 2001.

AIA (American Insurance Association), *Hazard Survey of the Chemical and Allied Industries, Technical Survey No. 3*, New York, 1968.

Andrew, Steve, *Demolition Man: Expert observations of demolition dangers and how to avoid them*, The Chemical Engineer, Issue 920, Feb 2018.

ANSI (American National Standards Institute), *Safety and Health Program Requirements for Demolition*, ANSI/ASSE A10.6-2006, Washington, DC, 2006.

API 2001a (American Petroleum Institute), *Recommended Practice for Design and Hazards Analysis for Offshore Production Facilities, 2nd edition*, API RP 14J, American Petroleum Institute, 2001.

API 2001b (American Petroleum Institute), *A Manager's Guide to Reducing Human Errors, Improving Human Performance in the Process Industries*, API Publication 770, 1st Edition, 2001.

API 2003a (American Petroleum Institute), *Security Vulnerability Assessment Methodology for the Petroleum and Petrochemical Industries*, American Petroleum Institute and National Petroleum Refiners Association (NPRA), 2003.

API 2003b (American Petroleum Institute), *Security for Offshore Oil and Natural Gas Operations, 1st Edition*, API RP-70, American Petroleum Institute, 2003 (revised 2010).

API 2004 (American Petroleum Institute), *Security for Worldwide Offshore Oil and Natural Gas Operations, 1st Edition*, API RP-70I, American Petroleum Institute, 2004 (revised 2012).

API 2006 (American Petroleum Institute), *Recommended Practice for the Design of Offshore Facilities Against Fire and Blast Loading*, API RP 2FB, American Petroleum Institute, 2006.

API 2007 (American Petroleum Institute), *Management of Hazards Associated With Location of Process Plant Portable Buildings*, API RP 753, American Petroleum Institute, Washington, DC, 2007.

API 2009 (American Petroleum Institute), *Management of Hazards Associated with Location of Process Plant Buildings*, 3rd edition, API RP 752, American Petroleum Institute, 2009.

API 2012 (American Petroleum Institute), *Recommended Practice for Classification of Locations for Electrical Installations at Petroleum Facilities Classified as Class I, Division I and Division 2, 3rd Edition*, API RP 500, American Petroleum Institute, 2012.

API 2014 (American Petroleum Institute), *Management of Hazards Associated with Location of Process Plant Tents,* API RP 756, American Petroleum Institute, Washington, DC, 2014.

API 2016(a) (American Petroleum Institute), *Guide to Reporting Process Safety Events*, Version 3.0, American Petroleum Institute, Washington D.C., 2016.

API 2016(b) (American Petroleum Institute), *Process Safety Performance Indicators for the Refining & Petrochemical Industries, Part 2: Tier 1 and 2 Process Safety Events,* API RP 754, 2nd edition, American Petroleum Institute, Washington D.C., 2016.

APM 2004, (Association of Project Management), *Project Risk Analysis and Management Guide,* 2nd edition, APM Publishing Ltd, 2004.

ARIA (Analysis, Research and Information on Accidents), *Leak and Ignition of Toluene, Manufacture of Basic Pharmaceutical Products,* N° 14500 (03/12/1998, Saint Vulbas, Ain, France), ARIA database; Ministry of Ecology, Sustainable Development and Energy, France.

ASME (The American Society of Mechanical Engineers), *Performance Test Codes*, New York, NY.
- PTC1 - 1999 - *General Instructions*
- PTC2 - 1985 - *Code on Definitions and Values*
- PTC4- 1998 - *Fired Steam Generators*
- PTC4.2 - 1997 - *Coal Pulverizes*
- PTC4.3 - 1991 - *Air Heaters*
- PTC4.4 - 1992 - *Gas Turbine Heat Recovery Steam Generators*
- PTC5 - 1949 - *Reciprocating Steam Engines*
- PTC6 - 1996 - *Steam Turbines*
- PTC6A - 2001 - *Test Code for Steam Turbines* - Appendix to PTC 6
- PTC6-Report - *Guidance for Evaluation of Measurement Uncertainty in Performance Tests of Steam Turbines*
- PTC6-S - *Procedures for Routine Performance Test of Steam Turbines*
- PTC7.1 - 1969 - *Displacement Pumps*
- PTC8.2 - 1990 - *Centrifugal Pumps*
- PTC9 - 1997 - *Displacement Compressors, Vacuum Pumps and Blowers*
- PTC10 - 1997- *Test Code on Compressors and Exhausters*
- PTC11 - 1995 - *Fans*
- PTC12.1 - 2000 - *Closed Feed Water Heaters*
- PTC12.2 - 1998 - *Steam Surface Condensers*
- PTC12.3 - 1997 - *Deaerators*

- PTC12.4 - 1997 - *Moisture Separator Reheaters*
- PTC17 - 1997 - *Reciprocating Internal-Combustion Engines*
- PTC18 - 1992 - *Hydraulic Prime Movers*
- PTC18.1 - 1984 - *Pumping Mode of Pump/Turbines*
- PTC19.1 - 1998 - *Measurement Uncertainty*
- PTC19.2 - 1998 - *Pressure Measurement*
- PTC19.3 - 1998 - *Temperature Measurement*
- PTC19.5 - 1972 - *Application, Part II of Fluid Meters: Interim Supplement on Instruments and Apparatus*
- PTC19.5.1 - 1964 - *Weighing Scales*
- PTC19.7 - 1988 - *Measurement of Shaft Power*
- PTC19.8 - 1985 - *Measurement of Indicated Power*
- PTC19.10 - 1981 - *Flue and Exhaust Gas Analyses*
- PTC19.11 - 1997 - *Steam and Water Sampling, Conditioning, and Analysis in the Power Cycle*
- PTC19.14 - 1958 - *Linear Measurements*
- PTC19.22 - 1998 - *Digital Systems Techniques*
- PTC19.23 - 1985 - *Guidance Manual for Model Testing*
- PTC20.1 - 1988 - *Speed and Load-Governing Systems for Steam Turbine-Generator Units*
- PTC20.2 - 1986 - *Over speed Trip Systems for Steam Turbine-Generator Units*
- PTC20.3 - 1986 - *Pressure Control Systems Used on Steam Turbine Generator Units*
- PTC21 - 1991 - *Particulate Matter Collection Equipment*
- PTC22 - 1997 - *Performance Test Code on Gas Turbines*
- PTC23 - 1997 - *Atmospheric Water Cooling Equipment*
- PTC24 - 1982 - *Ejectors*
- PTC25 - 1994 - *Pressure Relief Devices*
- PTC26 - 1962 - *Speed Governing Systems for Internal Combustion Engine Generator Units*
- PTC28 - 1985 - *Determining the Properties of Fine Particulate Matter*
- PTC29 - 1985 - *Speed-Governing Systems for Hydraulic Turbine-Generator Units*
- PTC30 - 1998 - *Air Cooled Heat Exchangers*
- PTC31 - 1991 - *Ion Exchange Equipment*
- PTC32.1 - 1992 - *Nuclear Steam Supply Systems*
- PTC33 - 1991 - *Large Incinerators*
- PTC33 - 1979 - *Large Incinerators Codes and Appendix Package*
- PTC33A - 1980 - *Abbreviated Incinerator Efficiency Test*
- PTC36 - 1998 - *Measurement of Industrial Sound*
- PTC38 - 1985 - *Determining the Concentration of Particulate Matter in a Gas Stream*

• PTC39.1 - 1991 - *Condensate Removal Devices for Steam Systems*
• PTC40 - 1991 - *Flue Gas Desulfurization Units*
• PTC42 - 1998 - *Wind Turbines*
• PTC46 - 1997 - *Overall Plant Performance*
• PTCPM - 1993 - *Performance Monitoring Guidelines for Steam Power Plants*

Broadribb, M. P., *What have we really learned? Twenty-five years after Piper Alpha.* Process Safety Progress, Volume 34: Issue 1: Pages 16–23. doi:10.1002/prs.11691, 2015.

Broadribb, M.P., Currie, M.R., *HAZOP/LOPA/SIL, Be Careful What You Ask For!,* Proceedings of the 6th Global Congress on Process Safety, San Antonio, Texas, 2010.

Broadribb, M.P., *Do You Feel Lucky? Or Do You Want to Identify and Manage Safety Critical Equipment?,* Proceedings of the 12th Global Congress on Process Safety, Houston, Texas, 2016.

BSI (British Standards Institute), *Code of Practice for Full and Partial Demolition,* BS 6187:2011, British Standards Institute, London, United Kingdom, 2011.

CCME, *National Guidelines for Decommissioning Industrial Sites,* CCME-TS/WM-TRE013E, Canadian Council of Ministers of the Environment, Canada, 1991.

CCPS 1989 (Center for Chemical Process Safety), *Workbook of Test Cases for Vapor Cloud Source Dispersion Models,* American Institute of Chemical Engineers, New York, NY, 1989.

CCPS 1994a (Center for Chemical Process Safety), *Guidelines for Evaluating the Characteristics of Vapor Cloud Explosions, Flash Fires, and BLEVEs,* American Institute of Chemical Engineers, New York, NY, 1994.

CCPS 1994b (Center for Chemical Process Safety), *Guidelines for Preventing Human Error in Process Safety,* American Institute of Chemical Engineers, New York, NY, 1994.

CCPS 1995(a) (Center for Chemical Process Safety), *Guidelines for Chemical Transportation Risk Assessment,* American Institute of Chemical Engineers, New York, NY, 1995.

CCPS 1995(b) (Center for Chemical Process Safety), *Guidelines for Process Safety Documentation,* American Institute of Chemical Engineers, New York, NY, 1995.

CCPS 1995(c) (Center for Chemical Process Safety), *Guidelines for Technical Planning for On-Site Emergencies,* American Institute of Chemical Engineers, New York, NY, 1995.

CCPS 1996(a) (Center for Chemical Process Safety), *Guidelines for Integrating Process Safety Management, Environment, Safety, Health, and Quality,* American Institute of Chemical Engineers, New York, NY, 1996.

CCPS 1996(b) (Center for Chemical Process Safety), *Guidelines for Use of Vapor Cloud Dispersion Models,* 2nd edition, American Institute of Chemical Engineers, New York, NY, 1996.

CCPS 1996(c) (Center for Chemical Process Safety), *Guidelines for Writing Effective Operating and Maintenance Procedures,* American Institute of Chemical Engineers, New York, NY, 1996.

CCPS 1998(a) (Center for Chemical Process Safety), *Estimating Flammable Mass of a Vapor Cloud,* American Institute of Chemical Engineers, New York, NY, 1998.

CCPS 1998(b) (Center for Chemical Process Safety), *Guidelines for Pressure Relief and Effluent Handling Systems,* American Institute of Chemical Engineers, New York, NY, 1998.

CCPS 1999 (Center for Chemical Process Safety), *Guidelines for Consequence Analysis of Chemical Releases,* American Institute of Chemical Engineers, New York, NY, 1999.

CCPS 2000 (Center for Chemical Process Safety), *Guidelines for Chemical Process Quantitative Risk Analysis, 2nd edition,* American Institute of Chemical Engineers, New York, NY, 2000.

CCPS 2001(a) (Center for Chemical Process Safety), *Layer of Protection Analysis: Simplified Process Risk Assessment,* American Institute of Chemical Engineers, New York, NY, 2001.

CCPS 2001(b) (Center for Chemical Process Safety), *Making EHS an Integral Part of Process Design,* American Institute of Chemical Engineers, New York, NY, 2001.

CCPS 2001(c) (Center for Chemical Process Safety), *Revalidating Process Hazard Analyses,* American Institute of Chemical Engineers, New York, NY, 2001.

CCPS 2002 (Center for Chemical Process Safety), *Wind Flow and Vapor Cloud Dispersion at Industrial and Urban Sites,* American Institute of Chemical Engineers, New York, NY, 2002.

CCPS 2003(a) (Center for Chemical Process Safety), *Guidelines for Analyzing and Managing the Security Vulnerabilities of Fixed Chemical Sites,* American Institute of Chemical Engineers, New York, NY, 2003.

CCPS 2003(b) (Center for Chemical Process Safety), *Guidelines for Fire Protection in Chemical, Petrochemical, and Hydrocarbon Processing Facilities,* American Institute of Chemical Engineers, New York, NY, 2003.

CCPS 2003(c) (Center for Chemical Process Safety), *Guidelines for Investigating Chemical Process Incidents, 2nd edition,* American Institute of Chemical Engineers, New York, NY, 2003.

CCPS 2003(d) (Center for Chemical Process Safety), *Understanding Explosions,* American Institute of Chemical Engineers, New York, NY, 2003.

CCPS 2004 (Center for Chemical Process Safety) *Guidelines for Safe Handling of Powders and Bulk Solids,* American Institute of Chemical Engineers, New York, NY, 2004.

CCPS 2005 (Center for Chemical Process Safety), *Building Process Safety Culture: Tools To Enhance Process Safety Performance,* American Institute of Chemical Engineers, ISBN #0-8169-0999-7,New York, NY, 2005.

CCPS 2006 (Center for Chemical Process Safety), *The Business Case for Process Safety*, Second Edition, American Institute of Chemical Engineers, New York, NY, 2006.

CCPS 2007(a) (Center for Chemical Process Safety), *Guidelines for Performing Effective Pre-Startup Safety Reviews,* American Institute of Chemical Engineers, New York, NY, 2007.

CCPS 2007(b) (Center for Chemical Process Safety), *Guidelines for Risk Based Process Safety,* American Institute of Chemical Engineers, New York, NY, 2007.

CCPS 2007(c) (Center for Chemical Process Safety) *Guidelines for Safe and Reliable Instrumented Protective Systems*, American Institute of Chemical Engineers, New York, NY, 2007

CCPS 2007(d) (Center for Chemical Process Safety), *Human Factors Methods for Improving Performance in the Process Industries,* American Institute of Chemical Engineers, New York, NY, 2007.

CCPS 2008(a) (Center for Chemical Process Safety), *Guidelines for Chemical Transportation Safety, Security, and Risk Management*, American Institute of Chemical Engineers, New York, NY, 2008).

CCPS 2008(b)   (Center for Chemical Process Safety), *Guidelines for Hazard Evaluation Procedures, 3rd edition,* American Institute of Chemical Engineers, New York, NY, 2008.

CCPS 2008(c) (Center for Chemical Process Safety), *Guidelines for the Management of Change for Process Safety,* American Institute of Chemical Engineers, New York, NY, 2008.

CCPS 2009(a) (Center for Chemical Process Safety), *Continuous Monitoring for Hazardous Material Releases* American Institute of Chemical Engineers, New York, NY, 2009.

CCPS 2009(b) (Center for Chemical Process Safety), *Guidelines for Developing Quantitative Safety Risk Criteria,* American Institute of Chemical Engineers, New York, NY, 2009.

CCPS 2009(c) (Center for Chemical Process Safety), *Guidelines for Process Safety Metrics*, American Institute of Chemical Engineers, New York, NY, 2009.

CCPS 2009(d) (Center for Chemical Process Safety), *Inherently Safer Chemical Processes: A Life Cycle Approach*, 2nd edition, American Institute of Chemical Engineers, New York, NY, 2009.

CCPS 2010(a) (Center for Chemical Process Safety), *A Practical Approach to Hazard Identification for Operations and Maintenance Workers,* American Institute of Chemical Engineers, New York, NY, 2010.

CCPS 2010(b) (Center for Chemical Process Safety), *Guidelines for Vapor Cloud Explosion, Pressure Vessel Burst, BLEVE and Flash Fire Hazards, 2nd edition,* American Institute of Chemical Engineers, New York, NY, 2010.

CCPS 2011(a) (Center for Chemical Process Safety), *Conduct of Operations and Operational Discipline: For Improving Process Safety in Industry,* American Institute of Chemical Engineers, New York, NY, 2011.

CCPS 2011(b) (Center for Chemical Process Safety), *Guidelines for Auditing Process Safety Management Systems*, American Institute of Chemical Engineers, New York, NY, 2011.

CCPS 2011(c) (Center for Chemical Process Safety), *Process Safety Leading and Lagging Metrics... You Don't Know What You Don't Measure,* American Institute of Chemical Engineers, New York, NY, 2011.

CCPS 2012(a) (Center for Chemical Process Safety), *Guidelines for Engineering Design for Process Safety, 2nd edition,* Center for Chemical Process Safety, American Institute of Chemical Engineers, New York, NY, 2012.

CCPS 2012(b) (Center for Chemical Process Safety), *Guidelines for Evaluating Process Plant Buildings for External Explosions, Fires and Toxic Releases,* 2nd edition, Center for Chemical Process Safety, American Institute of Chemical Engineers, New York, NY, 2012.

CCPS 2013(a) (Center for Chemical Process Safety), *Guidelines for Enabling Conditions and Conditional Modifiers in Layer of Protection Analysis,* American Institute of Chemical Engineers, New York, NY, 2013.

CCPS 2013(b) (Center for Chemical Process Safety), *Guidelines for Managing Process Safety Risks During Organizational Change,* American Institute of Chemical Engineers, New York, NY, 2013.

CCPS 2013(c) (Center for Chemical Process Safety), *Process Safety Leading Indicators Industry Survey,* American Institute of Chemical Engineers, New York, NY, 2013.

CCPS 2015(a) (Center for Chemical Process Safety), *Guidelines for Defining Process Safety Competency Requirements,* Center for Chemical Process Safety, American Institute of Chemical Engineers, New York, NY, 2015.

CCPS 2015(b) *Guidelines for Initiating Events and Independent Protection Layers in Layer of Protection Analysis,* American Institute of Chemical Engineers, New York, NY, 2015.

CCPS 2016(a) (Center for Chemical Process Safety), *Guidelines for Implementing Process Safety Management, 2nd edition,* American Institute of Chemical Engineers, New York, NY, 2016.

CCPS 2016(b) (Center for Chemical Process Safety), *Guidelines for Integrating Management Systems and Metrics to Improve Process Safety Performance,* American Institute of Chemical Engineers, New York, NY, 2016.

CCPS 2017(a) (Center for Chemical Process Safety), *Guidelines for Asset Integrity Management,* American Institute of Chemical Engineers, New York, NY, 2017.

CCPS 2017(b) (Center for Chemical Process Safety), *Guidelines for Safe Automation of Chemical Processes,* 2nd edition, American Institute of Chemical Engineers, New York, NY, 2017.

CCPS 2018a (Center for Chemical Process Safety), *Guidelines for Siting and Layout of Facilities, 2nd edition,* American Institute of Chemical Engineers, New York, NY, 2018.

CCPS 2018b (Center for Chemical Process Safety), *Process Safety During Transient Operations (Start-Up/Shutdown),* American Institute of Chemical Engineers, New York, NY, 2018.

CCPS 2018c (Center for Chemical Process Safety), *Bow Ties in Risk Management – A Concept Book for Process Safety)*, American Institute of Chemical Engineers, New York, NY, 2018.

CCPS 2018d (Center for Chemical Process Safety), *Dealing with Aging Process Facilities and Infrastructure*, American Institute of Chemical Engineers, New York, NY, 2018.

CCPS 2018e (Center for Chemical Process Safety), *Essential Practices for Creating, Strengthening, and Sustaining Process Safety Culture*, American Institute of Chemical Engineers, New York, NY, 2018.

CII 1996 (Construction Industry Institute), *Project Definition Rating Index (PDRI)*, RR113-11, Austin, TX, June 1996.

CII 1998 (Construction Industry Institute), *Planning for Startup*, Implementation Resource (IR) 121-2, Austin, TX, 1998.

CII 2004 (Construction Industry Institute), *Application of Lean Manufacturing Principles to Construction*, James E. Diekmann, Mark Krewedl, Joshua Balonick, Travis Stewart, and Spencer Won, Austin, TX, 2004.

CII 2006 (Construction Industry Institute), *Constructability Implementation Guide*, 2nd Edition, Special Publication SP34-1, Austin, TX, 2006.

CII 2010 (Construction Industry Institute), *Implementing and Improving Quality Management Systems in the Capital Facilities Delivery Industry*, IR254-2, Austin, TX, 2010.

CII 2010 (Construction Industry Institute), *Best Practices in Quality Management for the Capital Facilities Delivery Industry*, RS 254-1, Austin, TX, 2010

CII 2012 (Construction Industry Institute), *CII Best Practices Guide: Improving Project Performance*, Implementation Resource 166-3, Version 4.0, Austin, TX, February 2012.

CII 2014 (Construction Industry Institute), *Benchmarking & Metrics Summary Report*, BMM 2013-1, Austin, Texas, 2014.

CII 2015 (Construction Industry Institute), *Achieving Success in the Commissioning and Start-up of Capital Projects*, Implementation Resource 312-2, Austin, TX, 2015.

CMPT, *Guide to Quantitative Risk Assessment for Offshore Installations*, The Centre for Marine and Petroleum Technology, 1999.

CSA (Canadian Standards Association), *Land use planning for pipelines: A guideline for local authorities, developers, and pipeline operators*, Plus 663, CSA Group, Canada, 2004.

CSB (U.S. Chemical Safety and Hazard Investigation Board), T2 Laboratories, Inc., Runaway Reaction, Report Number 2008-3-I-FL, Washington, DC, 2009.

Deming, W. Edward, *Out of the Crisis*, Massachusetts Institute of Technology, Cambridge, MA, 1982.

Dixon-Jackson, K., *Lessons Learnt from Decommissioning a Top Tier COMAH Site*, Symposium Series No. 154, IChemE, Rugby, UK, 2008.

Duguid, I.M., *Analysis of Past Incidents in the Process Industries*, Symposium Series No. 154, Institution of Chemical Engineers, Rugby, UK, 2008.

EI 2007 (Energy Institute), *Guidelines for the Management of Safety Critical Elements, 2nd edition*, London, UK, 2007.

EI 2013 (Energy Institute), *Guidance on Meeting Expectations of EI Process Safety Management Framework*, London, UK, 2013.

EI 2015 (Energy Institute), *Model Code of Safe Practice Part 15: Area Classification Code for Installations Handling Flammable Fluids*, EI 15 (formerly IP 15), 4th edition, Energy Institute, London, UK, 2015.

EPA (U.S. Environmental Protection Agency), *Accidental Release Prevention Requirements: Risk Management Programs*, 49 CFR Part 68, Washington, DC, 1996.

Hicks, D.I., Crittenden, B.D., Warhurst, A.C., *Design for decommissioning: Addressing the future closure of chemical sites in the design of new plant*, Process Safety and Environmental Protection, 78 (6), pp. 465-467, 2000.

HM Government 1992, *The Offshore Installations (Safety Case) Regulations 1992*, Statutory Instruments, 1992 No. 2885, Health and Safety, UK, 1992. (Subsequently replaced by *The Offshore Installations (Safety Case) Regulations 2005*, (HM Government 2005)).

HM Government 1995, *The Offshore Installations (Prevention of Fire and Explosion, and Emergency Response) Regulations 1995*, Statutory Instruments, 1995 No.743, Health and Safety, UK, 1995.

HM Government 1996a, *Equipment and Protective Systems for Use in Potentially Explosive Atmospheres Regulations*, Statutory Instruments, 1996 No. 192, Health & Safety. UK, 1996.

HM Government 1996b, *The Pipelines Safety Regulations 1996*, Statutory Instruments, 1996 No. 825, Health and Safety, UK, 1996.

HM Government 2002, *The Dangerous Substances and Explosive Atmospheres Regulations*, Statutory Instruments, 2002 No. 2776, Health & Safety, UK, 2002.

HM Government 2005, *The Offshore Installations (Safety Case) Regulations 2005*, Statutory Instruments, 2005 No. 3117, Offshore Installations, UK, 2005.

HM Government 2015, *The Construction (Design and Management) Regulations (CDM, 2015)*, Statutory Instruments, 2015 No. 51, Health And Safety, UK, 2015.

HSE 1993 (Health & Safety Executive), *Offshore Gas Detector Siting Criterion Investigation of Detector Spacing* OTO 93 002, Health & Safety Executive, Bootle, UK, 1993.

HSE 1999 (Health & Safety Executive), *Reducing Error and Influencing Behaviour, 2nd Edition, HSG48*, Health & Safety Executive, Bootle, UK, 1999.

HSE 2006a (Health & Safety Executive), *Step-by-Step Guide to Developing Process Safety Indicators*, HSG 254, Bootle, UK, 2006.

HSE 2006b (Health & Safety Executive), *Guidance on Risk Assessment for Offshore Installations*, Offshore Information Sheet No. 3/2006, Health & Safety Executive, Bootle, UK, 2006.

HSE 2006c (Health & Safety Executive), *Plant Ageing, Management of Equipment Containing Hazardous Fluids or Pressure*, Research Report RR509, Health & Safety Executive, Bootle, UK, 2006.

HSE 2010a (Health & Safety Executive), *Human Factors & COMAH, A Gap Analysis Tool*, Health & Safety Executive, Bootle, UK, 2010.

HSE 2010b (Health & Safety Executive), *Managing Ageing Plant: A Summary Guide*, Research Report RR823, Health & Safety Executive, Bootle, UK, 2010.

HSE 2013 (Health & Safety Executive), *Modelling Smoke and Gas Ingress into Offshore Temporary Refuges*, Research Report RR997, Health & Safety Laboratory, Buxton, UK, 2013.

HSE 2016 (Health & Safety Executive), *Prevention of fire and explosion, and emergency response on offshore installations. Offshore Installations (Prevention of Fire and Explosion, and Emergency Response) Regulations, 1995. Approved Code and Practice and Guidance*, L65, 3rd Edition, Health and Safety Executive, Bootle, UK, 2016.

HSE 2017 (Health & Safety Executive), *COMAH Guidance, Technical Measures, Operating Procedures*, Bootle, UK, http://www.hse.gov.uk/comah/sragtech/techmeasoperatio.htm accessed October 2017.

ICC 2012, *International Fire Code (IFC)*, International Code Council, Washington, DC, 2012.

ICC 2018, *International Building Code (IBC)*, International Code Council, Washington, DC, 2018.

IChemE 2011 (Institution of Chemical Engineers), *Chemical and Process Plant Commissioning Handbook: A Practical Guide to Plant System and Equipment Installation and Commissioning*, 1st edition, Rugby, UK, 2012.

IChemE 2018 (Institution of Chemical Engineers), *Demolition Man, Expert Observations of Demolition Dangers and How to Avoid Them*, The Chemical Engineer, Issue 920, February 2018, Rugby, UK, 2018.

IEC 2015, *Explosive Atmospheres - Part 10-1: Classification of Areas - Explosive Gas Atmospheres, IEC 60079-10-1, 2nd edition*, International Electrotechnical Commission, Geneva, Switzerland, 2015.

IEC 2016, *Functional Safety: Safety Instrumented Systems for the Process Industry Sector - Part 1: Framework, definitions, system, hardware and application programming requirements*, IEC 61511, International Electrotechnical Commission, Geneva, Switzerland, 2016.

IGEM, *Hazardous area classification of Natural Gas installations, 2nd edition*, SR25, Communication 1748, Institution of Gas Engineers & Managers, Safety Recommendations, UK, 2010.

IOGP 2006, *Human Factors ... a means of improving HSE performance*, International Association of Oil and Gas Producers, London, UK, 2006.

IOGP 2010, *HSE Management – Guidelines for Working Together in a Contract Environment*, Report No. 423, International Association of Oil & Gas Producers, London, UK, 2010.

IRI, *Plant layout and Spacing for Oil and Chemical Plants*, IRInformation Manual IM 2.5.2, Industrial Risk Insurers, Hartford, Connecticut, 1991.

ISA, *Functional Safety: Safety Instrumented Systems for the Process Industry Sector*, ANSI/ISA 84.00.01 (IEC 61511 modified), Durham, NC, 2004.

ISA 2010, *Guidance on the Evaluation of Fire and Gas System Effectiveness*, TR84.00.07, International Society of Automation, Durham, NC, 2010.

ISA 2015, *Performance-Based Fire and Gas Systems Engineering Handbook*, International Society of Automation, Durham, NC, 2015.

ISO 1998 (International Organization for Standardization), *Petroleum and Natural Gas Industries - Control and Mitigation of Fires and Explosions on Offshore Production Platforms - Requirements and Guidelines*, ISO/FDIS 13702, Geneva, Switzerland, 1998.

ISO 1999 (International Organization for Standardization), *Petroleum and Natural Gas Industries – Offshore Production Installations – Guidelines on Tools and Techniques for Identification and Assessment of Hazardous Events*, ISO/DIS 17776, Geneva, Switzerland, 1999.

ISO 2009 (International Organization for Standardization), ISO 9004:2009, *Managing for the sustained success of an organization - A quality management approach*, Geneva, Switzerland, 2009.

ISO 2010 (International Organization for Standardization), ISO/TS 29001:2010, *Petroleum, petrochemical and natural gas industries - Sector-specific quality management systems - Requirements for product and service supply organizations*, Geneva, Switzerland, 2010.

ISO 2011 (International Organization for Standardization), ISO 19011:2011, *Guidelines for auditing management systems*, Geneva, Switzerland, 2011.

ISO 2015a (International Organization for Standardization), *Quality Management Principles*, Geneva, Switzerland, 2015.

ISO 2015b (International Organization for Standardization), ISO 9001:2015, *Quality management systems – Requirements*, Geneva, Switzerland, 2015.

Kidam, K. and Hurme, M., *Analysis of Equipment Failures as Contributors to Chemical Process Accidents*, Process Safety and Environmental Protection, 91, 61–78, 2013.

Killcross M., *Chemical and Process Plant Commissioning Handbook*, Institution of Chemical Engineers, Rugby, UK, 2011.

Kletz, T.A. and Amyotte, P., *Process Plants: A Handbook for Inherently Safer Design*, 2nd Edition, 2010.

MFB, *A Best Practice Approach to Shelter-in-Place for Victoria*, Metropolitan Fire and Emergency Services Board, Victoria, Australia, 2011.

MTI (Materials Technology Institute), Twigg, R.J., *Guidelines for Mothballing of Process Plants*, MTI Publication No. 34, The Materials Technology Institute of the Chemical Process Industries, Inc., St. Louis, MO, 1989.

NFPA 2014, *Standard on Explosion Prevention Systems*, NFPA 69, National Fire Protection Association, Quincy, MA, 2014.

NFPA 2015a, *Flammable And Combustible Liquids Code*, NFPA 30, National Fire Protection Association, Quincy, MA, 2015.

NFPA 2015b, *Uniform Fire Code (UFC)*, NFPA 1, National Fire Protection Association, Quincy, MA, 2015.

NIST, *Airtightness Evaluation of Shelter-in-Place Spaces for Protection Against Airborne Chemical and Biological Releases*, NISTIR 7546, National Institute of Standards & Technology, U.S. Dept. of Commerce, Gaithersburg, MD, 2009.

OSHA 1972 (U.S. Occupational Safety and Health Administration), *Explosives and Blasting Agents*, 29 CFR 1910.109, Washington, DC, 1972.

OSHA 1992 (U.S. Occupational Safety and Health Administration), *Process Safety Management of Highly Hazardous Chemicals*, 29 CFR 1910.119, Washington, DC, 1992.

Ostrowski, S.W. and Keim, K.K., *Tame Your Transient Operations; Use a Special Method to Identify and Address Potential Hazards*, Chemical Processing, June, 2010.

PMI (Project Management Institute), *A Guide to the Project Management Body of Knowledge (PMBOK Guide)*, 5th Edition, Newtown Square, PA, 2013.

Sanders, R.E., *Chemical Process Safety, Learning from Case Histories*, 4th edition, IChemE (Elsevier), Rugby, UK, 2015.

Seveso, *Council Directive 82/501/EEC of 24 June 1982 on the Major-Accident Hazards of Certain Industrial Activities (Seveso Directive)*, Council of the European Union, 1982. This was subsequently amended by Council Directive 96/82/EEC of 9 December 1996 on the *Control of Major-Accident Hazards Involving Dangerous Substances* (Seveso II), Council of the European Union, 1996. This in turn was repealed and replaced by Directive 2012/18/EU of the European Parliament and of the Council of 4 July 2012 on the *Control of Major-Accident Hazards Involving Dangerous Substances* (Seveso III), European Parliament, Council of the European Union, 2012.

Stapelberg, R.F., *Handbook of Reliability, Availability, Maintainability and Safety in Engineering Design*, 2009.

UKOOA (UK Offshore Operators Association), *Guidelines for Fire and Explosion Hazard Management*, UKOOA, 1995.

U.S. Army, *Product Assurance, Reliability, Availability, and Maintainability*, Army Regulation 702–19, Headquarters, Department of the Army, Washington, DC, 2015.

USCG (U.S. Coast Guard), 33 CFR 127 *Waterfront Facilities Handling Liquefied Natural Gas and Liquefied Hazardous Gas (*subpart 007 *Letter of Intent and Waterway Suitability Assessment)*, 2005.

Wintle, J. et al., *Plant Ageing: Management of Equipment Containing Hazardous Fluids or Pressure*, U.K. Health and Safety Executive Research Report RR509, HSE Books, 2006, www.hse.gov.uk.

Additional information can be found in several publications:

Amyotte, P.R., Goraya, A.U., Henershott, D.C. and Khan, F.I., *Incorporation of Inherent Safety Principles in Process Safety Management*, Process Safety Progress, Vol. 26, No.4, 2007.

Anles, M.K., Miri, M.F., and Flamberg, S.A., *Selection and Design of Cost-Effective Risk Reduction Systems*, Process Safety Progress, Vol. 20, No.3, 2001.

API, *Guide For Pressure-Relieving And Depressurizing Systems*, RP 521, 6th Edition, American Petroleum Institute, Washington, DC, 2014.

API, *Fire Protection in Refineries*, RP 2001, 9th Edition, American Petroleum Institute, Washington, DC, 2012.

API, *Protection Against Ignitions Arising Out of Static, Lightning, and Stray Currents*, RP 2003, 8th Edition, American Petroleum Institute, Washington, DC, 2015.

API, *Application of Fixed Water Spray Systems for Fire Protection in the Petroleum and Petrochemical Industries*, RP 2030, 4th Edition, American Petroleum Institute, Washington, DC, 2014.

API, *Fireproofing Practices in Petroleum and Petrochemical Processing Plants*, RP 2218, 3rd Edition, American Petroleum Institute, Washington, DC, 2013.

Corvaro, F., Giacchetta, G., Marchetti, B. and Recanati, M., *Reliability, Availability, Maintainability (RAM) study, on reciprocating compressors API 618*, Petroleum 3, 266-272, 2017.

Cox, A.W., Lees, F.P. and Ang, M.L., *Classification of Hazardous Locations*, 1993.

Dow, *Fire & Explosion Index, Hazard Classification Guide*, Sixth edition, Dow Chemical Company, available from the American Institute of Chemical Engineers, New York, NY, 2010.

EEMUA, *A practitioner's handbook - Electrical installation and maintenance in potentially explosive atmospheres*, Publication No. 186, 7th Edition, The Engineering Equipment and Materials Users Association, London, UK, 2016.

Gupta, V. and Borserlo, B., *Retrofit Experiences with a 32-Year Old Ammonia Plant*, Process Safety Progress, Vol. 21, No.3, 2002.

Hendershott, D.C., *An Overview of Inherently Safer Design*, Process Safety Progress, Vol. 25, No.2, 2006.

House F.F., *An Engineer's Guide To Process-Plant Layout*, Chemical Engineering, McGraw Hill, New York, NY, July 28, 1969.

Lees F. P., *Loss Prevention In The Process Industries*, 4th Edition, Volumes 1 & 2, Butterworths, Boston, MA, 2012.

Mecklenburgh J.C., *Process Plant Layout*, John Wiley & Sons, New York, NY, 1985.

NFPA, *Liquefied Petroleum Gas Code*, NFPA 58, National Fire Protection Association, Quincy, MA, 2017.

NFPA, *Hazardous Materials Code*, NFPA 400, National Fire Protection Association, Quincy, MA, 2016.

NFPA, *Standard for Purged And Pressurized Enclosures For Electrical Equipment*, NFPA 496, National Fire Protection Association, Quincy, MA, 2017.

NFPA, *Recommended Practice for the Classification of Flammable Liquids, Gases, or Vapors and of Hazardous (Classified) Locations for Electrical Installations in Chemical Process Areas*, NFPA 497, National Fire Protection Association, Quincy, MA, 2017.

NFPA, *Standard for the Prevention of Fire and Dust Explosions from the Manufacturing, Processing, and Handling of Combustible Particulate Solids*, NFPA 654, National Fire Protection Association, Quincy, MA, 2017.

Slye O. M. Jr., *Loss Prevention Fundamentals For The Process Industry*, Loss Prevention Symposium, American Institute of Chemical Engineers, New York, NY, March 1988.

Tam, V., Moros, T., Webb, S., Allinson, J., Lee, R. and Bilimoria, E. *Application of ALARP to the Design of the BP Andrew platform against Smoke and Gas Ingress and Gas Explosion*, Journal of Loss Prevention in the Process Industries, 9, 317-322, 1996.

INDEX

Action tracking *42, 54, 79, 122, 135, 142, 181, 236, 258, 287*
Asset integrity management .... *119, 139, 168, 230, 250, 282, 306, 350, See* Quality Management
inspection, testing and preventive maintenance *96*
maintenance management system *183, 239, 251*
Auditing *173, 234, 257, 285*
Basic Engineering Package *58*
Basis of design *17, 119*
Blowdown and depressurization study *76, 117*
Brownfield projects *22, 121*
Case for safety *125, 140, 141, 174, 238*
Change management *42, 80, 122, 142, 170, 283*
design change notice *165*
Commissioning
planning *148, 210, 213*
preparation *210*
procedures *180, 208*
Commissioning and startup *10, 148, 207, 302*
equipment testing *222*
planning *211*
preparation *123, 147, 183, 211, 226, 299*
procedures *180, 183, 224*
safety *214*

Compliance with standards *30, 31, 35, 72, 129, 159, 247, 306, 345*
Conduct of operations *160, 353*
Consequence analysis *79, 100*
Constructability *144*
Construction .... *10, 135, 143, 300, See* Quality Management
constructability *144*
construction planning *146*
execution *144, 154, 159*
installation *164, 178, 201*
management *151, 152, 172, 174, 180, 181*
mechanical completion *179*
mobilization *36, 146, 155, 158*
planning *37, 147, 154, 155*
pre-mobilization *277*
preparation *123, 146, 299*
request for information *165*
Contractor management *160, 253, 277, 352*
contracting strategy *27, 33*
contractor oversight *38, 147*
contractor selection. *120, 145, 277*
Decommissioning *117, 262, 307*
checklist *266*
cleaning and decontamination *273*
deconstruction *274*
demolition *276*
disposal *264, 270, 272, 285*
engineering survey *266, 271, 276, 308*

hazards *265, 269, 276, 315, 332*
late-life operations *272*
management *263, 276*
planning *271, 274*
re-engineering *271*
remediation *272, 286*
Design change notice *147, 165*
Design Hazard Management *66, 84, 88, 131*
design safety measures.. *66, 88, 97*
Detailed design . *9, 72, 122, 128, 130*
Development options *27, 48, 54, 61, 62, 63, 65*
evaluation *66, 85*
selection *62, 65, 85*
Document management *180, 291*
Documentation.. *24, 36, 44, 291, 341*
commissioning and startup *148, 233, 298, 299*
construction *43, 146, 148, 180, 197, 204, 297, 298*
detailed design *135, 143, 298*
end of life *307*
FEL 1 .... *294*
FEL 2 .... *295*
FEL 3 .... *296*
handover *303*
operation *305*
Dropped object study *113*
Emergency management *354*
emergency response *233, 284*
Emergency Management
emergency response *255*
End of life *10, 262*
Environment, health and safety.... *48, 59, 84, 130, 153, 210, 244, 263*
Evacuation, Escape and Rescue study
.... *112*
Facility siting study *74, 89, 100*
Factory acceptance tests *159*
Fire and explosion analysis ... *74, 103*
Fire and gas detection *73, 77, 116*
Fire Hazard Analysis *77, 102*
Firewater analysis *77, 117*

Front End Engineering Design *8, 83, 86, 126, 128*
Front End Loading *7, 8*
FEL-1 .... *9, 47*
FEL-2 .... *9, 58*
FEL-3 .... *9, 83*
Functional safety *66, 131, 133*
assessment *136, 184, 217, 251*
safety instrumented function . *107, 109*
safety instrumented system *66, 98, 108, 136*
safety requirements specification
.... *251*
Greenfield projects *22*
Handover *43, 123, 129, 148, 237, 280, 293, 299, 303*
Hazard and risk register *40, 54, 79, 121, 142, 181, 236, 258, 287, 316*
Hazard Identification and Risk
Analysis *14, 50, 53, 61, 71, 79, 99, 118, 134, 136, 228, 248, 269, 279, 347*
Hazardous area classification *108, 136*
Human factors analysis *115, 136, 189*
Inherently safer design *51, 71*
approach *5*
optimization *88, 132*
review *51, 61*
Inspection, testing, and preventive
maintenance *140, 148, 149*
ITPM *251*
See Quality Control
Major accident hazards *61, 66, 88*
Major accident risks *90*
Management of Change *24, 130, 232, 254, 353*
Management review and continuous
improvement *161, 257*
Materials handling *162*
Measurement and metrics *174, 235, 256*

Mechanical completion *123*, *134*, *179*
Operation *10*, *37*, *124*, *239*, *242*, *272*, *305*
  conduct of operations..............*255*
  post-operational review..........*260*
  technical support....................*259*
Operational readiness..*215*, *254*, *353*
  pre-startup safety review.*183*, *217*
  review*44*, *124*, *148*, *183*, *217*, *283*
Performance measurement..*173*, *235*
  benchmarking and metrics......*173*
Performance standards *85*, *88*, *98*, *134*, *193*
Performance test run...........*236*, *259*
  performance guarantee test run *237*
Pre-commissioning.........*8*, *175*, *207*
  preparation..............................*147*
Pre-mobilization *146*, *155*
Pre-startup safety review *44*
Process knowledge management *248*, *345*
  documentation.................*294*, *305*
  process safety information......*294*
Process safety *48*, *59*, *84*, *130*, *153*, *210*, *244*, *264*
  activities *1*, *13*, *66*, *85*, *119*, *140*, *153*, *211*, *244*, *264*
  competency......................*247*, *345*
  information *80*, *122*, *143*, *180*
  plan *54*, *79*, *142*, *181*, *236*, *258*, *270*, *287*, *312*
  procedures *167*, *183*, *230*, *239*, *243*, *258*, *281*
  safety checklist........................*318*
  studies...........*60*, *85*, *87*, *130*, *310*
Process safety management system *2*, *118*, *245*
Procurement *36*, *59*, *123*, *137*, *159*, *196*
Project budget...............................*18*
Project close-out *184*, *239*, *261*
  activities..................................*44*
  evaluation...............................*240*

lessons learned.........................*44*
report *240*
Project controls.............................*41*
  budget.......................................*41*
  cost control *41*
  cost estimate *41*
  planning....................................*41*
  reporting *42*
Project execution *128*, *151*, *207*
  plan *18*, *122*, *154*
Project implementation strategy ...*32*
Project life cycle......................*7*, *19*
Project management *16*, *20*
Project management team *27*, *30*, *47*, *58*, *83*, *129*, *152*, *210*, *243*, *263*
Project organization.....................*26*
  roles and responsibilities *29*
  teams..*26*
Project risk....................................*39*
  hazard & risk register *315*
  risk assessment cycle...............*40*
Project Risk Assessment...............*39*
Project scope.................................*17*
  scope creep *20*
Project types *22*
Punch-listing...............................*178*
Quality Assurance *195*
Quality Control... *120*, *137*, *192*, *195*
Quality management..... *44*, *120*, *188*
  construction and installation...*201*
  design / engineering................*193*
  documentation *204*
  fabrication..............................*197*
  operation................................*203*
  procurement............................*196*
  quality assurance *191*
  quality control........................*192*
  storage and retrieval *200*
Quality management plan *120*, *137*, *159*
Reliability, availability, and maintainability study *111*, *139*
Relief, blowdown and flare study *117*, *136*
Risk analysis

concept risk analysis *53, 62*
layer of protection analysis *107*
preliminary *71*
quantitative risk analysis *104*
Risk management *4, 38, 316*
Risk-based process safety *245*
elements *11, 245*
Safe work practices *165, 229, 250, 279, 349*
Safety assessments *107, 135*
Safety checklist *318*
Safety critical equipment *98, 109, 111, 136*
vulnerability *75, 111, 136*
Safety culture *153, 160, 247, 278, 345*
Safety Integrity Level
determination *107, 108, 136*
Security Vulnerability Analysis ... *78, 113*
Shelter-in-place *75, 90, 94, 103, 136, 172*
Simultaneous operations study .. *114, 136, 155, 163, 175, 202, 228, 250, 279*
Site acceptance test *134, 223*
Site layout *89, 133*
buildings *94*

confinement & congestion *93*
drainage & containment *92*
spacing *92*
storage *93*
utility routing & locations *95*
Smoke and gas ingress analysis ... *75, 103*
Stage gate review *19, 40, 45, 55, 80, 125, 149, 184, 216, 259, 288*
protocol *356*
Stakeholder outreach *84, 118, 160, 248, 278, 345*
Startup *207, 208, 226*
preparation *147, 226*
startup with process chemicals *227*
Start-Up Efficiency Review *220*
Statement of requirements *16*
Statement of work *16*
Temporary refuge impairment *75, 95, 136*
assessment *112*
Training and competence assurance *168, 231, 281*
Training and performance assurance *352*
Transportation studies .. *75, 106, 161*
Workforce involvement *160, 248, 278, 345*